The Art of Assembly Language Programming Using PIC® Technology

The Art of Assembly Language Programming Using PIC® Technology

Core Fundamentals

Theresa Schousek

ELSEVIER

Newnes
An imprint of Elsevier

Newnes is an imprint of Elsevier
The Boulevard, Langford Lane, Kidlington, Oxford OX5 1GB, United Kingdom
50 Hampshire Street, 5th Floor, Cambridge, MA 02139, United States

Library of Congress Cataloging-in-Publication Data
A catalog record for this book is available from the Library of Congress

British Library Cataloguing-in-Publication Data
A catalogue record for this book is available from the British Library

ISBN: 978-0-12-812617-2

For information on all Newnes publications
visit our website at https://www.elsevier.com/books-and-journals

Publisher: Mara Conner
Acquisition Editor: Tim Pitts
Editorial Project Manager: Charlotte Kent
Production Project Manager: Vijay Bharath R.
Cover Designer: Mark Rogers

Typeset by SPi Global, India

"To Brian Walter, Elizabeth Margaret, and Victoria Rose"
"In Memory of Zachary James and Nathaniel Howard"
"With special thanks to Keith Curtis at Microchip Technologies for his assistance throughout this book"

Contents

Preface

PURPOSE AND SCOPE

The purpose of this textbook is to give a thorough introduction to Microchip Assembly Language Programming for a wide range of users. Microchip's most basic 33 instructions are used heavily throughout the text. It is recommended that the new user, regardless of 8-bit processor selection, self-restrict their use to the basic 33 instructions until these are second nature. This will prove to be the best approach to learning assembly language. This book provides a basic introduction to get the reader up to speed quickly. It provides beginning code development chapters on such things as basic data transfers, addition, and subtraction. Application notes for more complex mathematical routines, multiplication and division are presented for use as a practical solution. For the novice and returning programmers, there is also a chapter on Mathematical Numbering Systems. The chapters on arithmetic are reserved for the end of the book so as to not overwhelm the newest of users. Practical use and selection of specific evaluation boards, recommended by Microchip, is briefly presented with a cost, and feature, analysis.

Fundamentals of good design practice are presented using tried and true approaches of flowcharting, Warnier-Orr diagrams, and State Machines. In addition, a separate chapter is dedicated to SysML, by way of introduction, with a thorough case example. A new text on the market is recommended for those expecting to use SysML daily in their business.

The hardware interface is presented in a chapter on Embedded Control. Focus is on basic "electronic glue" used on 8-bit processors. Most of the interfaces can be used directly from this text. With only a modicum of electronic hardware changes, such as actual values of components, that will be required.

PROGRAM CONVERSION PROCESS; FROM PROGRAM TO INDIVIDUAL BITS

Complexity of computer programs is shown in Fig. 1. This diagram starts with the most fundamental element of a computer; the two states of 5 and 0V. These are interpreted by the computer as a logical 1, for 5V, and a logical 0, for 0V or ground. The logical bits are then grouped by sequences of 8 bits, known as bytes. The binary strings are then grouped as clusters of 8 bits, or 1 byte; each byte defined by two hexadecimal digits from 0 through F. These binary strings are organized in groups of two nybbles; known as Machine Language. Each successive bubble, in Fig. 1, then introduces a higher level of programming. Assembly Language occupies the sweet spot between Machine Language and High level Language where knowledge of the embedded device is required and optimal efficient code can be produced.

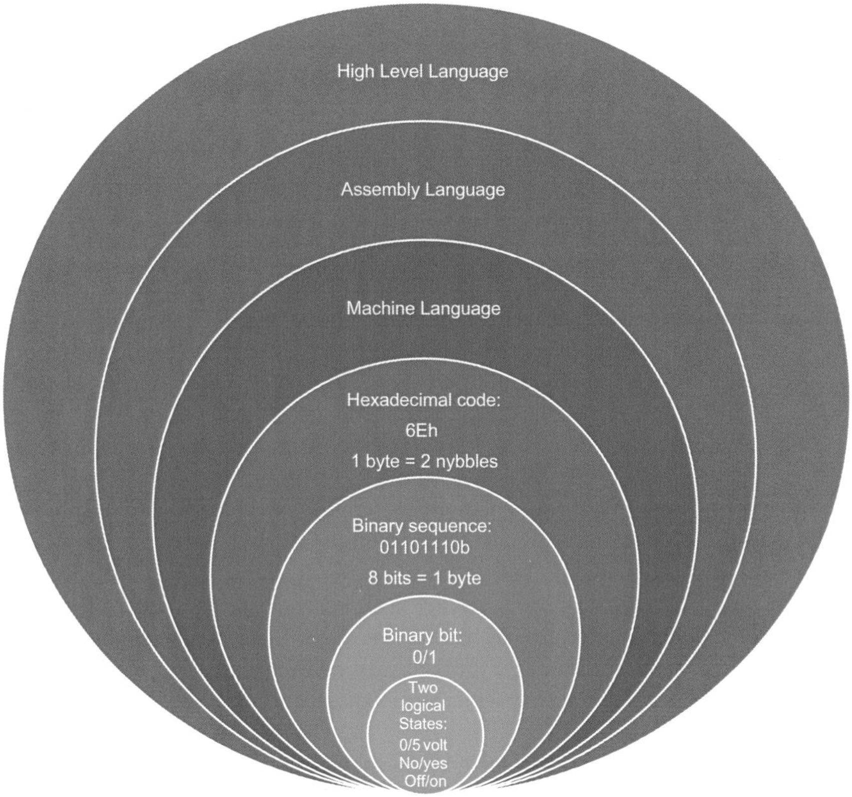

FIG. 1

The circle of languages and their position on the complexity scale.

The compiler converts a high level language such as C, C++, FORTRAN, and Python into machine language. Each level of programming then requires "translation" ultimately to Machine Language, as shown in Fig. 2. This begins with Assembly Language, which is converted to Machine Language by the use of an assembler and linker. The linker is not a separate language, per se. It is a necessary step in the process of converting the program, as written by the user, into nybbles to be read by the computer.

AUTHOR'S NOTE

Congratulations, you have arrived! This is the greeting that I anticipated when first starting my journey into Assembly Language. Those who came before me learned assembly as their first language. Assembly was a rite of passage. I started programming

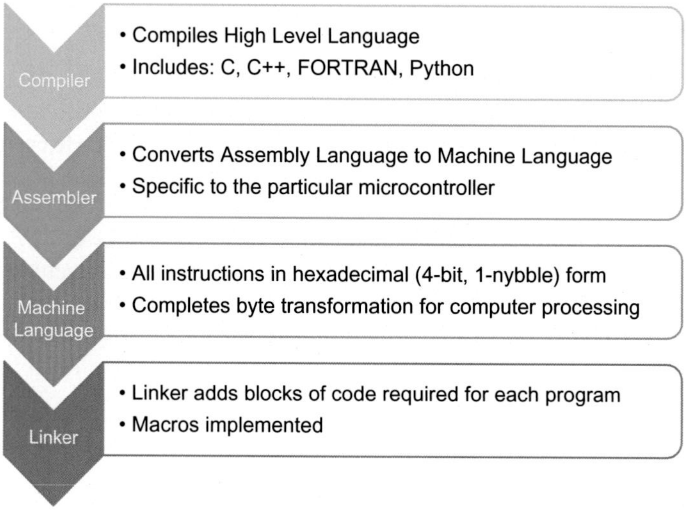

FIG. 2

Relationship between compiler, assembler, machine language, and linker.

with Basic, for the Commodore 64. Data Processing Class taught me FORTRAN, RPG II, COBOL, and Pascal.

There was a thirst for assembly, as some of us undertook manipulation of bits. I inherited a survey program. I was to expand the program so they could ask more questions. However, after pouring over the details in the program, I learned that the original author had taken the Yes/No questions, treated the answers as bit responses and made a single byte hold eight questions worth of binary answers. There would be no easy means of "just adding a couple more questions."

I would not be given the opportunity to learn assembly in detail until my 4th year as a double major in Computer Engineering and Electrical Engineering. We emulated the PDP 11 with our programs. We learned the details of how a microprocessor works. Within this class, we wrote in assembly as we could gain unfettered access to control our micro. When you are seeking control of your processor, there is no better way to control it than to use Assembly Language.

However you come by this text, you will learn a new way of thinking before you are done. Assembly requires some nominal electrical engineering knowledge, in that you are intimately meshing your code with the operation of the processor. Electronic glue is introduced and fundamental electrical interface circuitry is presented. Usually, this takes the form of communication with a sensor or input from a switch. This is the heart of the microcontrollers' functionality.

My approach is to be as thorough as possible, presuming no prior knowledge. Feel free to jump ahead if you have more background than the absolute beginner.

DARE WE CALL IT FUN?

A few months back, I had the opportunity to talk with a neurologist. We happened upon a brief discussion of this text when I shared information on my work. He chuckled and then said he had programmed in assembly many years back. He even recalled the use of a rotate function to complete a multiply by two. Wistfully, he said "That was fun, really fun stuff."

In recollection of this exchange, I have pondered this notion. Dare we call it fun? Throughout college engineering courses, I was so wrapped up in getting the next project done. I do not believe that I ever stopped to think of just how much fun I had in this course. I also really enjoyed Digital Logic; a course that I would later teach at a community college. Always remember, a course can be both difficult and fun.

Enjoy!
Theresa Schousek

CHAPTER

Introduction

1

CHAPTER OUTLINE

PRACTICAL APPLICATIONS

Common applications for embedded control are devices or systems that require simple repeatable programs, with limited interaction from the user, such as switch inputs or light emitting diodes (LED) outputs. Generally, once programmed, the device continues to operate for years without any necessary changes, unless there is a desire to change the performance. The most common occurrence requiring rewriting of the program is, generally, the need to add features or performance improvements in coming years. In this text, the conscientious design of the microcontroller program

The Art of Assembly Language Programming Using PIC® Technology. https://doi.org/10.1016/B978-0-12-812617-2.00001-8

is laid out in a manner to maintain upward compatibility. If you follow through with using this approach in your own designs, it will pay back in the future.

Some practical applications, suggested by Microchip, include:

- Identification tags
- Drug tester
- Electronic lock
- Electronic chime
- Pressure sensor
- Water consumption gauge
- Medication dispenser
- LED flashlight
- Intelligent power switch
- Light dimmer
- Fan controller
- Smoke/CO alarm
- Engine governor
- Protocol handler
- Flat iron temperature control
- Capacitive switch
- Irrigation controller
- Security monitor

Beginning with the use of Sets and *operators* (such as *op*, *opcode*, or *operation*) the relationship between real-world applications and program representation is introduced. The set S is a relationship between a set S and itself, where x ρ y is a relation "on S" that represents x and y.

Our program example will show how the elements, coins, and pushbuttons are part of a vending machine program. In this example, x is a coin which will be received prior to the selection pushbutton. They could represent a list of coins that are acceptable. Then y is a list of pushbuttons that may be selected.

WHY ASSEMBLY?

There are numerous reasons to program in Assembly Language. Generally, Assembly Language is simply more efficient. One of two explanations will be appropriate: the code will take less time to execute and it will take less memory space to execute.

> From Jan Hext, in his book Programming Structures, an example of simply accessing memory to illustrate how assembly code can use registers to accomplish the task of interchanging the values of A and B. With a high level language, an additional memory location will be necessitated. A total of six memory accesses is required.
>
> temp := A; Memory-to-memory requires 2 accesses
> A := B; Memory-to-memory requires 2 accesses
> B := temp; Memory-to-memory requires 2 accesses

With Assembly Language, registers are used which are much faster and do not require a memory access. A total of four memory accesses is required.

R ← A; Memory-to-register requires 1 access
A ← B; Memory-to-memory requires 2 accesses
B ← R; Register-to-memory requires 1 access

This code may seem only minimally better than the high level language version. However, four (4) accesses versus six (6) is a savings of 33%. Some compilers do optimize and use registers. However, this is not always the case.

In addition, when some form of a loop structure is used, the net effect is to multiply the number of accesses by the number of control loops. This is where it will clearly show an unhelpful multiplication effect. For example, if the loop was "FOR $T=1$ to 10, NEXT," the total accesses on the high level language side would be 60 (plus 10 for the loop counter) compared with the assembly language approach of 40 (plus 0 for the loop counter executed with the use of a register.) This would result in a savings of (70–40)/70 or approximately 43%.

CORE FAMILIES ("BASELINE," "MIDRANGE," "ENHANCED MIDRANGE," "HIGH PERFORMANCE")

This text is intended to give the reader thorough information on using Microchip's 8-bit Core Family Systems. Microchip groups their 8-bit products into three families: Baseline, Midrange, and Midrange Enhanced. These family systems have similar architecture, peripherals, and programming. Embedded control devices, or *controllers*, differ from *processors* as the controller user is not concerned, nor aware, with the use of devices to communicate with the controller. For example, the processor will have a keyboard and a monitor through which the user will communicate with the processor. In embedded control, the user will typically interface with the controller by way of switches and LEDs. The controller user will not necessarily even be aware of the controller's clock speed, instruction cycles, and the like.

BASELINE

Baseline families include the PIC10F, PIC12X5, PIC16X5X. The Baseline products have an 8-bit data memory access and a 12-bit wide instruction set. Paging is used to access four banks of program memory. These products have the smallest footprint of all products. There is even a small 6-pin SOT-23 in the 10F2xx series along with the 8-pin PDIP and 8-pin DFN. Low voltage operation, down to 2V, is available in the PIC16C5x and PIC10F2xx series. Features of the smallest PIC10F2xx include 33 single-word instructions; two-level deep stack; direct, indirect, and relative addressing modes for data and instructions; 8-bit wide data path; eight special function hardware registers; and an operating speed of 4 MHz, or 1 μs instruction cycle, generated by an internal clock.

Peripheral features include four input/output (I/O) pins: three with individual direction control and one input-only pin. Another feature is an 8-bit Real-Time clock/counter (TMR0) with 8-bit programmable prescalar and one comparator (on the PIC10F204/206 version) with an internal absolute voltage reference.

Mark McComb, a Technical Training Engineer, presented in his Webinar, the following summary of Baseline products. "Our 8-bit MCUs, including the PIC10, PIC12, PIC16, and PIC18 families, offer the designer a range of choices with variations in performance, memory and pin count. Baseline Products offer a 12-bit instruction set covering 6-40-Pins and up to 3 kByte program memory or 2k instructions with a basic peripheral set including comparators and Analog-to-Digital Converters."

MIDRANGE

Mark McComb further elaborates upon Microchip's "existing Midrange family offers a 14-bit instruction set covering 8-64-Pins with up to 14 kByte program memory or 8k instructions. With up to 5 MIPS performance, a more advanced peripheral set is offered including Serial Communications, Capture/Compare/PWM, LCD, EEPROM, 10-bit A/D and, more recently, our Capacitive mTouch Sensing peripherals."

McComb comments: "Recognizing the demand for increased performance and peripherals within the 8-bit MCU market, Microchip has recently revisited and enhanced the Midrange Architecture."

ENHANCED MIDRANGE

"The new enhanced core builds upon the best elements of the existing Midrange core and provides additional performance, while maintaining compatibility with existing Midrange products for true product migration. The enhancements provide users with a boost of performance of up to 50% and code-size reductions of up to 40% for various algorithms and functions. Improvements such as more program and data memory including "C" efficiency optimizations, increased peripheral support, and reduced interrupt latency.

"Customers now have more options for applications that may not require the extended 16-bit program memory sizes offered by the PIC18 family but could benefit from more capabilities than currently offered by the existing Midrange architecture," states McComb.

HIGH PERFORMANCE

"Finally, McComb states that the high-performance PIC18 products offer a 16-bit instruction set covering 18-100-Pins with up to 128 kByte program memory and up to 16 Million Instructions Per Second (MIPS) performance with hardware multiply. This family offers a very advanced peripheral set including Advanced Serial Communications (CAN, USB), Capture/Compare/PWM, LCD, USB, Ethernet, 12-bit A/D and the new

Charge Time Measurement Peripheral developed specifically for capacitive sensing applications." The high performance devices are beyond the scope of this text.

DATA SHEET—WALK THROUGH

Herein, we will take a look through the incorporated PIC10F200/202/204/206 data sheet from Appendix E. Follow along in Appendix E as we seek out information from the data sheet. Note that "reading" a data sheet is a little different from "reading" a book. In a typical book or novel, one generally starts at the beginning and reads each page to the last page. Technical data sheets are all about knowing how to seek out the information you need at any given time. It is important that you know how to look through the data sheet for the item(s) that you need. This means that it is also equally important to know how to ignore the information that you do not need. One does not read from beginning to end but rather scans for the required information within the datasheet format, generally, common among processors from one manufacturer.

Let's explore this together. Let us presume that, today, we are looking for information on how to wire in inputs for a comparator. We are getting a printed circuit board laid out for use in our new project, before we start programming. As we gather information for our new project, **pertinent information is highlighted in bold type.**

What does the table of contents (TOC) show us? **The TOC is found on page 3.** First, however, we are getting a bit ahead of ourselves. Before we look at the TOC, we "backup" a little to first look at the summary page.

Glance through page 1 as this is a summary page. Page 1 shows a lot of information that you may need for quick reference. There is a lot of information, also, that does not help us with this specific issue. We find some information on the lower right side. **It tells us that the one comparator uses an internal voltage reference and that both comparator inputs, as well as the output, are all visible externally.** This is good information. This is found under "Peripheral Features" (PIC10F204/206.) We note that we **must use either the 204 or the 206 version of the 10F to have access to the comparator**.

After the summary page, the pinouts for the device are shown on page 2. Let's look through each of the options. After perusing the pin labels, we know that **we can use the 10F chip in any of the three packages: 6-pin SOT23, 8-pin PDIP, and 8-pin DFN**. For ease of testing a prototype circuit board, **we select the 8-pin PDIP** as we can place a socket in this location and easily change processors if we damage it during testing.

Finally, we are at the TOC. Looking through it, we should peruse the General Description and the Device Varieties to ensure that we do not miss something of fundamental importance. Specifically, we need to understand information from the **I/O Port generally, on page 20, and the Comparator Module, on page 31**. We find, in the general description page, that there is a Table 1-1 that shows a comparison between the different versions of the chip. For the comparator, we know we cannot use the 200/202 version. We must use 204/206, as these are the only ones with the comparator. Looking at the Table, we note that the **primary difference**, between 204 and 206, is the **Flash Program Memory and the Data Memory**. As the pinouts are the same, we will note this information but not set our selection in stone as we have yet to write the software. Until the software has progressed, we really do not know whether we can use the 204 or need to step up to a 206.

Now, follow along as we move on to the I/O Port general information on page 20. We scan through this page. To scan, instead of "read," look at the subtitles and the first sentence of each paragraph. This approach will allow you to select which paragraphs are relevant to our information search.

At Section 5.0, I/O PORT, we learn that **upon Reset, all I/O ports are defined as inputs**. We must ensure that we redefine the registers, immediately after reset, to guarantee that the pins are set as the comparator option. At Section 5.1, General Purpose Input/Output (GPIO), notes that the **GPIO register low order bits of 0 through 3 are the only bits used**. At Section 5.2, Tri-State (TRIS) Registers, we learn that **additional registers are used to configure the GPIO pins and that GP3 is always configured as an input only**. At Table 5-1, we note that the priority order of our comparator inputs is top priority 1. The priority order of the COUT on GP2 is 2, after the "oscillator divide by 4." All other information on this page is more specific to software and registers and not needed at this time. This concludes our review of page 20.

Next, scanning the TOC, we move to page 31 Section 8.0, Comparator Module. This is the most important section for use of the Comparator Module. Scan the page, and the following pages through page 34. As we scan, we note that the focus is on the block diagram, Figure 8.1, on the second page, page 32. Return to the first page to see the **CMCON() Register, the primary control register for the comparator**. All the settings for the register are detailed on page 31.

Moving onto page 32, we return to the block diagram in Figure 8-1. The written information states that the **comparator pins are "steerable" with the CMCON0, OPTION, and TRIS registers.** We observe that there is a Table 8-1 that shows the **interaction between pins T0CS, ¬CMPT0CS, ¬COUTEN**, and the Source. We note that the block diagram shows the internal complexity of the comparator; similar to any external chip comparator with **feedback on CIN-**.

Again, we first peruse the information on page 33 and 34. Much of this material is not needed until it is time to write the software. Read through the first sentence of each paragraph. Some things we learn include:

- Figure 8-2 shows analog levels and the digital output results
- From Section 8.3, there is an internal reference signal dependent upon mode selection.
- From Section 8.4, … blah, blah, blah …. nothing much that we need here until software.
- From Section 8.5 Comparator Output **CMCON() is read-only** …. we expect this….
- From Sections 8.6, 8.7, 8.8, and 8.9 … blah, blah … in 8.7 note that **the comparator remains active when device is put into Sleep mode**. This may be really important.

- From page 34, Figure 8-3 Diagram shows how to interface with the input. These other new **external parts, Rs, and Va are important for our printed circuit board layout.**
- Finally, Table 8-2 shows **all registers** associated with the comparator. This will be important reference information during software development.

This concludes our walk through of the data sheet. It will become more intuitive as you work through your project. I think the toughest part is knowing what specific information to discard or note for later use. This is what I have attempted to convey in the Data Sheet Walk Through.

STRAIGHT LINE PROGRAM IMPLEMENTATION

Straight line programming is a basic form of program structure which is simply designed to pass control straight through each instruction once. We will use a "mind-reading" exercise and the outline of a program from beginning to end. We will follow the general approach used by Jan Hext, in Programming Structures, Section 0.2, with an algorithm from https://nrich.maths.org/1051

"The basic syntactic structure of a program is as follows:
Program heading
  Declarations
Begin
  Statements
End.

"The *program heading* interfaces the program with the environment. The *declaration* introduces the entities required by the statements. The *statements* specify the computations that are to be performed."
The algorithm is as follows:

- Think of a number.
- Double it.
- Add 10.
- Halve it.
- Take away your original number.
- Is your answer 5?

THINK OF A NUMBER

For this exercise, we presume that the algorithm is given a two digit decimal number, arbitrarily, between 10 and 16. We call this value THINK.

To receive a number through a keypad and displayed on an LED display with up to four digits, one could use Application Note (AN) 590 "A Clock Design Using the PIC16C54 for LED Displays and Switch Inputs" or AN 529 "Multiplexing LED Drive and a 4x4 Keypad Sampling."

DOUBLE IT

To double the input, we could multiply the value THINK by two. However, multiply by two is a special condition, in Assembly Language, and we can use a Rotate Left function to double the input.

ADD 10 (DECIMAL)

Add decimal value 10.

HALVE IT

To halve the value, we could divide the intermediate value (IMV) by two. Again, however, divide by two is a special condition, in Assembly Language, and we can use a Rotate Right function to halve the intermediate value.

SUBTRACT ORIGINAL NUMBER

Subtract THINK from the intermediate value.

COMPARE OUTPUT TO 5

Subtract 5 from the IMV and look for the Z flag to be set to 1.

Using pseudocode, the *straight line* program is as follows:

```
Program exercise (x: integer)
  THINK : integer := x;
  IMV : integer := x;
Begin
  ; assert 10<= THINK <= 16
  IMV := THINK
  RLF   IMV ;multiply by 2
  ADD   d'10'
  RRF   IMV ;divide by 2
  SUB   THINK
  SUB   5
  ; Zero Set?
End.
```

LOOPED CODE IMPLEMENTATION

For the practice exercise, we will consider the case of a very basic soda vending machine. Our basic vending machine has six offerings, with one pushbutton for each option. Our vending machine uses coins; quarters only. In this example, X is a coin. A and B represent the row and column of one pushbutton selected. We are going to wire our pushbuttons such that we only need row and column information. Our machine will only take quarters and requires just one quarter for a soda (see Figs 1.1 and 1.2).

Our first task is to define inputs and outputs. Next, we will design the electronic interface with our inputs and outputs. Using the assessment of inputs and outputs, we have selected the PIC 10F200.

This will accommodate our program space, which will be quite short. As we do not need the comparator, we will be using the GPIO pins for our input and output. GP3 is input only. We will assign the "Coin Entered" pin as this input on GP3. Row selected will be GP0, with column selected as GP1. GP2 will be our "Dispense Soda" output.

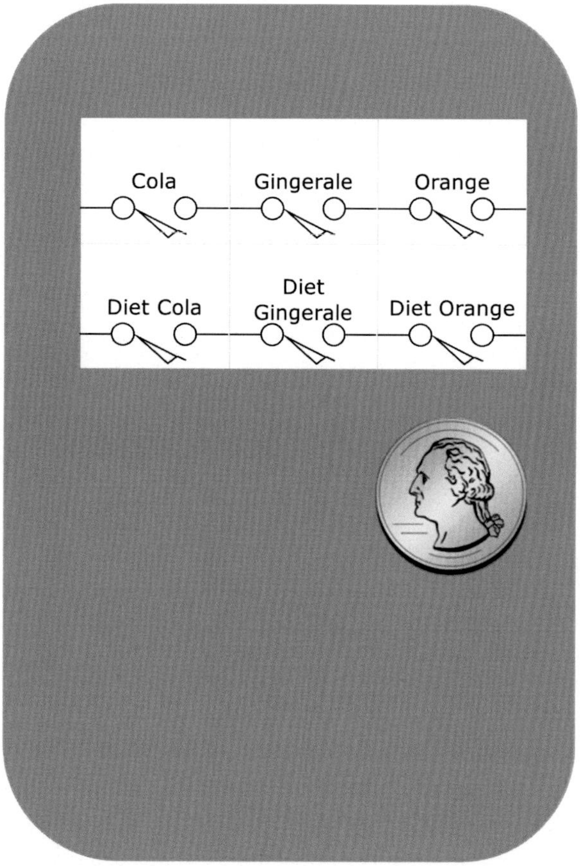

FIG. 1.1

Diagram of soda machine.

Inputs	Outputs
Row selected	Dispense soda
Column selected	
Coin entered	

FIG. 1.2

Define inputs and outputs.

```
Program exercise2 (x: integer)
   ROW : integer := x;
   COLUMN : integer :=y;
   DISPENSESODA : (yes, no);
   COINRECEIVED : (yes, no);
Begin
   ; assert 1<= ROW <= 2
   ; assert 1<= COLUMN <= 3
ENDLESSLOOP    Read ROW;
               Read COLUMN;
               Read COINRECEIVED;
               If COINRECEIVED = no,
               THEN    DISPENSESODA := no;
               ELSEIF   BEGIN
                        DISPENSESODA := yes;
                        CLOSE SWITCH Row, Column
                        END
               ENDIF
               GOTO    ENDLESSLOOP

End.
```

MICROCHIP TOOLS

Microchip's most basic tools are available for free on their website, www.microchip.com. To get started, download MPLAB X Integrated Design Environment (IDE). Initially, you will want to use the simulator. Use tutorials to ensure you have the tools set up correctly.

MICROCHIP MICROCONTROLLER COURSE INFORMATION

Microchip has an extensive network of field representatives that offer classes in each of the product families. To get up to speed quickly, look online at http://www.microchip.com for a local class on the particular family you are using. These courses are thorough, covering everything one needs to "fully develop and download applications" into Baseline, Midrange, and Midrange Enhanced microcontrollers (MCU).

The courses are also very hands on and, also, give you time to put your new knowledge to use with an experiment on a personal computer (PC).

Course number MCU0101 is "Getting Started with Microchip Baseline PIC MCU Architecture and Peripherals." This is a 2 class unit course.

Fundamentals of the Midrange are covered in three courses; MCU1101, MCU1102, MCU1121. MCU1101 (1 class unit) is entitled "Getting Started with Microchip's Midrange Microcontroller Family Architecture" and includes information on the "programmer's model, data and program memory organization, clocking structures, assembly language and special features of the devices." This is also a hands-on course with MPLAB X IDE.

MCU1102 (1 class unit) is entitled "Getting Started with Microchip's PIC16F1 Enhanced MCU Architecture." Topics include "enhancements to the data and program memory, clocking schemes, and special features of the MCU."

The final course, MCU1121 (2 class units) is entitled "Microchip Midrange PIC MCU Peripheral Configuration and Usage with C." This course provides details on peripherals including timers, A/D converters, UARTs, and Comparators. This course is also quite hands on, with the MPLAB X IDE, learning to program in assembly, using development tools including simulators.

JARGON NOTE—ITALICS

Assembly language uses quite a few new terms. When a new term is introduced, it is highlighted with *italics* type and the surrounding material provides the definition with explanation.

WORD SEARCH

```
S K N B N A Y Y E J D F E X I
N E M D W L I M Q T A J G Z N
S O T A I Z B C J N T Y M O E
P B I M R E P F P E A C C G E
E R A T D G K U D M S L C B C
Y F A D A G O O A P H O D J B
Y R E C V T C R G O E R P I L
C D F C T O N O P L E T V J D
I B Z J D I O E Z E T N U O K
H S Q U T P C W M V D O Q E O
X Y E T R G G A T E B C F G N
R S W A D J K U L D L E L V X
P A L G O R I T H M Z P A M Z
Y L B M E S S A C O R E M W B
K L A H D E C P V L O Q J I C
```

ALGORITHM
ASSEMBLY
CONTROL
CORE
DATASHEET
DEVELOPMENT
EMBEDDED
FAMILY
IMPLEMENTATION
PRACTICAL
PROGRAM
PSEUDOCODE

BINARY PUZZLE

Complete the grid so that each row and column contain four 0s and four 1s. The same number cannot appear in more than two consecutive squares in any row or column. Bonus: In the last column and the last row, write the hexadecimal value for each 8-bit column and row binary value (see Chapter 9).

			0					
						0		
	0							
0				0	0			
					0			
1				0				
0	1			1			1	
0								
								Hex

How to solve Binary

The puzzle is to be completed with 8-bit binary values, placing the hexadecimal value of each row in the right-most column and placing the hexadecimal value of each column in the bottom-most row. Complete the grid so that each row contains four 0s and four 1s. Each column may contain three, four, or five 1s, with the balance in 0s. The same number cannot appear in more than two consecutive squares in any row or column. In the finished puzzle, each row must have a different sequence of 0s and 1s to any other row, and likewise for each column.

FURTHER READING

A presentation on the "Why?" of assembly language is shown on page 76–77 in Programming Structures, Volume 1 Machines and Programs, by Jan Hext, Section 5.3 Registers.

Program structure "mind-reading" exercise and outline of program from beginning to end is presented in Programming Structures by Jan Hext from page 136 to 137, Section 7.2 Program Structures.

Mathematical puzzles can be found at https://nrich.maths.org/1051, including the specific "mind-reading" algorithm.

The Microchip Data Sheet used for the perusal of a Data Sheet is 10F2xx (DS40001239F) from page 45 to 52. In the Practice Exercise, references are made to Data Sheets 12(L)F1571/2 (DS40001723D) from page 247 to 258; and Data Sheets 16(L)F1454/5/9 (DS40001639B) pages 330–341.

Two slides are presented from Mark McComb's Technical Training Engineering Webinar from 2006, Slide 4 and Slide 6 of "Enhanced Mid-Range Architecture."

CHAPTER

Microchip 8-bit architecture

2

CHAPTER OUTLINE

THE MICROPROCESSOR

The microprocessor is an electronic device that is composed of a central processing unit, inputs and outputs, a system clock that executes instructions, a data bus for general information transfer, a program bus to facilitate instruction transfers, and specialized peripherals, both internal and external to the microprocessor. The diagram, see Fig. 2.1, shows the interrelationship between each component within a microprocessor. This diagram may also be found in the 10F Data sheet, provided in Appendix E.

SYSTEM CLOCK

The "heart" of the processor/controller is the system clock that functions much akin to a heartbeat. The system clock pulse is generated by either a crystal or an RC

The Art of Assembly Language Programming Using PIC® Technology. https://doi.org/10.1016/B978-0-12-812617-2.00002-X

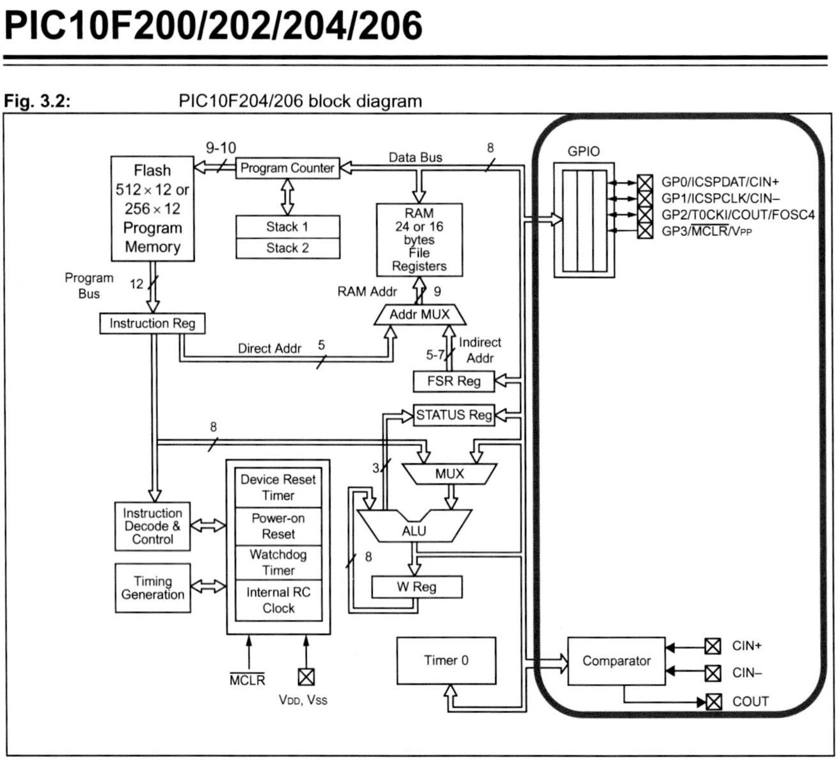

FIG. 2.1

Microprocessor 10F204/206 block diagram highlighting internal peripherals.

circuit. A crystal is, precisely what you think of, a natural element. There are two ways to supply an oscillator. First, you can use a crystal. The crystal responds by oscillating when electricity is applied to it. Second, you can create an oscillation with an RC circuit. An RC circuit contains one resistor and one capacitor. When electricity is applied, the capacitor charges up, with a rise in voltage, and then discharges through the resistor, and then repeats. When these two methods are compared, the crystal is generally more precise. However, the RC circuit is generally lower cost, but less consistent, part to part.

One of the nice features of many PIC® devices is that the oscillator is contained within the chip (see Fig. 2.2). Even the little 10Fxxx parts contain a 4 MHz crystal. The oscillator is a required part of any and all microprocessors. It serves as the "heart beat" of the processor and permits advances through each line of code, one by one. There may be a reason to override the internal crystal and increase the frequency to 20 MHz, for example, to speed up the processor. Always check the data sheet for any restrictions.

PIC10F200/202/204/206

Fig. 3.2: PIC10F204/206 block diagram

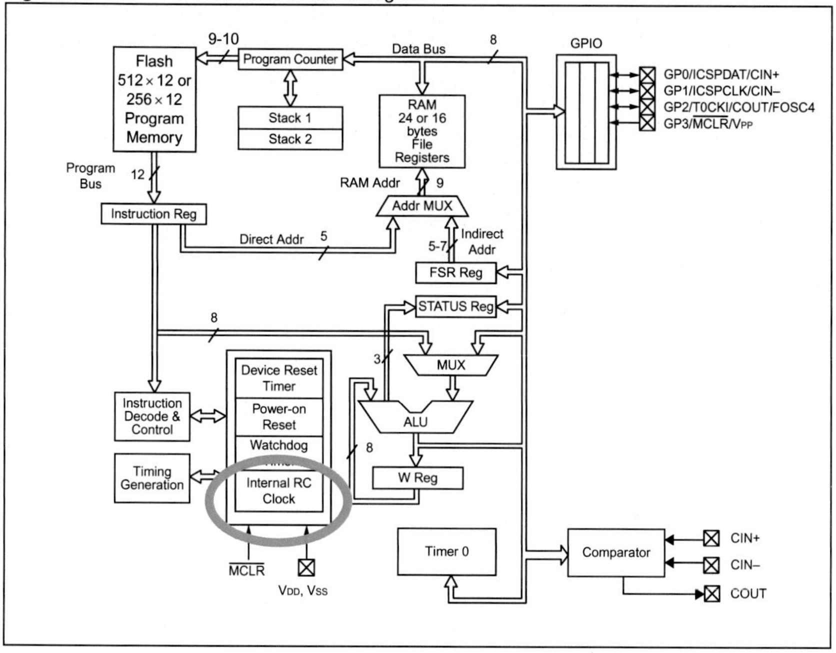

FIG. 2.2

System clock.

EXECUTION CYCLE

The system clock is necessary to step through the instructions. Each step, on a Microchip microcontroller, takes one execution cycle of four clock pulses for most instructions. For the 10Fxxx series, the internal clock speed is 4 MHz with an instruction cycle of 1 μs. Some more complicated instructions will require two execution cycles. In some instances, the next instruction step (the one after the step in play) is presumed to be the next instruction and the microprocessor attempts to begin the execution cycle early because there is space to begin in parallel with the current instruction. Most of the time, it is a safe assumption that the "next" instruction is next. However, sometimes this occurs when there is a "go to" or "call" routine. In this instance, the processor must be fed a NOP, or "no operation" instruction to clear the execution cycle before moving on to the actual next instruction.

INSTRUCTION FLOW PIPELINING

"The internal clock rate is divided by four, generating four nonoverlapping quadrature clocks, namely Q1, Q2, Q3, and Q4." See Appendix E 10Fxxx Data sheet, Fig. 3.3 Clock/Instruction Cycle. There is a Fetch instruction (program counter) followed by an execute instruction. These would ordinarily be sequential in execution. However, while the processor is fetching instruction 1, it is also executing instruction 0. This parallel process is known as pipelining (see Fig. 2.3).

BUS FOR DATA AND PROGRAM

The microchip microprocessor uses two separate buses, one for access to program memory (see Fig. 2.5) and one for access to data memory (see Fig. 2.4). The data bus is 8-bits wide, hence the "8-bit processor." The program memory may be 10-bit or 12-bit wide with up to four separate banks. *Paging* is the term for accessing four identical banks, with a 2-bit page number: page 00b, 01b, 10b, or 11b .

HARDWARE INTERFACES

The hardware interface uses four connections that may be applied to multiple uses: general purpose input/output pins, In-Circuit Serial Programming pins with Vpp programming voltage input pin, a clock input pin for Timer 0, an oscillator output, and a master clear reset. These are labeled as follows:

- GP0/ICSPDAT/CIN+
- GP1/ICSPCLK/CIN–

MOVLW 03H	Fetch 1	Execute 1				
MOVWF GPIO		Fetch 2	Execute 2			
CALL subroutine			Fetch 3	Execute 3		
(CALL 2nd cycle)				Fetch 4	Flush	
BSF GPIO Bit 1					Fetch subroutine	Execute subroutine

FIG. 2.3

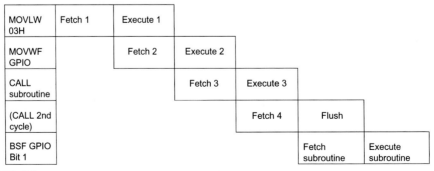

Instruction pipeline flow (see also Appendix E 10F Data sheet, Example 3.1).

PIC10F200/202/204/206

Fig. 3.2: PIC10F204/206 block diagram

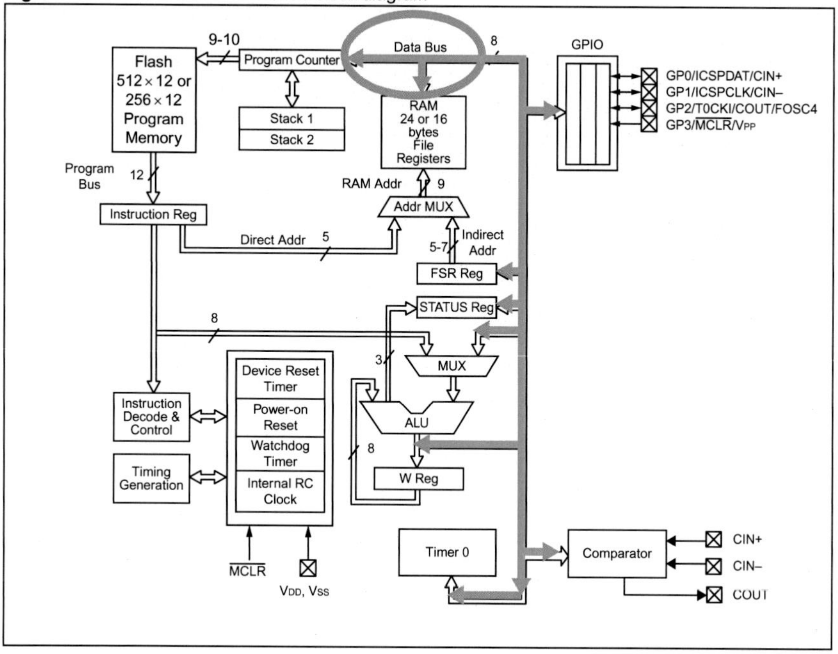

FIG. 2.4

Data bus.

- GP2/T0CK1/FOSC4/COUT
- GP3/¬MCLR/Vpp

Details on the pinout descriptions are shown in Appendix E 10F Data sheet Table 3.2, page 9.

INPUT/OUTPUT INTERFACE

There are four General Purpose (GP) Input/Output pins (GP0 to GP3). GP0 through GP2 are bidirectional and may be used as either input or output pins (see Fig. 2.6). GP3 may only be used as an input pin. GP0 and GP1 may be "software programmed for internal weak pull-up and wake from Sleep on pin change." The input type is TTL, while the output type is CMOS. Generally, TTL does not get damaged as easily as CMOS from static electricity on the input. In addition, TTL can take an input of up to 12v, whereas CMOS cannot exceed around 5v .

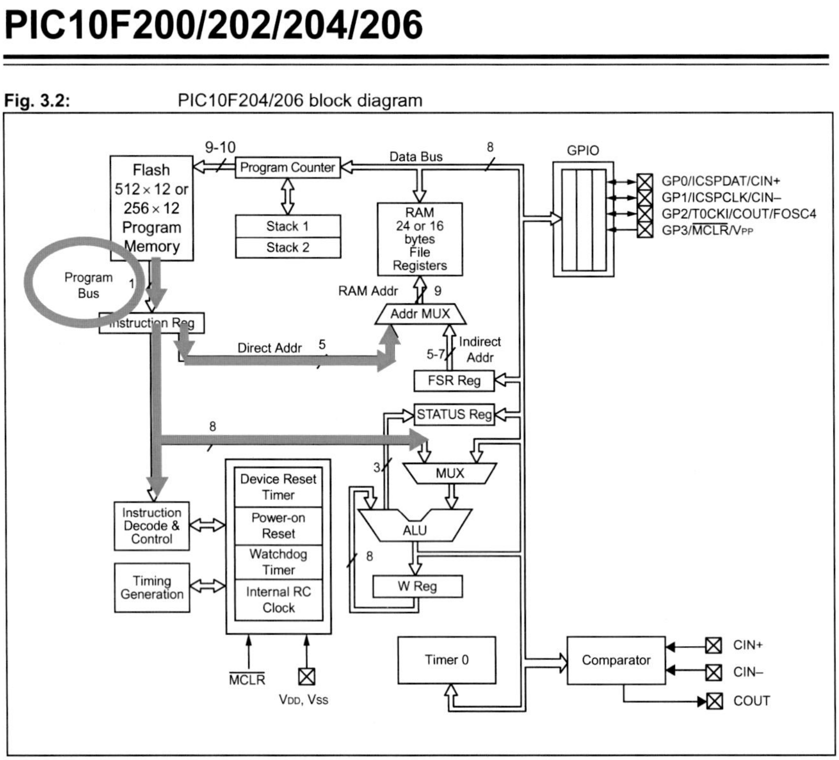

FIG. 2.5

Program bus (Appendix E Fig. 3.2 of Data sheet 10F200/202/204/206).

COMPARATOR INTERFACE

The comparator interface uses three pins: CIN+, CIN–, and COUT. These overlap in function with GP0 through GP2. The comparator is only available on PIC10F204/206 devices. Comparators are an abbreviated form of an opamp (see Fig. 2.7).

TIMER 0 INTERFACE

The Timer 0 interface is available on pin T0CK1; the same pin as GP2. This is the clock input pin to TMR0. The timer is 8 bits and connects through the data bus.

FLASH PROGRAMMING INTERFACE

The flash programming interface is found on pins ICSPDAT and ICSPCLK. These share pins with GP0 and GP1. As a flash part, the device can be programmed, erased, and reprogrammed hundreds of times without adversely affecting the device.

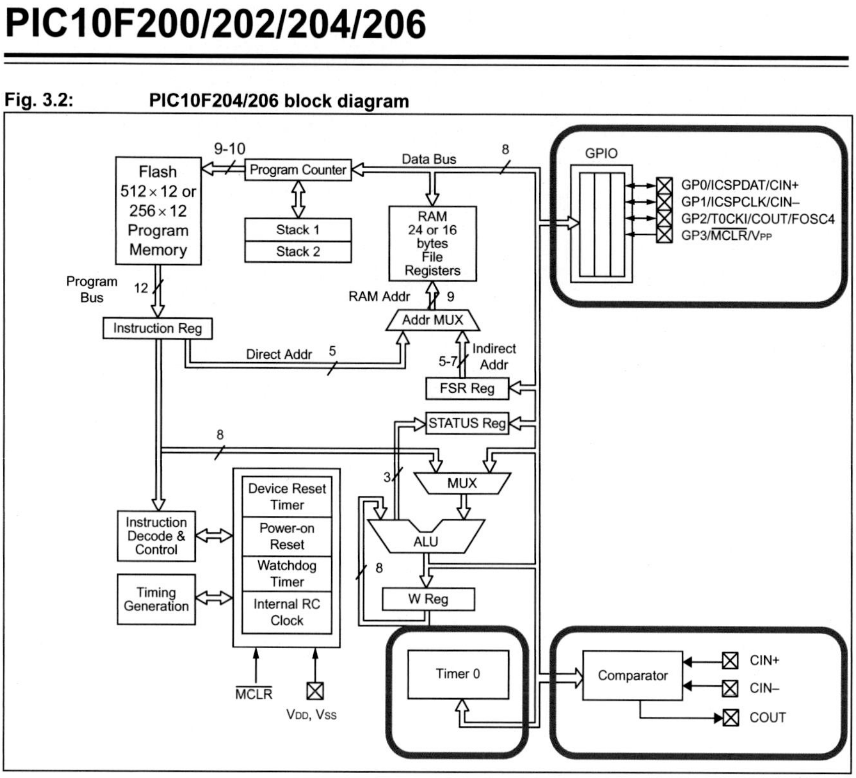

FIG. 2.6

Hardware interfaces.

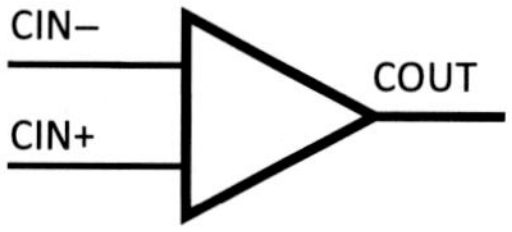

FIG. 2.7

Comparator pinout.

OSCILLATOR OUTPUT

The oscillator is a "divide by 4" output. This oscillator is an on-chip Resistor and Capacitor oscillator that is presented as "divide by 4" to sync with one pulse for each instruction execution; the exception being CALLs and GOTOs that require two instruction cycles.

MASTER CLEAR RESET

The Master Clear Reset is shared with the GP3 input. "When configured as NOT Master Clear (¬MCLR,), this pin is an active-low Reset to the device." It is important to not exceed the voltage on Vdd as the device will enter programming mode (see "Flash Programming Interface" section).

HARDWARE ORGANIZATION OF MEMORY ADDRESS SPACE

The program memory is a 12-bit program bus. The program bus is then split, internally, into an 8-bit program bus and a 5-bit direct address. The 8-bit bus proceeds to the instruction decode and control block. The 5-bit direct address is multiplexed with the 5–7 bit FSR indirect address to access file registers.

DATA MEMORY AND FILE REGISTERS

File Registers are Random Access Memory (RAM) bytes; 16 bytes for PIC10F200/202 and 24 bytes for PIC10F204/206. See Fig. 2.9 for an illustration of Data Memory and File Registers, shown highlighted in a *rounded rectangle*. Access to RAM is much quicker than accessing Data Memory. Therefore it is good practice to use the file registers for arithmetic calculations. The Program Counter and the 2 byte deep Stack are both shown in Figs. 2.8 and 2.9. In Fig. 2.8, one can directly compare the PIC10F200/204 to PIC10F202/206. Note the 8-bit Program Counter for the PIC10F200/202 versus the 9-bit Program Counter for the PIC10F202/206.

The diagram shown in Fig. 2.9 shows the RAM File Registers access off the data bus. Both the direct address from the Instruction Register and the indirect address from the FSR Register are compressed via the multiplexer into access to the RAM File Registers.

ARITHMETIC LOGIC UNIT (ALU) AND THE W REGISTER

The ALU is where the device performs arithmetic and logic instructions. See in Fig. 2.9 the ALU, Multiplexer, and the W register, all highlighted in an *ellipse*. Note that the multiplexer is used to compress inputs going to the ALU. The W register is a fundamental component of the ALU as it is used in most arithmetic or logic instructions. The W register, shown as an output of the ALU, also forms a feedback loop, as an input into the ALU. To access the contents of the registers, see Fig. 2.10, the file register map.

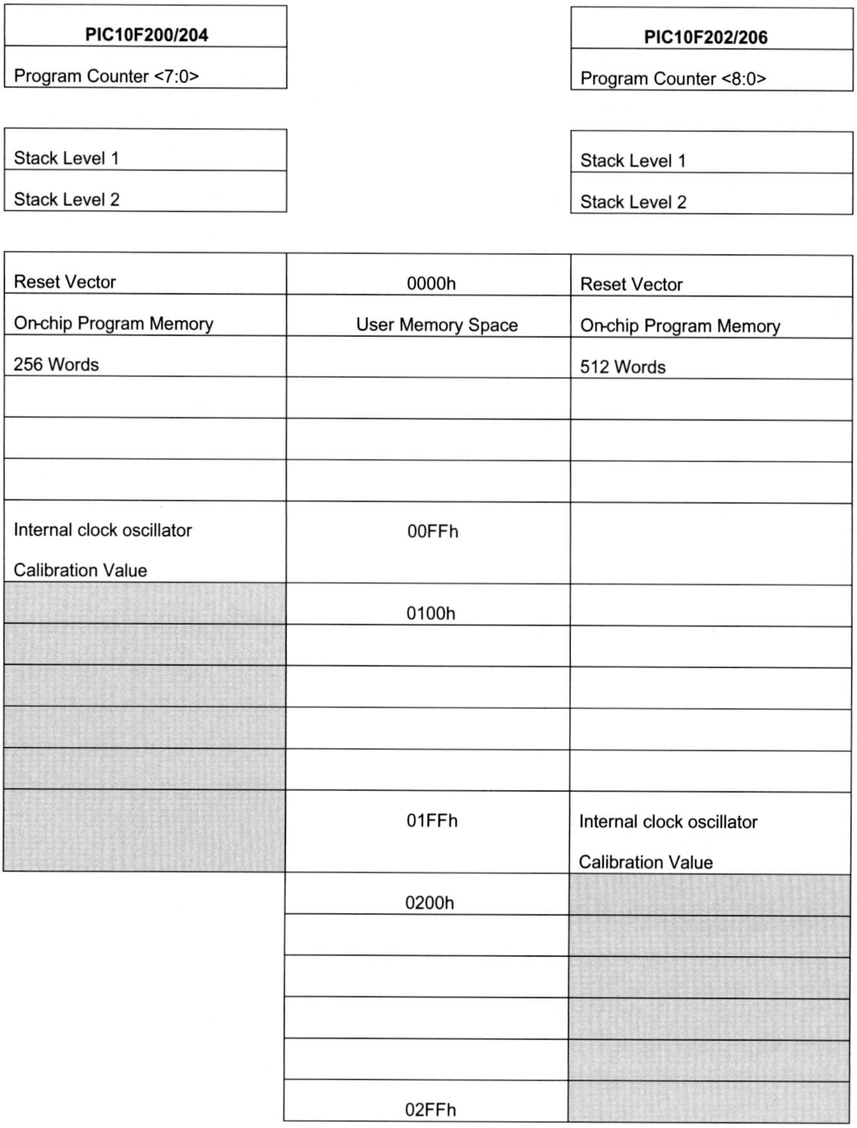

PIC10F200/204
Program Counter <7:0>
Stack Level 1
Stack Level 2
PIC10F202/206
Program Counter <8:0>
Stack Level 1
Stack Level 2
Reset Vector
On-chip Program Memory
256 Words
Internal clock oscillator
Calibration Value
0000h
User Memory Space
00FFh
0100h
01FFh
0200h
02FFh
Reset Vector
On-chip Program Memory
512 Words
Internal clock oscillator
Calibration Value

FIG. 2.8

Illustration of program memory address space.

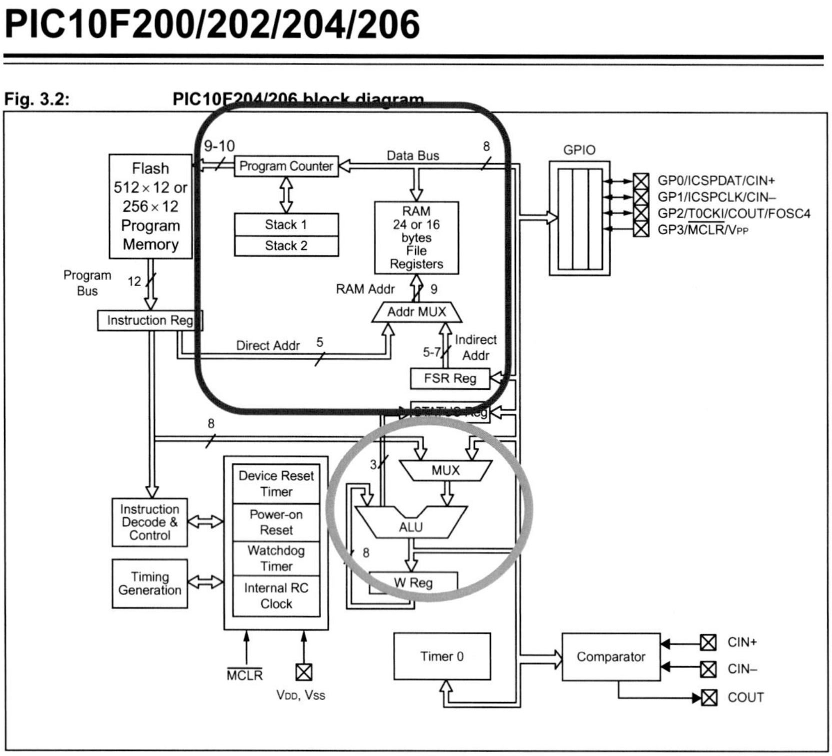

FIG. 2.9

Data memory, arithmetic logic unit, and registers.

PROTOTYPING

To set up a working prototype, there are a couple different approaches. Also, there are a couple of preferred distributors. I recommend using Mouser or Jameco, in the United States, and Farnell, in Great Britain.

Prototyping needs are shown in the following table. Mouser part numbers are supplied where applicable.

Always socket your processor so you can easily replace it if you damage it.

The processor can use an incoming supply as high as 5.5 V and as low as 2 V. Alkaline batteries are a good choice. At 1.5 V each, you can stack three for a 4.5-V supply.

For a simpler and solder-free option, kits are also available through Snap Circuits, an educational supplier for basic electronics. You will need to order a kit with an 8-pin Dual Inline Processor (DIP) socket. The processor of your choice would then be an 8-pin DIP microcontroller available through Microchip directly or as distributed by an electronics supplier such as Mouser Electronics (see Fig. 2.11, list of prototype components to order from an electronics supplier).

PIC10F200/204	Register File Map	PIC10F202/206
INDF	← 00h Not a physical register →	INDF
TMR0	01h	TMR0
PCL	02h	PCL
STATUS	03h	STATUS
FSR	04h	FSR
OSCCAL	05h	OSCCAL
GPIO	06h	GPIO
CMCON0 (204 only)	07h (Unimplemented 200/202)	CMCON0 (206 only)
Unimplemented	08h	General Purpose Registers
	0Fh	
General Purpose Registers	10h	
	1Fh	

FIG. 2.10

File register map.

WORD SEARCH

```
R A T E W Z S C W J D W J W R
R A N D W F W L P L R E L L F
F J B P B Z A R A L V T C D W
M T K N G T P W C Q E O E F D
F F W R O X F D G R D C M W D
R N O P T I E T X S F H I V A
L T F V O M T C Z S I N T O P
C C M W U L A P Z U C R R M L
Z S F C N I W B O F F Q T L P
I U I F I L M T B R C O M F F
D U F O D X H F S T F W R O I
A G R N O P F S L U F F N E R
R L A R G V C S F L B S L G U
W S L E E P B S N W A W C W V
E W Y V P D B L L N G C F P R
```

ADDWF
ANDLW
ANDWF
BCF
BSF
BTFSC
BTFSS
CALL
CLRF
CLRW
CLRWDT
COMF
DECF
DECFSZ
GOTO
INCF
INCFSZ
IORLW
IORWF
MOVF
MOVLW
MOVWF
NOP
OPTION
RETLW
RLF
RRF
SLEEP
SUBWF
SWAPF
TRIS
XORLW
XORWF

A single sided prototype board (requiring soldering) OR a prototyping plastic breadboard; TW-E40-1020	
A socket for the processor (8-pin DIP);	
The processor (10Fxxx) in 8-pin DIP	
Jumper wires for connections; 579-AC163029	
Battery connections and holder; 12BH331-C-GR	
Batteries; 658-LR6XWA	

FIG. 2.11

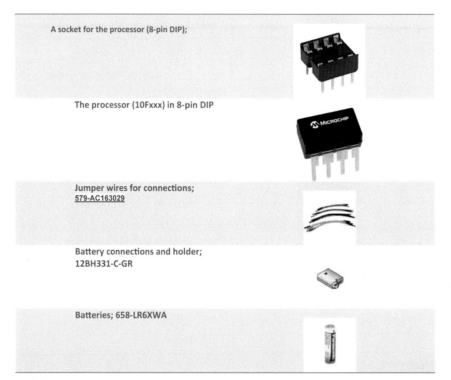

Prototype components.

PUZZLE: OSCILLATOR

Search and find the single instance of the word "Oscillator" spelled correctly.

```
O R Q V J R Y J J O R O Y A O
Z S O J L E O C C O K S P W S
V E C T K T A S T U F S R J C
I E P I I A I E C O J I Q S I
V W G M L L L Z K I D L U J L
Q R X F L L L O I P L L I V L
A O U A I I A A Z K Y A Z M T
U K T C D C Q T C V D T T S O
V O S W I S H B O S N O R O R
R O R S C O L G R R O R S N R
K I R O T A L L I C S A O U O
O S C A L L A T O R J E U B K
J Q T A R O T A L L I S O X C
F A Y D X Y K H B W Q U D J L
P E W F W Z A K M V Q H A B O
```

How to solve Oscillator

This is a word search puzzle with only one word. Search and find the single instance of the word "Oscillator" spelled correctly. Note there are 11 misspelled variants on the word "Oscillator."

FURTHER READING

For an in depth look at other processors, "The 80286 microprocessor, Hardware, Software, and Interfacing" by Walter A. Triebel and Avtar Singh is a good example of a detailed microcontroller textbook. Chapter 6 "The 80286 Microprocessor and Its Memory Interface" is the section most applicable to this chapter.

Again, note the data sheet for the Microchip 10F200/202/204/206 is provided in Appendix E. Peruse the contents of the data sheet, while reading this chapter, as it will help familiarize you with the standard format for Microchip Data sheets. You are also encouraged to print a copy of your specific device's data sheet to use as a reference.

CHAPTER

Instruction sets

3

CHAPTER OUTLINE

INSTRUCTION SET COMPARISON FOR PICMICRO 10F, 12F, 16F, 18F CONTROLLERS

Let us take a look at the various instructions for a small cross-section of the Microchip microcontroller family. There is a common base of 33 instructions that are used in most PICs. Note, see Fig. 3.1, a comparison of several PIC families with additional instructions available in the higher end microcontrollers. Throughout this chapter, the common 33 instructions will be our focus. Learning to manipulate operands with the common core 33 instructions will result in efficient coding and will prove beneficial regardless of the specific processor chosen for your particular project.

COMMON 33 INSTRUCTION SET FOR PICMICRO CONTROLLERS

In this chapter, focus will be on each logical subset of the instruction set. The first set will be the Move, Load, and Store functions. Followed by the Operation and Standby instructions. Next, Clear, Set and Skip; Add and Skip; Subtract and Skip;

The Art of Assembly Language Programming Using PIC® Technology. https://doi.org/10.1016/B978-0-12-812617-2.00003-1

10Fxxx	12Fxxx	16Fxxx	18Fxxx	10Fxxx	12Fxxx	16Fxxx	18Fxxx
		ADDLW	ADDLW	MOVLW	MOVLW	MOVLW	MOVLW
ADDWF	ADDWF	ADDWF	ADDWF	MOVWF	MOVWF	MOVWF	MOVWF
			ADDWFC				MULLW
ANDLW	ANDLW	ANDLW	ANDLW				MULWF
ANDWF	ANDWF	ANDWF	ANDWF				NEGF
			BC	NOP	NOP	NOP	NOP
BCF	BCF	BCF	BCF	OPTION	OPTION		
			BN				POP
			BNC				PUSH
			BNN				RCALL
			BNOV				RESET
			BNZ		RETFIE	RETFIE	RETFIE
			BOV	RETLW	RETLW	RETLW	RETLW
			BRA				
BSF	BSF	BSF	BSF	RETURN		RETURN	RETURN
BTFSC	BTFSC	BTFSC	BTFSC	RLCF			RLCF
BTFSS	BTFSS	BTFSS	BTFSS	RLF	RLF	RLF	
			BTG	RRF	RRF	RRF	
			BZ				RLNCF
CALL	CALL	CALL	CALL				RRCF
CLRF	CLRF	CLRF	CLRF				RRNCF
CLRW	CLRW	CLRW					SETF
CLRWDT	CLRWDT	CLRWDT	CLRWDT	SLEEP	SLEEP	SLEEP	SLEEP
COMF	COMF	COMF	COMF				SUBFWB
			CPFSEQ			SUBLW	SUBLW
			CPFSGT	SUBWF	SUBWF	SUBWF	SUBWF
			CPFSLT				SUBWFB
			DAW	SWAPF	SWAPF	SWAPF	SWAPF
			DCFSNZ				TBLRD*
DECF	DECF	DECF	DECF				TBLRD*-
DECFSZ	DECFSZ	DECFSZ	DECFSZ				TBLRD*+
GOTO	GOTO	GOTO	GOTO				TBLRD+*
INCF	INCF	INCF	INCF				TBLWT*
INCFSZ	INCFSZ	INCFSZ	INCFSZ				TBLWT*-
			INFSNZ				TBLWT*+
IORLW	IORLW	IORLW	IORLW				TBLWT+*
IORWF	IORWF	IORWF	IORWF	TRIS			
			LFSR		TRISGPIO		
MOVF	MOVF	MOVF	MOVF				TSTFSZ
			MOVFF	XORLW	XORLW	XORLW	XORLW
			MOVLB	XORWF	XORWF	XORWF	XORWF

FIG. 3.1

Comparison of 8-bit PICmicro controllers.

Logical Operators; Rotate, Complement, and Swap; Option and TRIS functions; and, finally, the Code Control functions.

INSTRUCTIONS

Instructions are called *Opcodes* for Operation Coding. Nearly all instructions are then followed by *Operands* for Registers and Memory locations upon which the Opcode will operate.

Move, Load, and Store Functions: MOVF, MOVWF, MOVLW

Opcode	Operands	Description	Status Affected
MOVF	f,d	Move f	Z
MOVWF	f	Move W to f	None
MOVLW	k	Move literal to W	None

The Move instructions are used to place values in a specific location or in the W or f registers. The use of k, noted as an operand, represents the use of a constant or literal value. These instructions are used very much akin to Load and Store functions in other processor families. If you need to get a value into a location, or the registers, **this is where you would start**. Often, this is one of the first commands in a program.

Operations and Standby: NOP and SLEEP

Opcode	Operands	Description	Status Affected
NOP	–	No Operation	None
SLEEP	–	Go into Standby mode	$\overline{TO}$, $\overline{PD}$

NOP, for No Operation, is used to effectively clear the *pipeline* from the prior instruction. The *pipeline* is the queue filled with instructions as the processor executes commands. Microchip products typically use one *instruction cycle*, in other words, four clock Q cycles, Q1, Q2, Q3, Q4, to execute an instruction. This is accomplished by first reading an instruction in Q1, followed by executing the prior instruction in Q2, Q3, and Q4. Note that upon reading the very first instruction, there is no prior instruction so the processor effectively executes a No Operation before continuing to then read the next instruction upon Q1. The Opcode NOP can be used during the program to clear the prior instruction, most specifically for use with jumps in the CALL, GOTO, and RETLW instructions; the next instruction will be within the called procedure and not a simple addition of one to the PROGRAM COUNTER.

SLEEP will place the device in Standby mode. The instruction SLEEP will zero the Watchdog Timer and will stop the oscillator.

Clear, Set, and Skip: CLRF, CLRW, CLRWDT, BCF, BSF, BTFSC, BTFSS

Opcode	Operands	Description	Status Affected
CLRF	f	Clear f	Z
CLRW	–	Clear W	Z
CLRWDT		Clear Watchdog Timer	$\overline{TO}$, $\overline{PD}$
BCF	f,b	Bit Clear f	None
BSF	f,b	Bit Set f	None
BTFSC	f,b	Bit Test f, skip if clear	None
BTFSS	f,b	Bit Test f, skip if set	None

CLRF and CLRW are byte-wise instructions. They will clear the affected register, W, or memory location f, and will affect the Z Status bit. CLRWDT will clear and effectively reset the *Watchdog Timer*. The *Watchdog Timer* is used to catch a runaway block of code. If the program is hung, such as with a Goto $, the watchdog will force a stop to the code.

BCF and BSF are bit-wise instructions that will clear or set only a specific bit, b, of the register specified. BTFSC and BTFSS are *bit-wise control operations*. If the specified bit, b, is clear, in BTFSC, then control passes to the instruction after the next instruction, Program Counter plus 2, skipping the next instruction. Likewise, if the specified bit, b, is set, in BTFSS, then control passes to the instruction after the next instruction, Program Counter plus 2, skipping the next instruction.

Note that use of the bit-wise instructions will reduce your code size and reduce inefficiencies. For example, setting a bit in a register is typically done by reading the register, ORing on the bit, and then storing the value back into the register. However, PIC16FXXXX parts all have a bit set instruction that allows you to set individual bits in a single clock cycle.

Add and Skip: ADDWF, INCF, and INCFSZ

Opcode	Operands	Description	Status Affected
ADDWF	f,d	Add W and f	C, DC, Z
INCF	f,d	Increment f	Z
INCFSZ	f,d	Increment f, Skip if 0	None

ADDWF works with the *working register* [W] and file memory location [f]; adding f to W and finally placing it in a *data memory location*, d, say 10h, for example. INCF increments location f by one. INCFSZ increments location f by one [f+1] and then transfers control [skip] to the second instruction following the program counter [PC +2] if f=0 after the increment. If f does not equal zero, then the program counter equals [PC+1] and control stays with the program's normal routine; effectively treating it as a *straight line* program. A straight line program is program that contains no loop structures and flows simply from start to end.

Subtract and Skip: SUBWF, DECF, and DECFSZ

Opcode	Operands	Description	Status Affected
SUBWF	f,d	Subtract W from f	C, DC, Z
DECF	f,d	Decrement f	Z
DECFSZ	f,d	Decrement f, Skip if 0	None

SUBWF works with the working register, W, and f; subtracting W from f and placing it in a data memory location, d. DECF decrements f by one and places the result in d. DECFSZ decrements f by one [f−1] and then transfers control [skip] to the second instruction following the program counter [PC+2] if f=0 after the decrement.

Logical Operators: ANDWF, IORWF, XORWF, ANDLW, IORLW, XORLW

Opcode	Operands	Description	Status Affected
ANDWF	f,d	AND W with f	Z
IORWF	f,d	Inclusive OR W with f	Z
XORWF	f,d	Exclusive OR W with f	Z
ANDLW	k	Bit Set f	Z
IORLW	k	Inclusive OR literal with W	Z
XORLW	k	Exclusive OR literal with W	Z

To understand how the *Logical Operators* work, we engage the use of *truth tables. Truth tables* show, in tabular form, the inputs on the left side columns and the output on the right side column of the table. We will take a look at the AND function, for ANDWF and ANDLW; the inclusive OR function, for IORWF and IORLW; and the XOR exclusive OR function for XORWF and XORLW.

Note that the truth tables, shown here, are set in a specific order that permits **only one bit to change** at a time. This order makes the inputs: 00, or zero; 01, or one; 11, or three; 10, or two. This technique has advantages in making more robust code. If implemented in hardware, this approach has other distinct advantages in preventing spurious output from electromechanical switches and in minimizing the effect of error in conversion of analog signals to digital. This technique is referred to as *reflective binary code (RBC)* or, simply, *Gray code*, after Frank Gray who discovered and patented this technique.

Follows is a detailed Truth Table with Gray Code that elaborates on each input and the value that has changed. The smaller tables group Input 1 and Input 2 into one column.

Changing Values	Input 1	Input 2	Output
	0	0	0
Input 2 changes	0	1	0
Input 1 changes	1	1	1
Input 2 changes	1	0	0

For the AND Truth Table, both inputs need to be one.

AND (Input 1 & 2)	Output
00	0
01	0
11	1
10	0

The AND functions show that, for the output to be one, both inputs (f and W for ANDWF) need to be one. The output needs both inputs to be one, for the output, *destination* d, to be one. For the ANDLW function, the two inputs are, first, the literal k and, second, the working register, W. Both inputs need to be one for the output of ANDLW, the working register, W, to be one. **The zero flag, Z, in the status register will be set to 1 when the output is zero.**

Opcode	Operands	Description	Status Affected
ANDWF	f,d	AND W with f	Z
ANDLW	k	AND literal/constant with W	Z

The "inclusive or" function, IOR, requires that **either Input 1 or Input 2, or both** inputs, must be 1 for the output to be 1.

IOR	Output
00	0
01	1
11	1
10	1

Again, the zero flag, Z, in the status register will be set to 1 when the output is zero.

Opcode	Operands	Description	Status Affected
IORWF	f,d	Inclusive OR W with f	Z
IORLW	k	Inclusive OR literal with W	Z

The "exclusive or" function, XOR, requires that either **Input 1 or Input 2, but NOT both**, must be 1 for the output to be 1.

XOR	Output
00	0
01	1
11	0
10	1

Again, the zero flag, Z, in the status register will be set to 1 when the output is zero.

Opcode	Operands	Description	Status Affected
XORWF	f,d	Exclusive OR W with f	Z
XORLW	k	Exclusive OR literal with W	Z

There are two additional functions that are not represented with an Opcode. For completeness sake, we will take a quick look at the NAND and NOR functions. NAND is simply a **NOT AND** function and the NOR is simply a **NOT IOR**. For the NAND, the output is one, where the AND is zero. For the NOR, the output is zero where the IOR is one. The truth tables are supplied.

Note that there is a unique special case for NAND. In digital logic, there are gates that may be implemented in electronic hardware. **A chain of NAND functions is the only function that can be used, through changes in the inputs, to construct any, or all, of the other gates.** This is left to the reader as a fun exercise. *Hint*: tie inputs together to form a NOT function.

NAND	Output
00	1
01	1
11	0
10	1
NOR	**Output**
00	1
01	0
11	0
10	0

Rotate, Complement, and Swap Words Functions: RLF, RRF, COMF, SWAPF

Opcode	Operands	Description	Status Affected
RLF	f,d	Rotate left f through Carry	C
RRF	f,d	Rotate right f through Carry	C
COMF	f,d	Complement f	Z
SWAPF	f,d	Swap f	None

The purpose of these commands, rotate, complement, and swap is to execute arithmetic functions. More coverage is found in Chapters 9 and 10.

The *rotate commands* specify a direction of rotation, bit by bit through the value. Further, the rotation includes the carry value. To rotate left through the carry, one would specify:

RLF f,d

The destination, d, may also be f.

Note that, in other processors, there are commands referred to as *shifts* instead of *rotates*. The difference is that a shift will not incorporate the Carry bit.

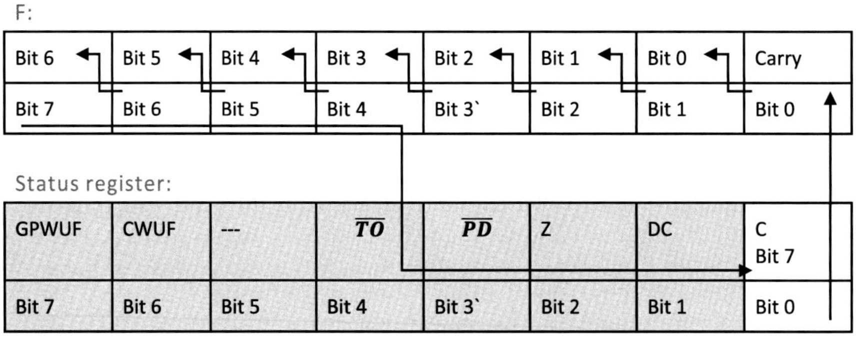

The *complement* of a set gives the elements not in the set. For the instruction,

```
COMF    f,d
```

f is complemented, where for every 0 there is a 1 and for every 1 there is a 0, and the result is then sent to destination d. If f were 11011001b, then the complement of f would be 00100110b. Note that the complement is an arithmetic operation and covered in more detail in Chapters 9 and 10.

The *swap* function causes the upper 4 bits, or upper *nybble*, to be swapped with the lower 4 bits, or lower nybble. If f were 11011001b, viewed as 1101b 1001b, then the result would be 10011101b, or 1001b 1101b.

Option and TRIS Functions: OPTION, TRIS

Opcode	Operands	Description	Status Affected
OPTION	–	Load OPTION_REG register with W	None
TRIS	f	Load TRIS register with f	None

The Option and TRIS functions set up timers, and tristate ports, respectively. The Option function is shown in detail in the following section on registers.

Code Control Functions: CALL, GOTO, RETLW

Opcode	Operands	Description	Status Affected
CALL	k	Call subroutine	None
GOTO	k	Unconditional branch	None
RETLW	k	Return, place literal in W	None

A GOTO instruction changes bits 8-0 of the PC provided by the GOTO Instruction word. The Program Counter Low (PCL) is mapped to PC<7..0>.

For a CALL Instruction, or any instruction where the PCL is the destination, bits 7:0 of the PC again are provided by the instruction word. However, PC<8> does not come from the instruction word, but is always cleared (0).

Instructions where the PCL is the destination, or modify PCL instructions, include MOVWF PC, ADDWF PC, and BSF PC, <5>.

Note: Because PC<8> is cleared in the CALL instruction or any modify PCL instruction, **all subroutine calls or computed jumps are limited to the first 256 locations** of any program memory page (512 words long.)

This information is also presented in the Program Counter Register information. Memory map for calls or jumps is detailed in the Program Counter Register section.

REGISTERS

A *Register* is any, typically 8 bit, easily accessible memory location that holds a value, or address, for use by the instructions. Registers hold addresses of important memory locations. Registers also provide information on how the controller is operating and information on the prior instructions' result, namely, whether zero was the result of the operation. Memory access to registers is generally faster access than general data or program memory. The most used register for the PICmicro is the Working Register, W. These are additional general-purpose registers used within the context of the instructions. The file register, generally located in data memory, is f. As part of the instructions, the destination is called d, but is not designated as a register. Other abbreviations follow. The bit is b. The constant, or literal, is k. The radix value is n.

The general-purpose registers are located at the file address 10h through 1Fh. Depending upon your choice of device, there are additional general-purpose registers from 08h through 0Fh, although use of these specific memory locations is highly discouraged as the use would preclude writing reusable code. Follow along on page 13 of Appendix E. This visually demonstrates the register file map, showing the unimplemented section from 08h to 10h on PIC10F200 and PIC10F204. Further note that INDF is allocated to address 00h, although it is not a physical register.

Note that one may use any memory location as a Register. This can be done in data memory. However, it is not advisable for program memory.

There are also *special function registers* that provide information on the current status of the processor and peripheral functions. These registers common to 8-bit PIC® controllers are as follows: STATUS, OPTION, OSCCAL, PROGRAM COUNTER, INDF, FSR, GPIO, TRIS, and TMR0. Material herein is briefly covered. Please follow along in the data sheet in Appendix E. Note that the presentation herein of the registers follows the organization of the 10F processor registers starting on page 15. Data sheets are structurally similar in presentation for all Microchip processors. If you are using a different processor, you may wish to print out or follow along in the specific processor's data sheet. Page 14 of the 10F data sheet summarizes the special function registers. Appendix F provides excerpts from a more advanced PIC® processor for your review. Look for the summarization of the special function registers.

Status Register

R	W	-	$\overline{T}$	$\overline{P}$	Z	D	C
GPWUF	CWUF	–	$\overline{\textbf{TO}}$	$\overline{\textbf{PD}}$	Z	DC	C
Bit 7	Bit 6	Bit 5	Bit 4	Bit 3	Bit 2	Bit 1	Bit 0

The Status register is also covered in detail on page 15 of the 10F data sheet in Appendix E.

Bit 7: GPWUF Reset GPIO bit
Bit 6: CWUF Wake-up Comparator on Change Flag bit; **Do not use**.
Bit 5: Reserved **Do not use.**
Bit 4: $\overline{TO}$: Time-out bit
Bit 3: $\overline{PD}$:Power-down bit
Bit 2: Z: Zero bit
Bit 1: DC: Digit Carry/$\overline{Borrow}$
Bit 0: C: Carry/$\overline{Borrow}$ *bit*

The *arithmetic logic unit* (ALU) sets (1) or clears (0) the Status register according to the device logic. Bit 5 is an unimplemented bit. To maintain upward code compatibility, you should not use Bit 5. In addition, bit 6 is used only on the PIC10F204/206. For code compatibility, you should not use Bit 6, either. Following along, as we examine the PIC10F data sheet, we find details on the Status register on page 15. Every detail that you will require on this register is available here. There are exceptions to every rule. However, the general gist of this is: **read this register.**

Option Register

$\overline{GPWU}$	$\overline{GPPU}$	TOCS	TOSE	PSA	PS2	PS1	PS0
Bit 7	Bit 6	Bit 5	Bit 4	Bit 3	Bit 2	Bit 1	Bit 0

The Option register is a **write-only** register used to select control bits to configure Timer0/WDT prescaler and Timer0.

Bit 7: $\overline{GPWU}$ Reset GPIO Bit
Bit 6: $\overline{GPPU}$ Enable Weak Pull-ups Bit
Bit 5: TOCS Timer0 Clock Source Select Bit

1 = Transition on TOCK1 pin (overrides TRIS on the TOCK1 pin)
0 = Transition on Internal Instruction Cycle Clock, Fosc/4

Bit 4: TOSE Timer 0 Source Edge Select Bit

1 = Increment on high-to-low transition on the TOCK1 Pin
0 = Increment on low-to-high transition on the TOCK1 Pin

Bit 3: *Prescaler* (PS) Assignment Bit 3

1 = Prescaler Assigned to the Watch Dog Timer (WDT): 0 to 128
0 = Prescaler Assigned to Timer0: 0 to 256

Bit 2: Prescaler Bit 2
Bit 1: Prescaler Bit 1
Bit 0: Prescaler Bit 0

The general gist of this is: **write this register**.

OSCCAL Register

CAL6	CAL5	CAL4	CAL3	CAL2	CAL1	CAL0	FOSC4
Bit 7	Bit 6	Bit 5	Bit 4	Bit 3	Bit 2	Bit 1	Bit 0

The OSCCAL register is a read-write register used to calibrate the internal precision 4-MHz oscillator. Read the OSCCAL register prior to resetting the device so the calibration can be rewritten back to OSCCAL after reset.

Bit 7-1: Oscillator Calibration Bits

0111111 = Maximum frequency
0000001
0000000 = Center frequency
1111111
1000000 = Minimum frequency

Bit 0: FOSC4: INTOSC/4 Output Enable Bit

1 = INTOSC/4 output onto GP2
0 = GPS/T0CK1/COUT applied to GP2

There are exceptions to every rule. However, the general gist of this is: **read this register**.

Program Counter

PC	PCL	PCL	PCL	PCL	PCL	PCL	PCL	PCL
Bit 8	Bit 7	Bit 6	Bit 5	Bit 4	Bit 3	Bit 2	Bit 1	Bit 0

The Program Counter (PC) contains the address of the next program instruction to be executed. The value is increased by one every instruction cycle, unless an instruction specifically changes the PC.

Again, GOTO instruction changes bits 8-0 of the PC provided by the GOTO Instruction word. The Program Counter Low (PCL) is mapped to PC<7..0>.

For a CALL Instruction, or any instruction where the PCL is the destination, bits 7:0 of the PC again are provided by the instruction word. However, PC<8> does not come from the instruction word but is always cleared (0).

Instructions where the PCL is the destination, or modify PCL instructions, include MOVWF PC, ADDWF PC, and BSF PC, 5.

Note: Because PC<8> is cleared in the CALL instruction or any modify PCL instruction, **all subroutine calls or computed jumps are limited to the first 256 locations** of any program memory page (512 words long.)

Memory map for calls or jumps:

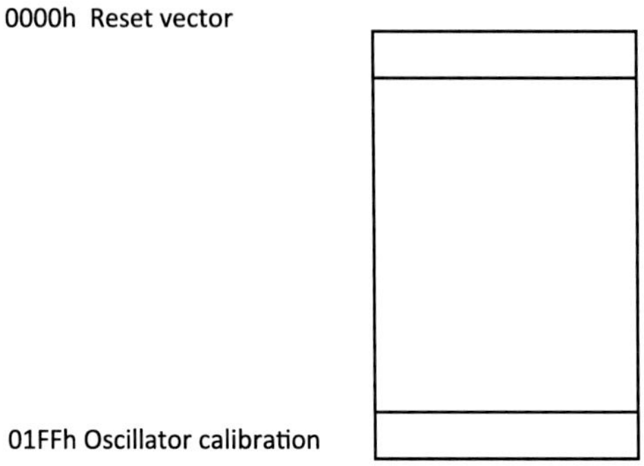

INDF Register

x	**x**	**x**	**x**	**x**	**x**	**x**	**x**
Bit 7	Bit 6	Bit 5	Bit 4	Bit 3	Bit 2	Bit 1	Bit 0

x = unknown.

Note that the INDF register and the FSR register function together. The INDF Register uses the FSR Register to obtain a *pointer* to address Data Memory. From the data sheet, page 19, "The INDF register is not a physical register. Addressing INDF actually addresses the register whose address is contained in the FSR register (FSR is a *pointer*.) This is *indirect addressing*."

FSR Register

1	**1**	**1**	**x**	**x**	**x**	**x**	**x**
Bit 7	Bit 6	Bit 5	Bit 4	Bit 3	Bit 2	Bit 1	Bit 0

x = unknown.

The FSR, File Select Register, and the INDF Register are used together to accomplish Indirect Addressing. The FSR Register is **only 5 bits long,** in the 10F family of processors, with the higher order bits 7 through 5 each set to 1. This is a device characteristic. Again, from the data sheet, "It is used in conjunction with the INDF register to indirectly address the data memory area." See section on addressing, following the registers section.

PORTS

The general-purpose input output (GPIO) pins are controlled by the 4 low-order bits of register GPIO. GPIO performs the function of a dedicated register for the port, such as PORTB, if there were multiple ports. Each port has a corresponding TRIS (for "*tristate*") register, such as TRISB, in our earlier example. TRISB determines the polarity of the port, whether the port is an input or output. In the case of the 10F, a second register, the TRIS register, is used to configure each pin of the port as either an input (logic 1) or an output (logic 0). On more complex processors, there are multiple ports, such as PORTB for reading input, with a correlated TRISB register dictating inputs and outputs.

The register configuration that follows shows GP3 and GP1 and GP0, configured as inputs, with GP2 as an output, in the TRIS register. Note that the PORT or GPIO show what values are active on the PORT, such as 1111, to show an input value of 1 for bit 3, an input of 1 for bit 1 and bit 0, and an output value of 1 for bit 2.

GPIO/PORT

				GP3	**GP2**	**GP1**	**GP0**
Bit 7	Bit 6	Bit 5	Bit 4	Bit 3	Bit 2	Bit 1	Bit 0

TRIS Register

				1	**0**	**1**	**1**
Bit 7	Bit 6	Bit 5	Bit 4	Bit 3	Bit 2	Bit 1	Bit 0

TIMERS

There are two internal timers in the 10F processor: one is the precision oscillator used to step through the program code, and the second is a rough, imprecise RC oscillator. Timer0 uses the TMR0 register to access the 8-bit Real-Time clock counter value. The imprecise oscillator is used by the Watchdog function to detect runaway code and is, therefore, dubbed to be the "Watchdog Timer." The Option register is used to configure the Watchdog prescaler. The Watchdog may be reset by a clear Watchdog function: CLRWDT, which also affects bits $\overline{TO}$ and $\overline{PD}$ of the Status register. A clear Watchdog function should be executed prior to executing any changes to the prescaler inputs of PS2, PS1, PS0 of the Option register. More details on the Watchdog Timer (WDT) may be found on page 40, section 9.6 of Appendix E.

One may also use an external timer. There are very specific details about synchronization with an external clock. Be sure to read Appendix E, section 6.1 on page 24 to obtain the details on synchronization and section 7.1 to use Timer0 with an external clock (available only on devices PIC10F204/206.)

TMR0 Register

Bit 7	Bit 6	Bit 5	Bit 4	Bit 3	Bit 2	Bit 1	Bit 0

Option Register

$\overline{\textbf{GPWU}}$	$\overline{\textbf{GPPU}}$	**T0CS**	**T0SE**	**PSA**	**PS2**	**PS1**	**PS0**
Bit 7	Bit 6	Bit 5	Bit 4	Bit 3	Bit 2	Bit 1	Bit 0

The Option register is a **write-only** register used to select control bits to configure Timer0/WDT prescaler and Timer0.

ADDRESSING

Addressing is simply a method of finding the location of the next instruction. (That's it. Read this three times.)

DIRECT

Direct addressing is where the **memory location's address** is specified as the instructions' operand. The **address** is given in the current instruction. For example, where d is the address of the destination of the instruction.

For example, in the GOTO instruction, the address of the next instruction is given as label NEXT.

```
GOTO    NEXT
```

INDIRECT

Indirect addressing is where the **memory location of the address** is stored in the location specified by the instruction. This is one step removed from direct addressing. Indirect addressing is where the memory location's address, not value, is then specified by the instruction's operand. This is referred to as one level of *indirection*. If the location of the address is further contained in the memory location's address, then this is considered two levels of indirection. In theory, this can be continued indefinitely.

For example, from page 19 in the data sheet, the code first initializes the pointer to 0x10.

```
MOVLW   0x10
MOVWF   FSR
```

Then, the INDF is cleared and the pointer is incremented for the next instruction in a loop.

```
CLRF INDF
INCF FSR,F
```

INDEXED/RELATIVE

Indexed indirect addressing is where the **memory location of the address** is stored in the location specified by the instruction, as with simple indirect. Indexed addressing uses the concept of a pointer. The pointer specifies the list start address with an offset value that indexes off the start value. The index is typically a variable that changes during the program, specifically during loops. This is useful in implementing a *jump table* where **the location is specified by the sum of the indirect address and the index**. In other words, the values are at the location given by [memory location + index].

CODE EXAMPLE

As an example of code, a *bit-sum* operation is shown in code as follows. A bit-sum is simply summing the number of bits that are 1s in an 8-bit value. The value A is used as the incoming variable containing the number of 1s. The value C contains the total number of 1s upon return to the prior procedure.

Label	Opcode	Operand	Comment
BITSUM:	MOVLW	0	;Clear variables B and C
	MOVWF	B	
	MOVWF	C	
	MOVLW	A	;Variable A comes from another process ;upon starting BITSUM:
	MOVWF	D	;Copy variable A to D to keep A's value ;intact during the procedure
LABEL1:	INCF	B	
	BTFSC	1, D	;Checking bit 1 of variable D, skip if clear
	INCF	C	;C holds total count of 1s in A, increment ;upon finding a 1 value
	DECF	D	;D is decremented upon each loop, ;decrement pulls 1s to the right
	BTFSS	4, B	;B holds total loops through the code, ;returns upon 8 (bit 4)
	GOTO	LABEL1	
LABEL2:	RETLW		

PAGING CONSIDERATION

Paging is a means of virtual memory programming. Pages are implemented in code blocks that work with a conceptual framework of pages that appear sequential but can be located, physically, in any order.

If you are working with a processor that uses pages, it is critical that you monitor precisely where your assembled code resides. This can get tricky. The diagram, Fig. 3.2, shows how the page structure works conceptually. In the older Microchip processors, 16C5x series, the programmer, not the assembler, is solely responsible for the location of the code in memory.

Most newer processors manage paging through the assembler. If you are watching, carefully, program memory usage, it may benefit you to watch for the overuse of page specification. (Review Appendix D for Best Practice recommendations from Microchip.) At times, the assembler may add a page specification that is already correct and unnecessary. The opposite may also occur, in other words, NOT specifying the page when it is necessary. The details, shown herein, apply only to the 16C5x series of processor. The concept, however, is important to understand as there are other processors that implement a slightly different mechanism for paging. Always consult the data sheet for your specific processor.

Note that the default page is 0. If you neglect to move the code execution to the next page in memory, you will likely rewrite over your first page of memory. As your code on page 0 expands, you will need to specify when your code is to move to page 1 and later to pages 2 and 3. If you are nearing the end of a page and you are furiously

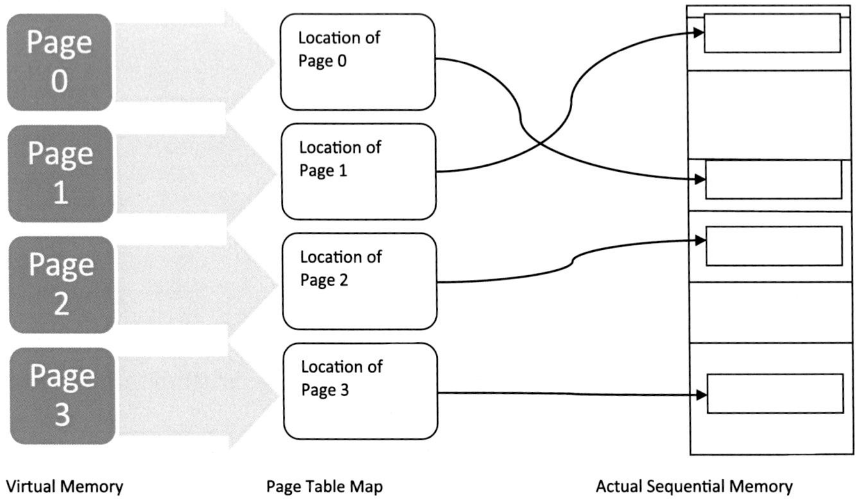

FIG. 3.2

Paging framework.

programming away, you may miss the fact that you just overran a page boundary. These types of errors are difficult to catch. This can also affect your variable locations. If you located some of your variables on page 0 and your program is on page 2, you will need to set your page back to 0, read variable, then set your page back to 2.

WORD SEARCH: INSTRUCTION SET

```
N O P A N D W F N W L Q J F Z
B S A K E H E N S O L L G C F
X T O C W Z P C X G I V A B B
F B F L S M O O F P J T O C T
L S D S W U R F G F U E P M F
Z N E L C L B F W R L E Y O S
A R T F W G L W Q L F V O M S
E E A K V O N F F C P X X X W
R I K J X T J W S L E E P D W
B C O M F O N V I F S R C R S
F W R O I S R O S W T I L I W
Z S F C N I R M I D F C R F A
F W R O X L F N L D R W W T P
M N L K W S C P F A R J D Q F
Y I W D B F T Y A V F O T K O
```

```
ADDWF   INCF    TRIS
ANDLW   INCFSZ  XORLW
ANDWF   IORLW   XORWF
BCF     IORWF
BSF     MOVF
BTFSC   MOVLW
BTFSS   MOVWF
CALL    NOP
CLRF    OPTION
CLRW    RETLW
CLRWDT  RLF
COMF    RRF
DECF    SLEEP
DECFSZ  SUBWF
GOTO    SWAPF
```

PUZZLE: INSTRUCTION SET

Across

5. Move W to f
7. Increment f, Skip if 0
8. Bit Test f, Skip if Set
9. Decrement f, Skip if 0
10. Bit Test f, Skip if Clear
12. Subtract W from f
13. Rotate Left f through Carry
15. Return with literal in W
17. Move f
18. Clear f
19. Exclusive OR W with f
21. Inclusive OR W with f
22. Unconditional Branch
25. AND literal with W
26. Inclusive OR literal with W

Down

1. Add W and f
2. Complement f
3. Increment f
4. Load TRIS Register
6. AND W with f
8. Bit Set f
9. Decrement f
10. Bit Clear f
11. Clear Watchdog Timer
12. Enter SLEEP Mode
14. Swap Nibbles in f
15. Rotate Right f through Carry
16. No Operation
18. Clear W
19. Exclusive OR literal with W
20. Move literal to W
23. Load OPTION Register
24. Subroutine Call

How to solve Crossword Instruction Set

Solve this Crossword by working in two directions. Follow the ACROSS clues and fill in the correct answers across, starting at each numbered box. Follow the DOWN clues and fill in the correct answers downward, starting at each numbered box. Both sets of clues will help you over the difficult spots. This Crossword is dedicated to the 33 Instructions in the Minimal Instruction Set. Each answer is an Instruction word.

FURTHER READING

Programming Structures, Vol. 1, Machines and Programs, 1990, by Jan Hext, takes a detailed mathematically proven approach to program registers and addressing. Hext uses examples in FORTRAN 77, Ada, Pascal, and C. This is an excellent book for historical information as well. It reaches back to the very beginnings of programming as a profession, including program and data memory storage on paper and tape. The basic principles are still relevant today.

Assembly Language, Step-By-Step, 1992, by Jeff Duntemann takes a very basic approach to assembly language programming with fun and interesting quips. Duntemann jumps right into the thick of the mathematical side of assembly in Chapter 1 and the hardware side of assembly in Chapter 2. Basics presented in Assembly Language, Step-By-Step, are more applicable to the Intel 8086.

CHAPTER

Beginning code

4

CHAPTER OUTLINE

SUBROUTINES

Subroutines are blocks of code that may be repeatedly called by the main program to serve a given function. Subroutines are executed during program run time. Following is a data transfer subroutine with the name of "loadAB," which loads the accumulators ACCaHI, ACCaLO, ACCbHI, and ACCbLO with 01FFh and 7FFF, their

The Art of Assembly Language Programming Using PIC® Technology. https://doi.org/10.1016/B978-0-12-812617-2.00004-3

respective values. This subroutine has a label of "loadAB" and ends with a "return." The "call" instruction places the PC+1 value in the stack. The "return" instruction pops the "PC+1" value off the stack and returns control to the main program.

Label	Opcode	Operands	Comments
loadAB	MOVLW	0x01	
	MOVWF	ACCaHI	
	MOVLW	OxFF	
	MOVWF	ACCaLO	;Loads ACCa = 01FF
			;
	MOVLW	0x7F	
	MOVWF	ACCbHI	
	MOVLW	OxFF	
	MOVWF	ACCbLO	;Loads ACCb = 7FFF
	return		

The code excerpt is from Application Note (AN) AN544, Math Utility Routines.

MACROS

Macros are similar to subroutines in that they may be used to execute blocks of code multiple times. Macros differ significantly from subroutines as they are used by the **assembler**, expanding out the code at each instance of the macro. Following is a macro that is called "multiply." This code is then repeated eight separate times in the final machine language code. Note that the only portion of the macro repeated is shown in **bold**. All other "code," shown in the macro, will be absent in the final machine code. Another difference from subroutines is that, if the commands indicate that the macro is to be repeated 0 times, there will be no instance of the macro in the final machine language implementation.

			;Define a macro for adding ;and right shifting
multiply	MACRO		
	variable i		;
	i=0		
.while	i<8		
	BTFSC	**mulplr,i**	
	ADDWF	**H_byte**	
	RRCF	**H_byte**	
	RRCF	**L_byte**	
	i=i+1		;
.endw			
ENDM			;End of macro

The code excerpt is from AN544, Math Utility Routines.

ELEMENTARY PROGRAMS

ACC CONVENTION

Within mathematical routines, there is a general consensus that data memory uses the convention of leading variable names with "ACC" to indicate accumulators. For example, ACCaLO and ACCaHI are used to indicate data memory uses "a" with two words indicated as "LO" or "HI" as LO byte and HI byte, respectively.

The following, allocation of variable memory locations, allows the programmer to use variable names instead of specific address names. The variables begin in location 0x40 or, interchangeably, 40h.

Label	Opcode	Operands	Comments
Start	CBLOCK	0x40	
		ACCaLO, ACCaHI, ACCbLO, ACCbHI	;Allocation of Variable ;Memory Locations
		ACCcLO, ACCcHI,	
		ACCdLO, ACCdHI	
End			

8-BIT DATA TRANSFER

Data transfer is the transfer of a byte of data; copying of a memory location to a second memory location. The MOVE instructions are used to move contents from one memory location to another. This is similar to other processors' Load and Store commands.

Problem: Copy the contents of memory location 40h to location 41h.
Test Case: (0040h) = 5B
Result: (0041h) = 5B

Solution:

Label	Opcode	Operands	Comments
	MOVF	40h, 0	;(40h)→W
	MOVWF	41h	;W→(41h)
	RETLW	00h	;Clear W register & return

Note that we are accessing base page 00 in memory, even though leading zeros have been omitted in the solution. MOVF 40h, 0 could more explicitly show this with MOVF 0040h, 0. The option 0, following the address, indicates that the W register is used. Your program may use "W" in place of 0 if it is declared at the beginning of the program.

The solution ends with RETLW, to return control back to the calling function. Alternatively, a GOTO $ command indicates you are to GOTO the same line as you are already on. The GOTO may be used to place the code in an infinite loop.

16-BIT DATA TRANSFER

Move the contents of 40h, most significant byte (MSB) and 41h, least significant byte (LSB) into 42h (MSB) and 43h (LSB). The MOVE instructions are used to move contents from one memory location to another.

Problem: Copy the contents of memory location 40h to location 42h. Copy the contents of memory location 41h to location 43h.

Test Case: (0040) = 5B
(0041) = 6E

Result: (0042) = 5B
(0043) = 6E

Solution:

Label	Opcode	Operands	Comments
	MOVF	40h, 0	;Get MSB in 40h→W
	MOVWF	42h	;W→42h
	MOVF	41h, 0	;Get LSB in 41h→W
	MOVWF	43h	;W→43h
	RETLW	00h	;Clear W register & return

8-BIT ADDITION

The Addition instructions are used to add two operands, in this problem, (ACCaLO) and (ACCbLO). The result is to store and, therefore, overwrite the second operand. Can you see the negative side to this approach? If you are testing the code, you will need to always reenter the initial value of ADDbLO.

Problem: Add the contents of two memory locations, 0040 and 0041. Place the result in memory location 0041.

Test Case: (0040h) = 27h
(0041h) = 1Ah

Result: (0041h) = 41h

Solution:

Label	Opcode	Operands	Comments
	MOVF	40h, 0	;(40h) →W
	MOVWF	41h	;W → (41h)
	RETLW	00h	;Clear W register & return

8-BIT SUBTRACTION

The Subtraction instruction is used to subtract one operand from another. In this problem, (ACCaLO) is subtracted from (ACCbLO). The result is in memory location ACCbLO.

Problem: Subtract the contents of memory location ACCaLO from location ACCbLO. Place the result in memory location ACCbLO.
Test Case: (ACCaLO) = 27h
(ACCbLO) = 1Ah
Result: (ACCbLO) = 0Dh

Solution:

Label	Opcode	Operands	Comments
	MOVF	ACCaLO,0	;ACCaLO →W
	SUBWF	ACCbLO,1	;Subtract W from ACCbLO and store in ACCbLO
	RETLW	00h	;Clear W register & return

8-BIT MULTIPLY BY TWO (ROTATE LEFT 1 BIT)

This example illustrates the rotation of ACCbLO as a one-byte operation. Rotate left through Carry places the most significant bit in the Carry position. Also, a zero is placed in the least significant bit position.

Problem: Multiply ACCbLO by two.
Test Case: (ACCbLO) = 5Eh = 01011110b
Result: (ACCbLO) = BCh = 10111100b Carry remains 0.

Solution:

Label	Opcode	Operands	Comments
	RLF	ACCbLO,1	;Rotate Left Thru Carry
	RETLW	00h	;Clear W register & return

8-BIT MULTIPLY BY FOUR (SHIFT LEFT 2 BITS)

This example illustrates the rotation of ACCbLO as a one-byte operation. Rotate left through Carry places the most significant bit in the Carry position. Also, a zero is placed in the least significant bit position.

Problem: Multiply ACCbLO by four.
Test Case: (ACCbLO) = 5Eh = 01011110b
Result: (ACCbLO) = 78h = 01111000b Carry is 1.

Solution:

Label	Opcode	Operands	Comments
	RLF	ACCbLO,1	;Rotate Left Thru Carry
	RLF	ACCbLO,1	;Rotate Left Again Thru Carry (MSB, in Carry ;from first Opcode, will be truncated)
	RETLW	00h	;Clear W register & return

8-BIT DIVIDE BY TWO (ROTATE RIGHT 1 BIT)

This example illustrates the right rotation of ACCbHI to perform a divide by two. The least significant bit is truncated.

Problem: Divide ACCbHI by two.
Test Case: (ACCbHI) = 5Eh = 01011110b
Result: (ACCbHI) = 2Fh = 00101111b

Solution:

Label	Opcode	Operands	Comments
	RRF	ACCbHI,1	;Rotate Right Thru Carry and store result in file register ;ACCbHI, Note Carry is pulled into the MSB position
	RETLW	00h	;Clear W register & return

MASK OFF MOST SIGNIFICANT 4 BITS

Problem: Mask off most significant 4 bits
Test Case: (ACCbHI) = 5Eh = 01011110b
Result: (ACCbHI) = 0Eh = 00001110b

Solution:

Label	Opcode	Operands	Comments
	MOVF	40h,0	;Get data from the content of (40h) into W register
	ANDWF	‘00001111’b,0	;Mask 4 most significant bits and return solution ;to file register W
	MOVWF	40h	;Return value to location 40h
	RETLW	00h	;Clear W register & return

MASK OFF LEAST SIGNIFICANT 4 BITS

Problem: Mask off least significant 4 bits
Test Case: (ACCbHI) = 5Eh = 01011110b
Result: (ACCbHI) = 50h = 01010000b

Solution:

Label	Opcode	Operands	Comments
	MOVF	40h,0	;Get data from the content of 40h into W register
	ANDWF	‘11110000’b,0	;Mask 4 least significant bits and return solution ;to working register W
	MOVWF	40h	;Return value to location 40h
	RETLW	00h	;Clear W register & return

CLEAR A MEMORY LOCATION

Problem: Clear memory location
Test Case: (40h) = 5Eh = 01011110b
Result: (40h) = 00h = 00000000b

Solution:

Label	Opcode	Operands	Comments
	CLRF	40h	;Clear register f and set zero bit Z, in Status ;register, to 1
	RETLW	00h	;Clear W register & return

SET A MEMORY LOCATION TO ALL ONES

Problem: Set a memory location to all ones
Test Case: (40h) = 5Eh = 01011110b
Result: (40h) = 00h = 00000000b

Solution:

Label	Opcode	Operands	Comments
	MOVF	40h,0	;Get data from the content of 40h into W register
	IORLW	'11111111'b	;Inclusive OR with FFh to set Working register to ;all ones
	MOVWF	40h	;Reset 40h to all ones
	RETLW	FFh	;Return with all ones

FIND LARGER OF TWO NUMBERS

Problem: Find the larger of two numbers
Test Case: (40h) = 5Eh = 01011110b
(41h) = 6Ch = 01101101b
Result: (41h) holds the larger number

Solution:

Label	Opcode	Operands	Comments
	MOVF	40h,0	;Get data from the content of 40h into W register
	SUBWF	41h,0	;
	BTFSS	STATUS,0	;N=1 Negative
	RETLW	40h	;Contents of 40h is the Larger Number
	BTFSS	STATUS,2	;Z=1 Zero
	RETLW	00h	;0 for numbers that are equal
	RETLW	41h	;Contents of 41h is the Larger Number

FIND SMALLER OF TWO NUMBERS

Problem: Find the smaller of two numbers
Test Case: (40h) = 5Eh = 01011110b
(41h) = 6Ch = 01101101b
Result: (40h) holds the smaller number

Solution:

Label	Opcode	Operands	Comments
	MOVF	40h,0	;Get data from the content of 40h into ;W register
	SUBWF	41h,0	;Subtraction sets Status Bits
	BTFSS	STATUS, 0	;N=1 Negative
	RETLW	41h	;Contents of 41h is the Smaller Number
	BTFSS	STATUS, 2	;Bit Test Z=1 Zero
	RETLW	00h	;0 for equality of numbers
	RETLW	40h	;Contents of 40h is the Smaller Number

PROGRAM COUNTER ADDRESSING

Simply, the Program Counter (PC) stores the address of the next instruction. A standard within microchip processors includes the ability to alter the Program Counter. Using this ability, one can implement subroutines, interrupts, stacks, and jump tables.

APPLICATION NOTES

Applications Notes (AN) cover numerous software techniques for any given microcontroller. Microchip's website has a search function where one can locate ANs on various topics. Herein, Application Notes are used to explain interrupts and jump tables. Following use of an AN, the note number is given.

INTERRUPTS AND JUMP TABLES (AN514)

"To create the interrupt subroutine jump table, each I/0 condition may have its own unique subroutine to respond to changes on the interrupt lines. The interrupt conditions are determined by detecting changes on the I/O lines that have been selected to be the interrupt lines. Direct access to these routines is achieved by using the microcontrollers' ability to change the program counter under software control. Here is an example of how two I/O lines may be polled:" [1]

Label	Opcode	Operands	Comments
	MOVF	CONDTN, W	;Load I/O condition into W
	ANDLW	3	;Mask off top 6 bits
	ADDWF	2,1	;Add input to PC
			;to create jump table
	GOTO	MAIN	;For no change go to Main
	GOTO	INT1	;For change in Bit0 GOTO INT1
	GOTO	INT2	;For change in Bit 1GOTO INT2
	GOTO	INT3	;For both change GOTO INT3

CONDTN is moved into the W register. To focus on the lower two bits, the ANDLW and 3 with the W register (CONDTN), masking off the top six bits. The two lower bits determine the jump address. Using ADDWF, add the interrupt input to the Program Counter. The PC value is then, for 0, MAIN; for 1, INT1; for 2, INT2; for 3, INT3.

Jump tables are a very efficient way to route control, using the PC, to different code locations.

STACKS (AN534)

We are going to digress for a moment. Since the 10F has only a two deep stack, we will be using a PIC17C42 that has a 16-level deep stack to illustrate the use of PUSH and POP instructions. Note that another Application Note AN527, Software Stack Management, implements a "software stack" that is five levels deep on a processor, one like PIC10F, that has a hardware stack that is only two deep.

Label	Opcode	Operands	Comments
Main_prog	SETF	FSR0	;Initialize & dedicate FSR0 as stack pointer
	BCF	ALUSTA, FS0	
	BCF	ALUSTA, FS1	;Set-up FSR0 for auto-dec
PUSH	MACRO		
	BCF	ALUSTA, FS0	;Set-up FSR0 for auto-dec
	MOVFP	ALUSTA, INDO	;Save ALUSTA first
	MOVFP	BSR, INDO	;
	MOVFP	W, INDO	;
	MOVFP	RAM_x, INDO	;Now save general purpose registers
	MOVFP	RAM_y, INDO	;
	ENDM		;
POP	MACRO		;
	BSF	ALUSTA, FS0	;Set-up for auto-inc
	INCF	FSR0	;

Continued

Label	Opcode	Operands	Comments
	MOVPF	INDO, RAM_Y	;
	MOVPF	INDO, RAM_X	;
	MOVPF	INDO, W	;
	MOVPF	INDO, BSR	;
	MOVFP	INDO, ALUSTA	;Restore ALUSTA last
	DECF	FSR0	;Adjust stack pointer
	ENDM		;
Interrupt_ routine	PUSH		;Save registers
;Main body			;interrupt service
	POP		;Restore status
	RETFIE		;Return

The code excerpt is from AN534, Saving and Restoring Status on Interrupt.

CALCULATE TIMING

Code execution timing is calculated in units called Million Instructions per Second, or MIPS. Timing is calculated with the following expression:

$$\begin{aligned}\text{CPU execution (time in seconds)} &= (\text{instructions / program}) \\ &\times (\text{clock cycles / instruction}) \\ &\times (\text{seconds / clock cycle})\end{aligned}$$

WORD SEARCH: BEGINNING CODE

S	H	S	P	A	R	E	N	T	H	E	S	E	S	P
F	R	H	N	X	M	S	X	Y	A	G	A	T	C	R
E	S	O	G	O	A	I	L	S	A	K	I	N	I	O
N	X	N	S	M	I	B	N	D	E	N	C	E	T	G
O	T	A	P	S	M	T	D	I	D	B	T	R	A	R
I	N	L	M	E	E	R	N	I	M	L	D	E	M	A
T	E	O	S	P	E	C	R	E	U	U	X	F	E	M
U	V	S	M	S	L	E	O	S	V	W	M	F	H	M
L	A	B	S	M	C	E	E	R	J	N	E	I	T	I
O	S	I	V	T	O	R	S	T	P	L	O	D	A	N
S	N	O	I	T	A	C	I	L	P	P	A	C	M	G
G	Y	R	A	T	N	E	M	E	L	E	N	Y	U	L
L	M	R	N	H	C	O	N	T	E	N	T	S	B	O
T	E	S	T	C	A	S	E	S	M	E	T	S	Y	S
D	Z	E	G	A	U	G	N	A	L	M	G	O	N	W

ADDRESSING
APPLICATIONS
ASSEMBLY
COMMON
CONTENTS
CONVENTIONS
DIFFERENT
ELEMENTARY
EXAMPLES
INDIRECT
LANGUAGE
MATHEMATICS
MINIMUM
PARENTHESES
PROCESSORS
PROGRAMMING
RESULT
SAMPLE
SOLUTION
SYSTEMS
TESTCASE

PUZZLE: NUMBER SQUARE

Use order of operations in decimal

7		5		1	35
9		4		3	8
2		8		6	50
61		17		−8	

How to solve Number Square

The Number Square is to be filled in, in each empty square, with an operator. The operator may be any one of the following four: +, for addition; -, for subtraction; x, for multiplication; or, ÷, for division. The puzzle is given in decimal numbers. The rules for mathematical Order of Operations must be followed to complete the Number Square correctly.

FURTHER READING

6502 Assembly Language Programming and 6809 Assembly Language Programming both by Lance A. Leventhal.

Application Notes from Microchip's Embedded Control Handbook 1994/95:

AN527 Stack Management

AN534 Saving & Restoring Status on Interrupt

AN581 Long Calls

AN544 Math Utility Routines

Computer Organization & Design by David A. Patterson and John L. Hennessy

REFERENCE

[1] Anon, AN514 Software Interrupt Techniques, n.d.

CHAPTER

Looping code

5

CHAPTER OUTLINE

LOOPS INTRODUCTION

A *program loop* is a program structure that causes a segment of code to repeat a fixed number of times. There are four basic segments of code to each loop structure. See program loop flowcharts: Fig. 5.1 (minimum one loop) and Fig. 5.2 (minimum zero loops).

1. Initialize
2. Process
3. Control
4. Conclude

Initialization is where the setup values are established. These include the start values of counters, pointers, indexes, and other variables. Counters allow the program to step through the Process section a fixed number of times. Pointers and Indexes allow the program to step through sections of addressed space, essentially counting a fixed number of address locations, searching for the end position.

Process is the block of instructions that are to be repeated in the *Program Loop*. An *iteration* describes one loop of the Process block. There will be a maximum number of iterations. However, depending on how the program loop is structured, the minimum number of loops may be either one or zero. This is based upon whether the test value (used to count iterations) is either at the beginning of the process structure or at the end.

Loop Control is the block of instructions that test whether and when the program loop should repeat. Loop Control monitors and updates the counters or pointers in the program loop.

The Art of Assembly Language Programming Using PIC® Technology. https://doi.org/10.1016/B978-0-12-812617-2.00005-5

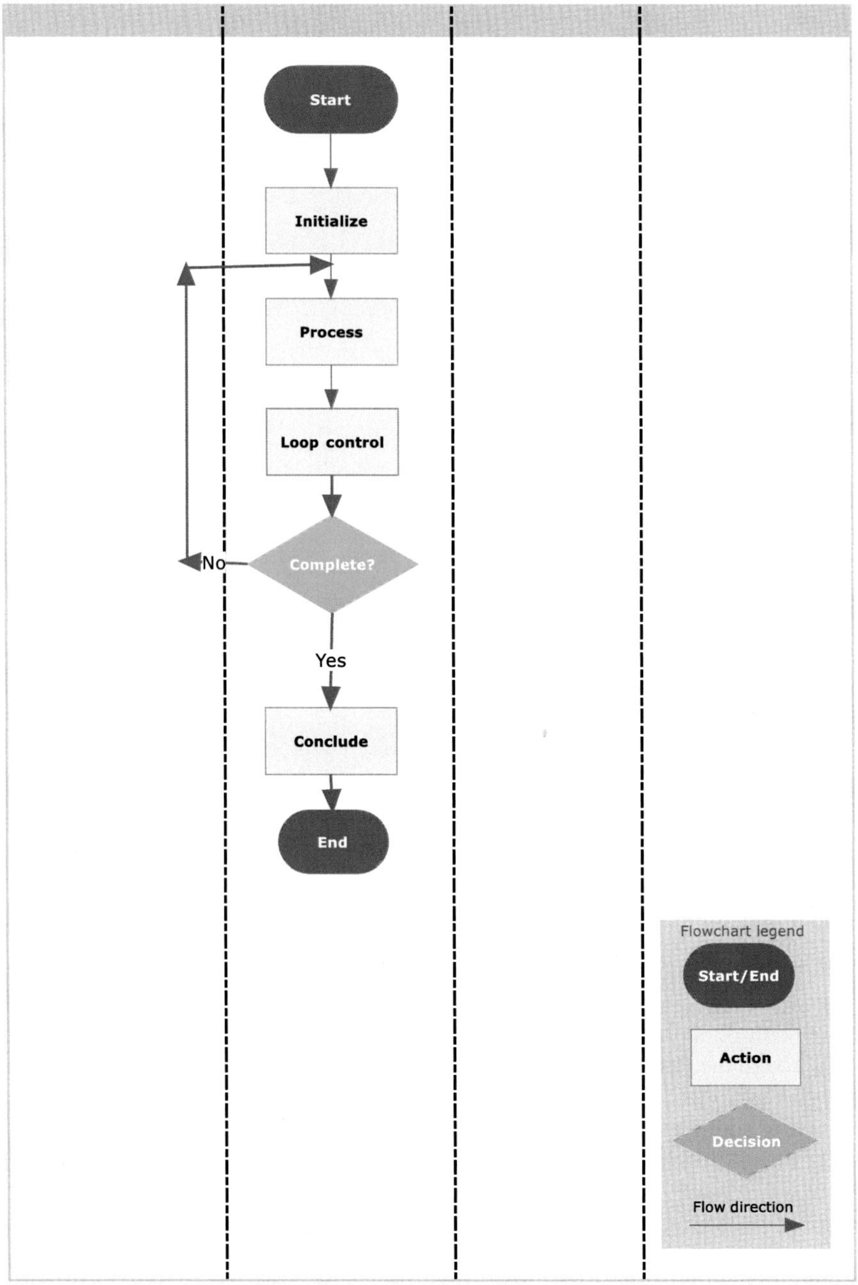

FIG. 5.1

Program loop (with minimum one loop) flowchart.

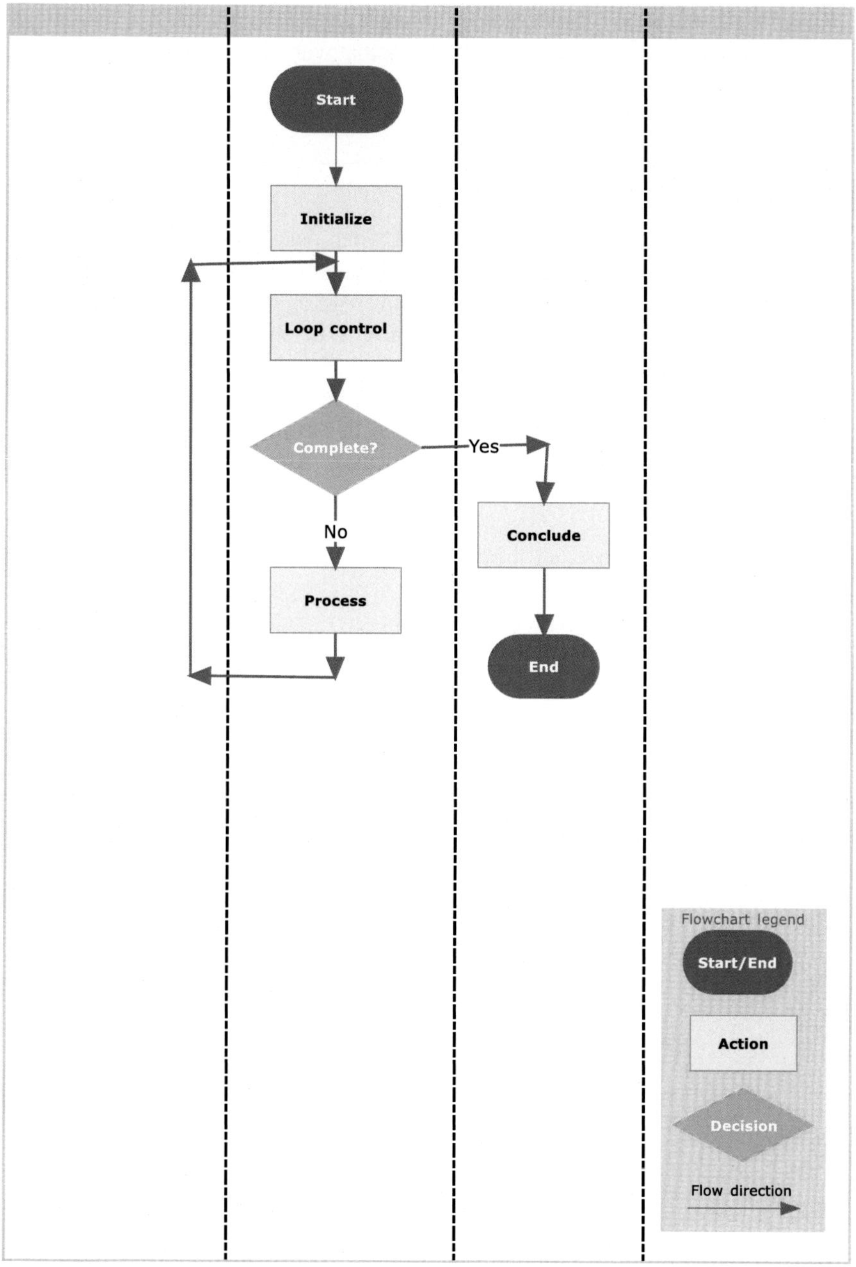

FIG. 5.2

Program loop (minimum zero processes) flowchart.

Conclude is the final section of code in a program loop. Conclude contains any analysis on the data or any storage of variables that were obtained from the program loop iterations.

8-BIT SUM OF DATA

The 8-bit sum of data routine may be useful when one requires a calculation of a group of data for any financial reason, that is, for computing hours worked in a week, or for determining calculation values for an average of sensor inputs, of which the sum of data is a first step followed by division to determine the mean value. See Fig. 5.3 for sum of data routine.

Problem:

Calculate the sum of a series of numbers. In this example, the array is five units long. The series' length is given in memory location 0041. The series starts in location 0042. Store the calculated sum in 0040.

Limitations:

Presume that the sum is an 8-bit, unsigned, hexadecimal number so that you can ignore carries. Further, the length of the series is greater than or equal to one. Note that a value contained within parenthesis refers to the *contents of* the memory location addressed by the value. Recall, on Microchip 8-bit processors, the value addressed may be either a 12-bit or 16-bit memory location, whereas the *contents of* the memory location are an 8-bit value.

Test Case and Result:

Memory Location Value	Contents of Location	Comment
(0×10)	=E7h	Sum (=05h+38h+25h+16h+23h+51h)
(0×11)	=05h	Number of data bytes (5)
(0×12)	=38h	Data byte 1
(0×13)	=25h	Data byte 2
(0×14)	=16h	Data byte 3
(0×15)	=23h	Data byte 4
(0×16)	=51h	Data byte 5

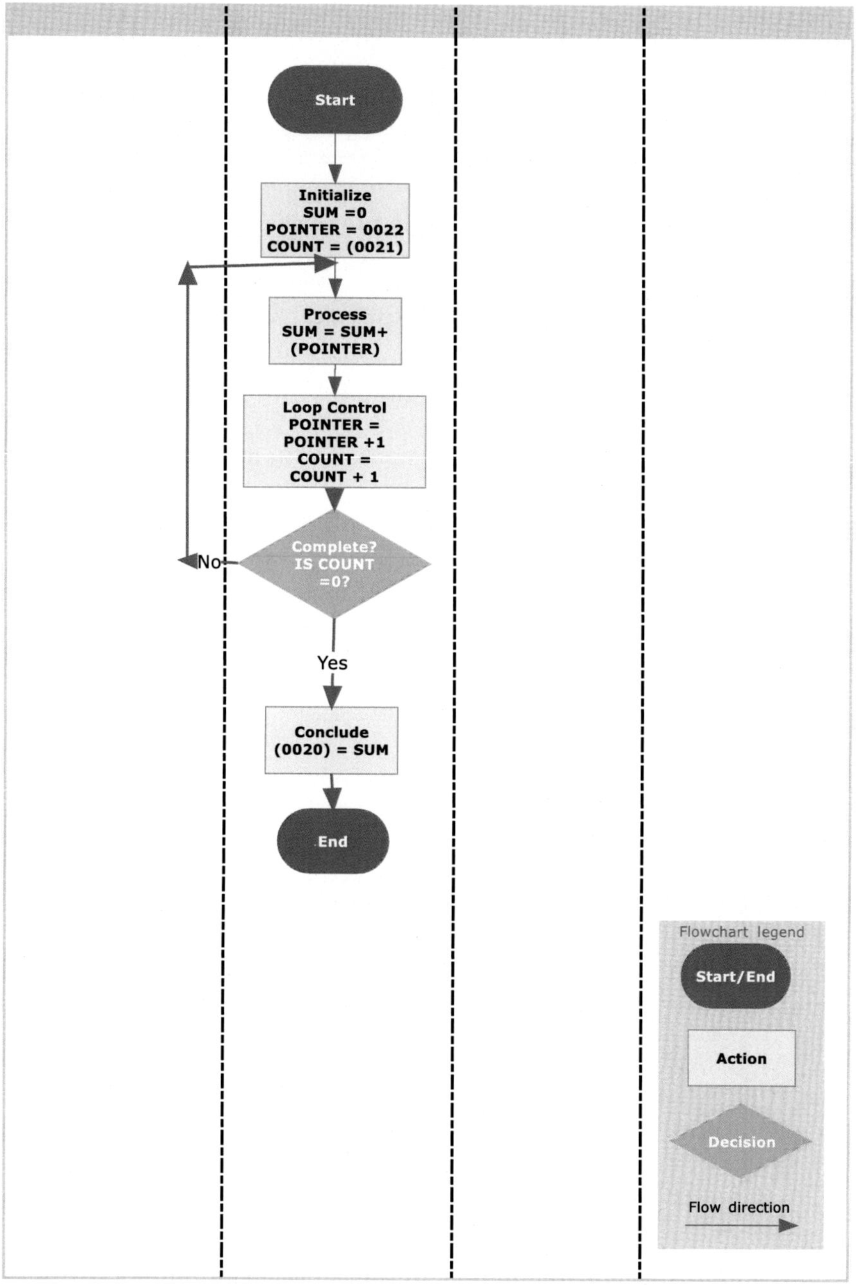

FIG. 5.3

Program loop, sum of data, flowchart.

Solution:

```
          #include   p10f200.inc
res_vect                          0x0000    ;processor reset vector
          GOTO       start        ;go to beginning code vector
count     EQU        0x11
sum       EQU        0x10
                                  ;main_prog code
```

Label	Opcode	Operand	Comment
start			;array elements placed
	MOVLW	0x00	;sum
	MOVWF	0x10	
	MOVLW	0x05	;number of data bytes (5)
	MOVWF	0x11	
	MOVLW	0x38	;data byte 1
	MOVWF	0x12	
	MOVLW	0x25	;data byte 2
	MOVWF	0x13	
	MOVLW	0x16	;data byte 3
	MOVWF	0x14	
	MOVLW	0x23	;data byte 4
	MOVWF	0x15	
	MOVLW	0x51	;data byte 5
	MOVWF	0x16	
	CLRF	sum	;clear sum
	MOVLW	12h	;start of array
	MOVWF	FSR	;pointer fsr is a 12h, start of loop
loop	MOVF	indf, 0	;move element of array into w
	ADDWF	sum, 1	;add array element to sum
	INCF	fsr, 1	;increment pointer
	DECF	count, 1	;decrement count & store back into count
	BTFSS	status, 2	;zero bit (at bit location 2) is 1 if count is 0
	GOTO	loop	;if not zero then loop in routine.
	GOTO	$	;loop forever onto itself
	END		

LOOPED CODE EXAMPLE

Follows is an example of looped code. There is a major loop defined as loop16a. There is also a call to a subroutine, shown in blue typeset, that is invoked several times within loop16a.

```
;*************************************************************
;                    Binary to BCD Conversion Routine (16 bit)
;                    Looped Version
;                    This routine converts a 16 Bit binary number to a 5 BCD
                     number.
;                    Performance:
;                                        Program Memory: 32
;                                        Clock Cycles: 750
;*************************************************************
```

Label	Opcode	Operand	Comment
			;B2_BCD_Looped
	BCF	_fs0	
	BSF	_fs1	
	BCF	_carry	
	CLRF	count	
	BSF	count, 4	
	CLRF	R0	
	CLRF	R1	
	CLRF	R2	
			;
			;loop16a
	RLCF	Lbyte	
	RLCF	Hbyte	
	RLCF	R2	
	RLCF	R1	
	RLCF	R0	
	DCFSNZ	count	
	RETURN		
			;
			;adjDEC
	MOVLW	R2	
	MOVWF	fsr0	
	CALL	adjBCD	
			;
	INCF	fsr0	
	CALL	adjBCD	
			;
	INCF	fsr0	
	CALL	loop16a	
			;Subroutine
			;adjBCD
	MOVFP	indf0,wreg	
	ADDLW	0x03	

Continued

Label	Opcode	Operand	Comment
	BTFSC	wreg, 3	
	MOVWF	indf0	
	MOVFP	indf0, wreg	
	ADDLW	0x30	
	BTFSC	wreg, 7	
	MOVWF	indf0	
	RETURN		

PARTIAL STRAIGHT LINE CODE EXAMPLE

Label	Opcode	Operand	Comment
			;B2_BCD_Straight
	BSF	_fs0	
	BSF	_fs1	
			;
	BCF	_carry	
	CLRF	count	
	BSF	count, 4	
	CLRF	R0	
	CLRF	R1	
	CLRF	R2	
			;loop16b
	RLCF	Lbyte	
	RLCF	Hbyte	
	RLCF	R2	
	RLCF	R1	
	RLCF	R0	
			;
	DCFSNZ	count	
	RETURN	;DONE	
			;
	MOVLW	R2	;load R2 as indirect addr
	MOVWF	fsr0	
			;adjustBCD
	MOVFP	indf0, wreg	
	ADDLW	0x03	
	BTFSC	wreg, 3	;test if result > 7
	MOVWF	indf0	
	MOVFP	indf0, wreg	
	ADDLW	0x30	

Label	Opcode	Operand	Comment
	BTFSC	wreg, 3	;test if result > 7
	MOVWF	indf0	;save as MSD
			;
	INCF	fsr0	
			;adjustBCD
	MOVFP	indf0, wreg	
	ADDLW	0x03	
	BTFSC	wreg, 3	;test if result > 7
	MOVWF	indf0	
	MOVFP	indf0, wreg	
	ADDLW	0x30	
	BTFSC	wreg, 3	;test if result > 7
	MOVWF	indf0	;save as MSD
			;
	INCF	fsr0	
			;adjustBCD
	MOVFP	indf0, wreg	
	ADDLW	0x03	
	BTFSC	wreg, 3	;test if result > 7
	MOVWF	indf0	
	MOVFP	indf0, wreg	
	ADDLW	0x30	
	BTFSC	wreg, 3	;test if result > 7
	MOVWF	indf0	;save as MSD
			;
	GOTO	loop16b	

In comparison to the fully looped code, the outer loop of the prior segment of code is maintained here as well. However, the three instances of adjustBCD are no longer called. The adjustBCD loops are repeated three times. Instead of calling as a subroutine, the instructions are simply repeated which makes a clear difference in the number of clock cycles. This *partial-straight-line* code reduces the clock cycles from 750 to 572! Note that, instead of a "partial" straight line code, you could feasibly create a fully straight line code by copying the entire loop contents (loop16b) the full number of executions, creating a "long" program that would execute more quickly.

MACROS FOR PAGE AND BANK SWITCHING (AN586)

There are two application notes that explain in detail macros for page and bank switching (AN586) and implementing long calls (AN581). These apply only to an older series of processors known as 16C5x series. See Chapter 3, Fig. 3.2, for details on paging consideration.

LONG CONDITIONAL BRANCH VECTORS (AN581)

There are special rules for dealing with multiple pages and in dealing with calling subroutines. Again, the concern is specifically for the older series 16C5x. Subroutines, on these specific devices, may only be called in the first 256 bytes of the program page. Long conditional branch vectors are one way to deal with these device limitations.

Recall from Chapter 3, in Fig. 3.2, Paging Framework, there are special rules for dealing with multiple pages and in dealing with calling subroutines. Subroutines may only be called in the first 256 bytes of the program page. Long conditional branches are one way to deal with these device limitations.

There are two application notes that explain in detail **macros for page and bank switching (AN586)** and implementing long calls (AN581). I recommend that you look up these application notes and follow along. Details, specifically from AN581, follow. The use of vectors allows the subroutine to be anywhere within the program memory page, not just the first 256 bytes.

Three concepts used in implementing long calls are as follows:

1. A CALL loads the entire PC on the stack.
2. A GOTO does not affect stack.
3. A GOTO can branch to any location in a program memory page.

To select the desired page, RP1 and RP0 are employed from the Status register, bit 6 and 5 (Status<6:5>). These two bits need to be programmed accordingly. See Fig. 5.4 for program counter structure. They do not get loaded into A10:A9 of the Program Counter (PC) until one of the following occurs:

1. A CALL instruction.
2. A GOTO instruction.
3. An instruction modifies the PC register PC<:7:0> such as ADDWF PC, F or RETLW instruction POPs the stack, or the stack contains the PUSHed PC from the CALL instruction.

The program developer places "call vectors" at the first 256 words of each page (see Fig. 5.5).

A10	A9	A8	A7:A0
From RP1	From RP0	See below	From PC or instruction

1. Carry from PC
2. From GOTO instruction
3. Forced to 0 by CALL instruction or Instruction with PC as destination

FIG. 5.4

Program counter structure.

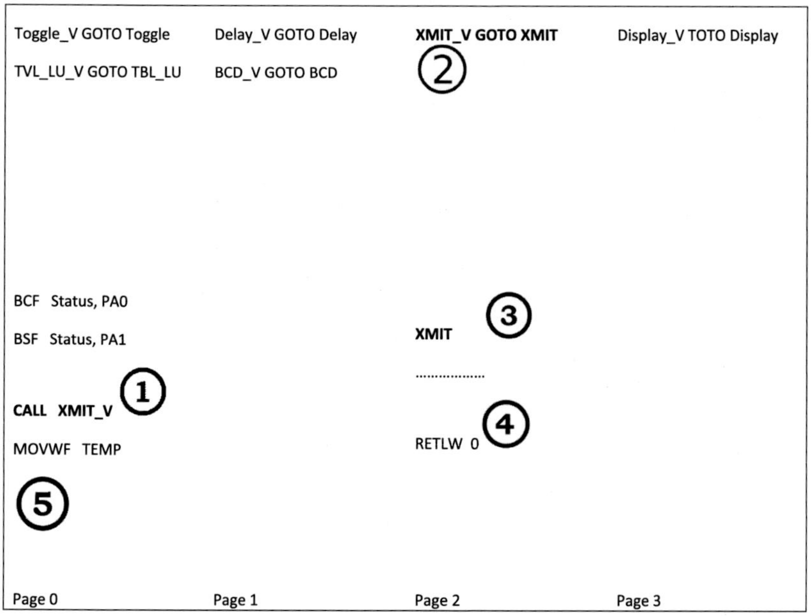

FIG. 5.5

Page vectors.

WORD SEARCH: LOOPS CODE

M	J	Q	W	A	S	J	V	R	F	H	S	M	D	A
U	A	C	M	S	P	M	V	A	I	T	E	U	H	F
L	G	R	I	J	N	P	W	H	R	J	E	S	D	T
N	O	L	G	I	J	C	L	U	R	D	W	A	J	B
T	B	O	B	O	K	N	C	I	K	B	H	T	S	V
E	E	L	P	J	R	T	J	F	C	I	C	A	R	E
N	O	T	E	S	U	P	G	L	P	A	G	D	E	D
P	R	O	G	R	A	M	C	O	U	N	T	E	R	B
M	J	C	E	Y	P	T	V	W	A	P	C	I	A	W
B	L	S	F	X	S	Z	Z	C	Z	S	K	N	O	E
D	K	F	G	O	N	R	J	H	G	P	K	P	N	N
W	U	B	R	D	H	B	W	A	A	E	E	A	J	I
A	N	C	P	I	H	C	O	R	C	I	M	G	V	Q
G	A	R	C	S	N	K	N	T	U	K	V	E	H	T
M	E	A	R	Y	T	U	X	S	B	J	P	B	D	R

APPLICATION
BANK
DATASUM
FLOWCHARTS
LOOP
MACROS
MICROCHIP
NOTES
PAGE
PROGRAM
PROGRAMCOUNTER
STRUCTURES

PUZZLE: LOOPS CODE

addressing
code
data
Instruction
interrupt
jump
location
loops
memory
numerals
oscillator
stack
table

How to solve Word Fit

Fill the grid with the words provided, each word is used once. Word Fits are like reverse crosswords. The words are given and you just have to find where they go. Try to solve the puzzle in your head, without writing the actual words into the grid.

FURTHER READING

6809 Assembly Language Programming by Lance Leventhal.

Much of the structure of Chapter 5 and the introduction to program loops was obtained from Lance Leventhal's 6809 Assembly Language Programming in Section II. Hat tip to Lance. This was the assembly language book that I used as a student nearly 27 years ago. He has a thoughtful approach to program examples that walks the reader through programs that quite gradually bring the reader up to speed, introducing code for the novice. This approach has been imitated here, with additional chapters explaining core fundamentals, between chapters that provide code examples.

Embedded Control Handbook by Microchip, 1994–1995.

AN586 Macros for page and bank switching.

AN581 Implementing long calls.

Binary to BCD Conversion Routines (Looped and Partial Straightline) p. 4-99–4-100.

PIC10F200/202/204/206 Data Sheet.

CHAPTER

Embedded control fundamentals

6

CHAPTER OUTLINE

EMBEDDED CONTROL

Embedded control refers to electronic devices that have user programmable space for software and connections to external circuitry through hardware. Embedded control devices have a very specific set of tasks, or states, that they are designed specifically for accomplishing. The user only interacts with specific inputs, like switches, and receives specific responses in output, such as light emitting diodes (LEDs). Generally, they are not used with a keyboard or mouse, and they typically do not provide any provision for user programmability.

ELECTRICAL WIRING DIAGRAMS

Nodes are points of electrical value where the electrical characteristics are identical across the connections. Herein, the impedance and capacitance of the wire, and connections to that wire, are similar enough across it as to be considered of negligible difference.

Looking at Fig. 6.1, there are three nodes to consider; Node 1 at 3.3V, Node 2 between resistors, and Node 3 at ground. Node 2 is calculated as a ratio of the two resistors; R1 and R2. Ohm's law, $V = IR$; where V is voltage in kilovolts, I is current

The Art of Assembly Language Programming Using PIC® Technology. https://doi.org/10.1016/B978-0-12-812617-2.00006-7

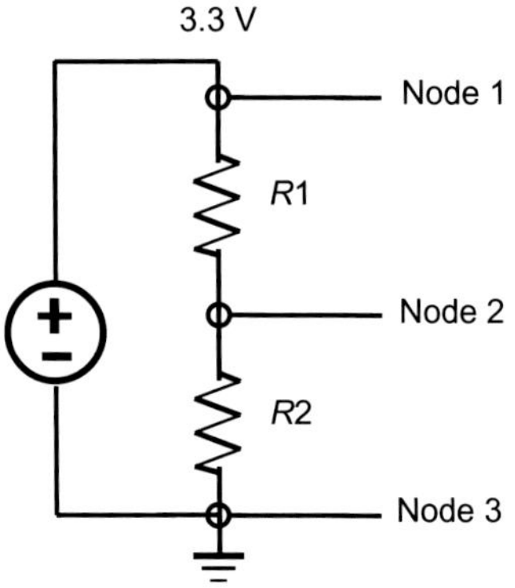

FIG. 6.1

Three nodes.

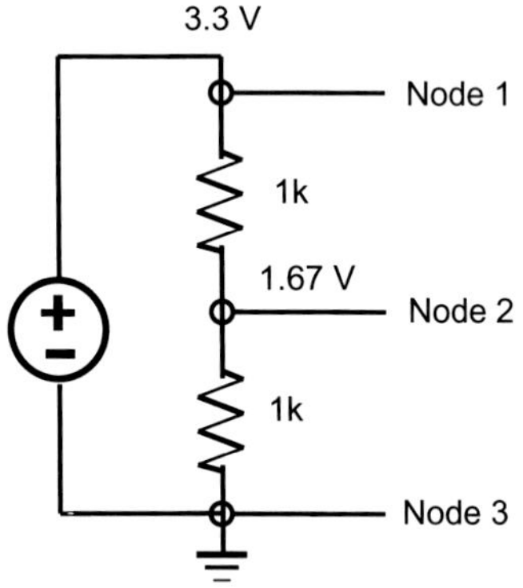

FIG. 6.2

Voltage divider.

in amperes, and R is resistance in kilohms. Rewritten as $R = V/I$. If Node 2 is of interest to us, we can use the resistor ratio to calculate the voltage at Node 2. $V_{\text{node2}} = (R2/(R1+R2)) * 3.3\text{V} = (1\text{k}/2\text{k}) * 3.3\text{V} = 1.67\text{V}$ (Fig. 6.2).

BREADBOARDS

Breadboards are temporary work boards for electronic circuits. The general shape of a breadboard is shown in Fig. 6.3. Compatible with most breadboards, 24-gauge wire is used to connect circuits; solid wire, not stranded. Sometimes, kits may be available with various colors of fixed lengths to specifically fit breadboards. These are a nice convenience.

A bare board, Fig. 6.3, already has some connections. The horizontal rows are connected throughout the row and may make a complete row with the addition of a

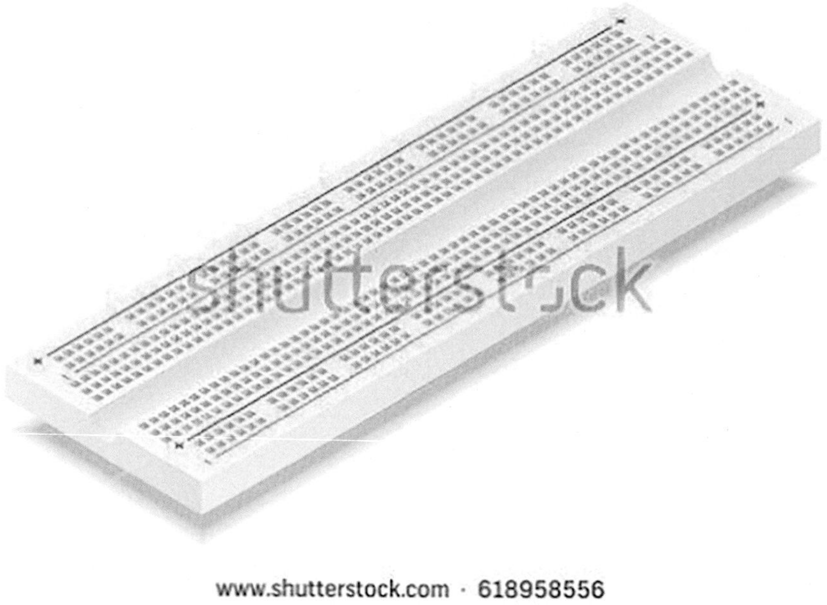

FIG. 6.3

Pictorial image of breadboard.

simple jumper at the center point. These rows are noted with red and blue or black markings (Fig. 6.4).

Similarly, the vertical columns are connected vertically down half the width of the board. A jumper may be inserted at the middle location to connect the full width of the breadboard. The horizontal rows are typically used for power and ground, and, with the middle NOT connected, one can apply more than one power level, such as 12V, 3.3V, and ground. Fig. 6.4 shows the orientation of the board for our purposes. As noted in Fig. 6.4, vertical columns are for connections to each leg of your integrated circuit (IC). Like the three integrated circuits shown in Fig. 6.5, your IC should straddle the center valley.

BASIC INPUT AND OUTPUT (I/O)

Some very basic I/O, commonly referred to as "electronic glue," is used to setup a PIC® controller. A PIC® 10F comes in an 8-pin Dual Inline Package (DIP), although only 6 of the 8 pins are active. This package style is ideal for use with a breadboard which accommodates the DIP package to straddle the center. The three DIP packages are already placed into this double width breadboard. Note that these packages arrive with their legs splayed at an angle of about 60 degrees. Simply lay the package on its side and rock the DIP to bend the legs to a clean 90 degrees.

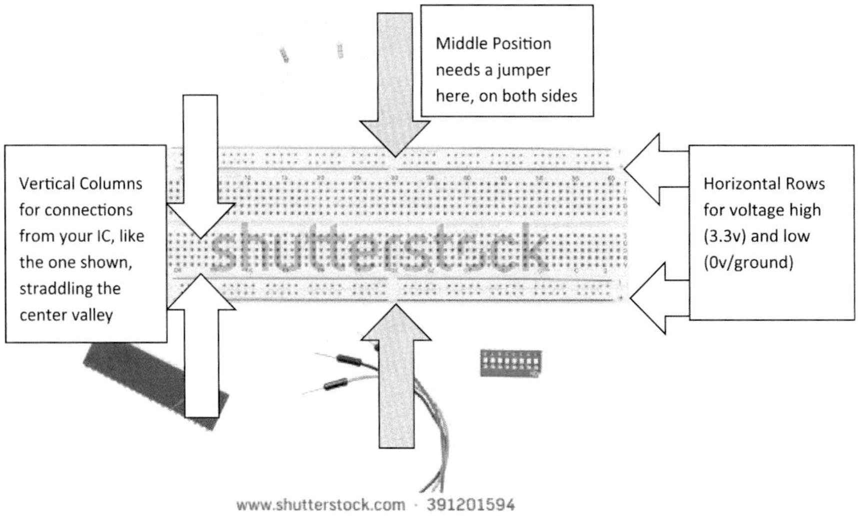

FIG. 6.4

Breadboard horizontal row and vertical column connections.

FIG. 6.5

Breadboard with integrated circuits (ICs) straddling the center valley.

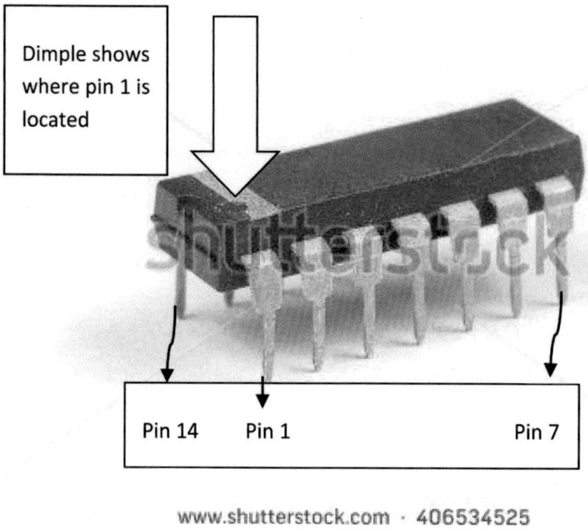

FIG. 6.6

Pin numbering conventions.

There is a slight dimple on the top of the device, where pin 1 is identified. Pin numbering then continues around the device; see Fig. 6.6. This pin numbering pattern is similar to all DIP packages. Note the difference between PIC10F204/206 and 10F200/202 is a comparator on the former 204/206 device; CIN+ on pin 5, CIN– on pin 4, and COUT on pin 3.

The data sheet shows the pinouts on page 2. The numbering conventions are shown in Fig. 6.6. The pinouts start on the lower left side of the 14-pin Dual Inline Package (DIP) pinout. Pins are numbered 1–7, from left to right, with pins 8 through 14 starting on the top side, from right to left, as shown in Fig. 6.6.

On the PIC10F200/202, four *general-purpose input/output (GPIO)* pins are shown as GP0 to GP3. Basic Inputs and Outputs (I/O) are I/O that are routinely found and available on all microprocessors. Inputs are typically 5- or 3.3-V inputs that read 0.0 V, or "ground," when inactive and 5 V, or "high," when active.

Other I/O may include TRIS, or Tristate, inputs. These are similar to standard I/O, except there is a third state "high impedance," where the pin is pulled high with external high impedance circuitry, thereby offering a third state. Outputs are low, or "ground," and "high," 5 V. This is using internal microprocessor circuitry.

Another common means of executing other states is by pulsewidth modulation, or PWM. This output is made by alternating quickly between high and ground, and leaving high up for a known percentage of time. For example, if one were to pulse a high for three periods and then pulse a ground for one period, this would be read as 3/4 of high, or 5 V multiplied by 3/4, or 15/4, or 3.75 V. PWM is a good means to measure, or create, analog signals which use the entire range of 0–5 V.

If more basic I/O is required than is available on the PIC® controller, one can use a multiplexer provided that the I/O does not switch or change faster than the multiplexer. For every one multiplexer, there can be four or more multiplexed I/O.

SPECIALIZED INPUT/OUTPUT

Specialized Input and Output (I/O) encompasses a wide range of unique features offered on PIC® microcontrollers. Specialized I/O often include I/O that support unique features that are supplied internal to the microprocessor at a lower cost than if it were to be implemented external to the microcontroller. Some of these features are implemented "on top" of other features, while all taking place on the same pin.

To take full advantage of these more unique features, carefully read through the PIC® controller's data sheet. These micros are well documented in their data sheets. Many of these specialized I/O are interfaced for additional sensor inputs, such as an onboard temperature signal or the onboard crystal for timing signals. These are typically options that are available at a slightly higher cost. This is one way to keep your options open to upgrades on the same pinout. This will position your code well as the features may be needed, after time, for a modest additional cost and would not necessarily require changing the circuit board and external hardware.

TIMER MODULES AND REGISTERS

There are three main ways to achieve a time delay of known duration. These are hardware alone, mixed hardware and software, and software alone. These different methods all yield different levels of accuracy. This requires knowledge of how long each cycle of the processor takes to complete individual instructions. PIC® controllers, 8-bit processors, contain crystals that create a known oscillation frequency. Some older processors require hardware connection to either an external crystal or a resistor capacitor (RC) network to create the pulse. Think of the timers as heartbeats for the processors. Every circuit will have a known pulse frequency although it may not be the same as another unit. RCs are the least accurate and least consistent of the choices. Typically, you can measure your processor's own "heartbeat" and, with a known pulse rate, you can compensate for inaccuracies. Note that compensation must be calculated each time the processor operates as variables like temperature will dramatically affect the pulse rate. Let's take a look at the three options to use to achieve time "intervals."

HARDWARE ALONE

One-shots and monostable multivibrators can produce time delays in hardware. These are among the least accurate methods but may be preferred if you do not want to have the processor tied up with "doing nothing" but counting.

HARDWARE AND SOFTWARE

A 555 CMOS timer can also be used to achieve a hardware pulse but with more consistency and accuracy. The 555 can be configured with microcontroller software to create a pulse of known frequency and pulse width. This configuration is more reliable than the hardware only process.

SOFTWARE ALONE

Software alone can be the best option, especially with PIC® controllers with an internal crystal. This is achieved first with a measurement of frequency and then a series of two or more nested loops to "waste time."

DEVELOPMENT BOARDS

There are a variety of development and evaluation boards available through Microchip. Reading through each processor's data sheet to find the ideal IC and development or evaluation board. Follows are some of the available boards.

Automotive Networking Development Board
ADM00716 $110.00
CAN & LIN network related
Curiosity Development Board
DM164137 $20.00
www.microchip.com/Curiosity
Curiosity HPC Development Board
DM164136 $32.00
Versatile for 8-bit 6-pin to 8-bit 100pin
Explorer 8 Development Kit
DM160228 $75.00
www.microchip.com/Explorer8
Includes options for external sensors, off-board communication and human interface
PICDEM Lab II Development Platform
DM163046 $100.00

Development and teaching platform with a large prototyping breadboard to easily experiment with different values and configuration

WORD SEARCH: EMBEDDED CONTROL

S	R	G	L	Z	X	F	S	S	R	S	D	P	S	C
M	N	N	W	W	O	I	O	O	N	N	Y	I	L	J
S	V	I	D	U	Y	F	O	O	U	R	A	X	A	P
A	C	R	R	I	T	S	I	O	U	T	F	Z	T	O
S	N	I	L	W	A	T	R	D	G	H	D	G	N	S
U	K	W	A	P	C	G	E	M	B	E	D	D	E	D
O	X	R	N	E	D	N	R	H	Q	O	M	E	M	S
K	E	W	N	M	N	B	A	A	W	C	I	L	A	J
T	M	N	Y	A	L	O	L	R	M	S	Y	L	D	Y
L	O	R	T	N	O	C	C	D	T	S	T	R	N	T
C	Q	O	B	B	G	A	Z	W	U	U	Y	K	U	M
L	U	Y	B	K	A	C	J	A	O	Q	O	G	F	H
Z	P	O	W	E	R	Z	O	R	J	Y	V	N	E	H
E	M	O	S	M	D	D	K	E	I	X	E	O	I	B
J	A	I	Z	D	J	R	A	A	Y	M	J	Y	K	P

CONNECTIONS
CONTROL
DIAGRAMS
EMBEDDED
FUNDAMENTALS
GROUND
HARDWARE
PINOUT
POWER
SOFTWARE
WIRING

PUZZLE: FLUSTERED

To solve the puzzle, place each of the words into the puzzle square. Letters may be shared horizontally and vertically, but not diagonally. Words are grouped by number of letters. You are given one letter's location, shown as X. To start, find a five letter word containing X, place it on the grid, and continue with words that may share letters with your first word.

				X

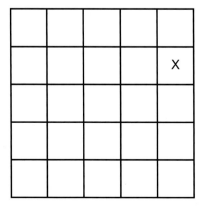

OR
AND
BIN
DEC
HEX
DATA
LOOP
CARRY
DESIGN
MACRO
MPLAB
RADIX
MPLAB
RANDOM

How to solve Flustered

In the familiar word game Fluster, players are given 16 letters (some of which may be duplicates) arranged in a four-by-four grid. Players try to form words by moving from letter to touching letter – horizontally, vertically, or diagonally. All letters of a given word must be found in different squares. Things are turned around a bit here. In the following puzzle, a list of words from a game of Fluster is provided, and you are to determine the correct location for each letter. One letter is given to get you started.

FURTHER READING

PIC10F200/202/204/206 Data sheet, see Appendix E.
www.Microchip.com.

CHAPTER

Fundamentals of good practice

7

CHAPTER OUTLINE

STRUCTURED PROGRAMMING

When one thinks of software design, the mind may conjure up a vision of huge programming efforts written by teams of software writers, not programming for small PIC® devices. However, we need to set our minds to think one step ahead and to *design* software code from the beginning. There are some pretty basic principles that are outlined here. Religiously following these recommendations shall increase the quality of your code to empower the next person, who could be you, too, to best utilize your completed work in new designs. This will result in tremendous savings in cost, quality, and time. Further, it will spare you the frustration in rewriting or attempting to follow poorly written code. In this chapter, guidelines are presented first with tools to follow. Tool usage and selection are often a personal preference. Many programmers use a combination of tools, such as Warnier-Orr for data structure, while using flowcharts for decision and repetition structure. Likewise, SysML users frequently use a modified version of state machines within the SysML model. Give your effort the best tool for your situation.

The Art of Assembly Language Programming Using PIC® Technology. https://doi.org/10.1016/B978-0-12-812617-2.00007-9

SOFTWARE DESIGN PATTERNS

Software Design Patterns are loosely defined as guidelines that are followed within a company, or any group of programmers, to format programs and software to a "common denominator." Design Patterns may dictate the order of variable declaration; the method of calling subroutines; the naming of routines, variables, and constants; and the use of a specific library of arithmetic routines. Design patterns are *not* algorithms and, as such, it is not a complete design. These patterns are "formalized best practices" that programmers can use when designing an application or system. At the end of this chapter, the SysML modeling language is one implementation of a Software Design Pattern that attempts to establish common elements within the larger context of software design. Wiki notes "that there are many types of design patterns for instance.[1,2] Follows are five general guidelines to review and consider in designing your code to be reusable in later designs. Please also review Appendix D Best Practices.

1. Algorithm strategy patterns
   Addressing concerns related to high-level strategies describing how to exploit application characteristics on a computing platform.
2. Computational design patterns
   Addressing concerns related to key computation identification.[3,4]
3. Execution patterns
   Addressing issues related to lower-level support of application execution, including strategies for executing streams of tasks and for the definition of building blocks to support task synchronization.
4. Implementation strategy patterns
   Addressing concerns related to implementing source code to support program organization, and the common data structures specific to parallel programming.
5. Structural design patterns
   Addressing concerns related to global structures of applications being developed."[1]

This is, by no means, a comprehensive list. This is a fluid, changing definition incorporating new sets of design patterns. SysML is one example of a comprehensive language built around design patterns. The key patterns in SysML include Behavior Diagrams, Requirement Diagrams, and Structure Diagrams. These are explored in detail at the end of this chapter.

REUSABLE CODE GUIDELINES

1. Comments
   Carefully consider your code before writing comments. Do not just state the obvious. You need to comment on overall structure and note the purpose of the code section, not just reiterate the code itself.

[1] https://sourcemaking.com/design_patterns
[2] https://pmsware.wordpress.com/tag/computational-design-patterns/
[3] Category: Computational Thinking Patterns—Scalable Game Design Wiki, sgd.cs.colorado.edu, (Retrieved 2015-12-26).
[4] Introduction to Software Engineering/Architecture/Design Patterns—Wikibooks, Open Books for an Open World, en.wikibooks.org (Retrieved 2015-12-26).

2. Graphics

   It is easy to spend a lot of time on graphics. Do not get hung up with pretty graphics. You are looking for content and tools that support the way you work. Carefully consider your objective. Do you need to produce a graphics based image of your code, for use by a large department? Or will well-designed handwritten work suffice? It is all in the content and the audience for whom you are writing.
3. Subroutines

   Make everything conceivable a function or a procedure. Document well the variables passing from the program to and from the function or procedure. Note register changes as well.
4. Modularity

   Structure your code as a library of routines to "include" for use with other programs. Using include files to create reusable code. Use #define as a tool for compatibility. This will improve your quality of code, as well as building a resource for you and your team. Note at the top of each function or procedure any architecture restrictions.
5. Upward compatibility

   Microchip data sheets provide a thorough accounting of where there need to be compromises for upward compatibility. Thoroughly consider where you need to account for changes. In the eventuality that a need arises, you will then be able to make a cautious transition, in the form of an upward move, for more memory or additional architecture features.

ORDERED LISTS

To build a program structure, first construct an ordered list of elements. Build a thorough list of program components, by type, in the following order.

a. Inputs (read)
b. Preparation of branches
c. Branches
d. Preparation of calculations
e. Calculations
f. Preparation of outputs
g. Outputs
h. Subroutine calls

ORDERED LISTS EXAMPLE

Examples are shown using the development of a front loading washing machine. This set of examples will be drawn out in all five methods of structured programming: flowcharts, Warnier-Orr diagrams, state machine diagrams, pseudocode, and SysML. The ordered list of elements, for this example, shows a Table of Values, Physical Switches, and Motor Control.

a. **Inputs (read):**

   On/Off switch

   Start/Pause switch

Potentiometer with settings of:
Whites
Colors/Normal
Stain Wash
Easy Care
Active Wear
Delicates
Hand Wash
Speed Wash
Drain and Spin
Rinse and Spin
Two state pushbutton switches:
Extra Rinse
Pre-Wash
Signal (audible)

b. Preparation of branches

Table of Values:

Descriptor	Soil Level	Spin Speed	Wash Temp	Agitation Duration
Whites	N	H	H	1:21
Colors/Normal	N	H	C	1:14
Stain Wash	H	H	H	2:44
Easy Care	N	M	W	1:01
Active Wear	H	M	C	1:08
Delicates	N	M	C	0:58
Hand Wash	N	L	C	0:52
Speed Wash	X	H	W	0:38
Drain and Spin	Off	H	Off	0:13
Rinse and Spin	Off	H	Off	0:18

c. Branches
Soil level: Heavy, Normal, Light, Extra Light, Off
Spin Speed: High, Medium, Low, No Spin/Off
Wash Temp: Hot, Warm, Cold, Off

d. Preparation of Calculations
Water valve solenoid for Hot water
Water valve solenoid for Cold water
Speed/Motor 3 Speeds and Off
Timer setting is calculated from soil level

e. Calculations

For Spin Speed, motor is set as:

Spin Speed	Motor Speed
Speed 3: High	100%
Speed 2: Medium	66%
Speed 1: Low	33%
Speed 0: No spin, Off	0%

For Wash Temp, solenoids are set at:

Wash Temp	Hot Valve	Cold Valve
Hot	100%	0%
Warm	100%	100%
Cold	0%	100%
Off	0%	0%

For Soil Level, timer is set as Agitation Duration:

f. Preparation of Outputs

On LED lit
Time display
Start LED lit
Wash Temp set
Motor Speed
Soil Level
Upon signal, Beep
Latch/Unlatch door

g. Outputs

LEDs (On, Start)
Timer
Motor Output
Audible Beep
Door Latch

h. Subroutine calls

STRUCTURED SYSTEMS DEVELOPMENT (SSD): FOUR BASIC STRUCTURES

A structured process is one that explicitly organizes functions in one of four logical control structures: **function**, **sequence**, **repetition**, and **alternation**. Bohm and Jacopini, 1966, have shown in Communications of ECM 9, pp. 366–371, article "Flow diagrams, turing machines and languages with only two formation rules" that any program can be developed using some combination of these **four logical control structures**. Herein is shown how to represent each of these entity types in each of the following four, omitting SysML, software design models: **Flowcharts**, **Warnier-Orr Diagrams**, **State Diagrams**, and **Pseudocode**.

PROGRAMMING FLOWCHARTS

Flowcharts are constructed top-down in a vertical fashion. Templates are available for doing flowcharting by hand. Computer rendering is also available in Microsoft Word. Word has shapes for all flowchart symbols. Within Word, use the menu sequence: Insert, Shapes, Flowchart. There are also other dedicated program packages that provide support for flowcharting, such as Smart Draw by Microsoft.

The four logical control structures are shown in flowcharting as function, see Fig. 7.1, alternation, see Fig. 7.2, repetition, see Fig. 7.3, and sequence, see Fig. 7.4.

WARNIER-ORR DIAGRAMS

Structured Systems Development, with Warnier-Orr diagrams, is designed to start the design at the end of the program, with the Outputs and Subroutine Calls. The construction of the Warnier-Orr diagrams is to be worked backward and horizontally; bottom up. These diagrams can be generated quite quickly, when using handwritten diagrams. On templates, such as those provided for flowcharting by IBM, there are braces for Warnier-Orr diagrams on the right outer edge of the template. The author is

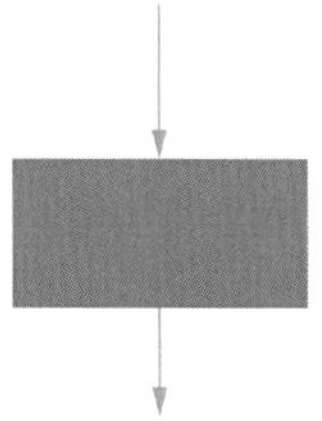

FIG. 7.1

Flowchart function.

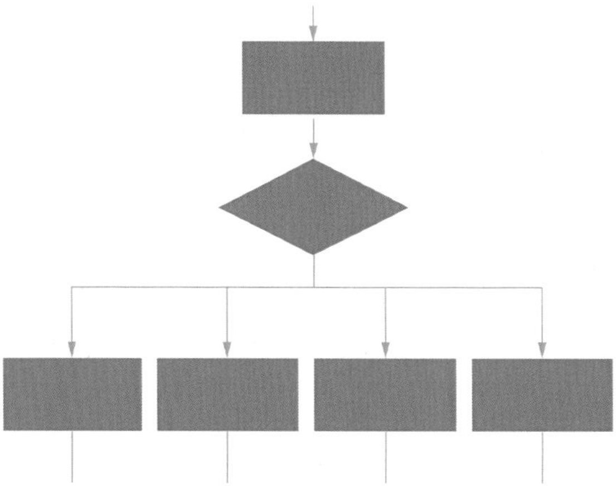

FIG. 7.2

Flowchart alternation.

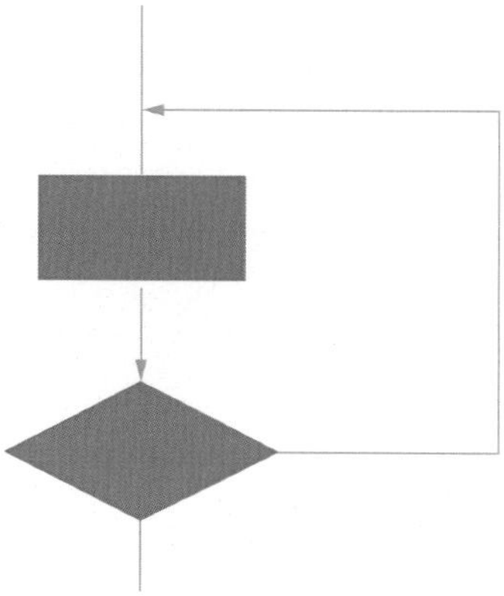

FIG. 7.3

Flowchart repetition.

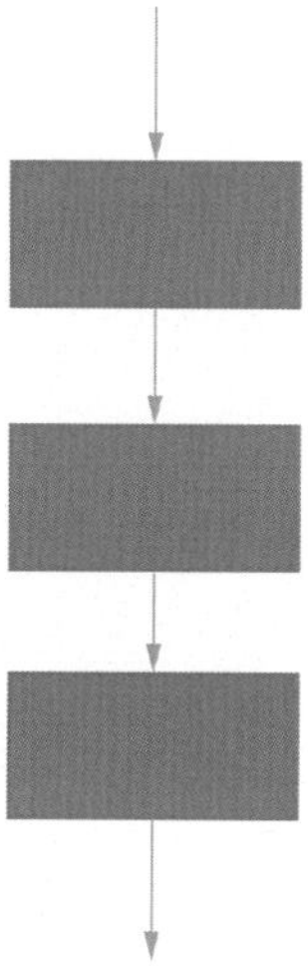

FIG. 7.4

Flowchart sequence.

not familiar with any adequate tool used to construct these diagrams in a computerized form. Smart draw does have some very basic provisions for Warnier-Orr diagrams.

Generate Warnier-Orr diagrams, as with flowcharts, by working backward from the Outputs and Subroutine calls for each ordered list generated (Figs. 7.5–7.8).

STATE MACHINE DIAGRAMS

State machine diagrams are *event-driven diagrams*. Each state is identified by a circle with an Sx number; where x is the number of the state. Each event follows an arrow indicating an *event*, change, or movement to another state (see Figs. 7.9–7.12).

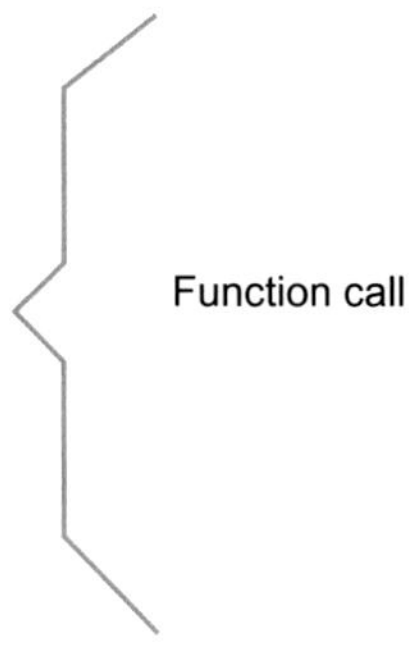

FIG. 7.5

Warnier-Orr function.

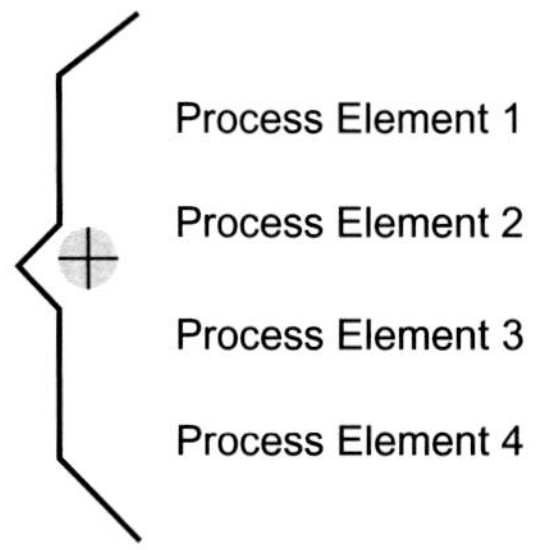

FIG. 7.6

Warnier-Orr alternation.

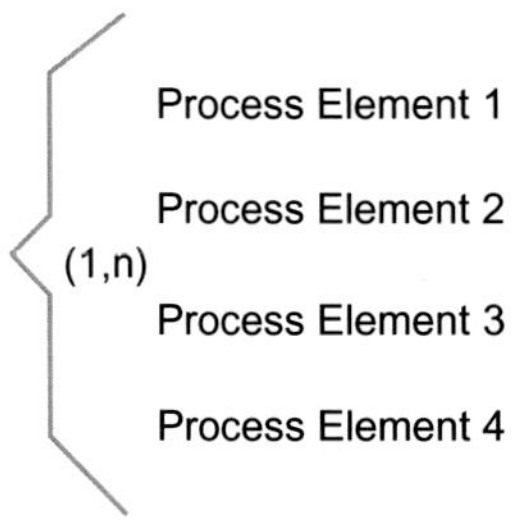

FIG. 7.7

Warnier-Orr repetition.

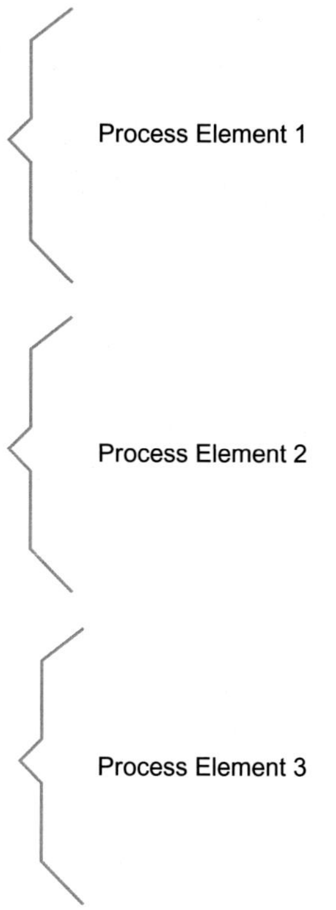

FIG. 7.8

Warnier-Orr sequence.

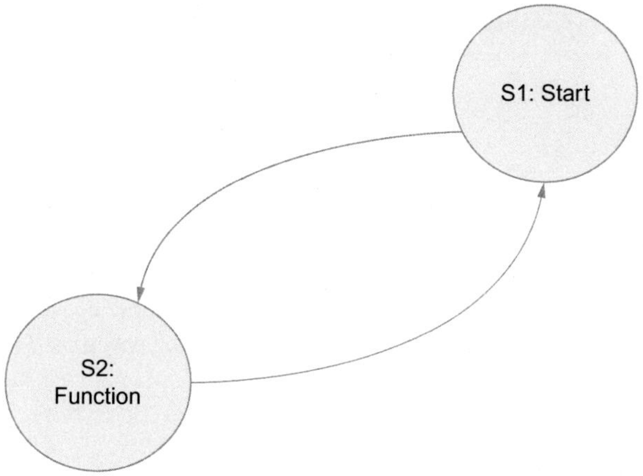

FIG. 7.9

State machine function.

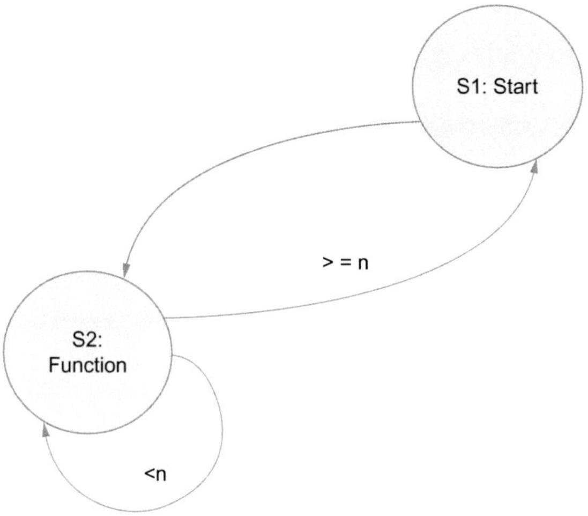

FIG. 7.10

State machine repetition.

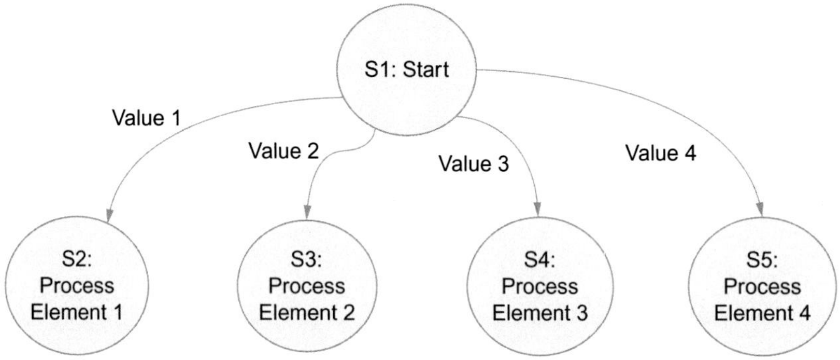

FIG. 7.11

State machine alternation.

PSEUDOCODE

Pseudocode provides a language-independent, generalized block of code. The details of the language are not an issue. This is mainly a way of showing, or illustrating, how we can begin to move from the diagrams into programming specifics (Figs. 7.13–7.15).

SysML

SysML Standards and Tools

SysML is *System Modeling Language*. This graphical language is extreme overkill for most Microchip software design and documentation. An overview is arranged

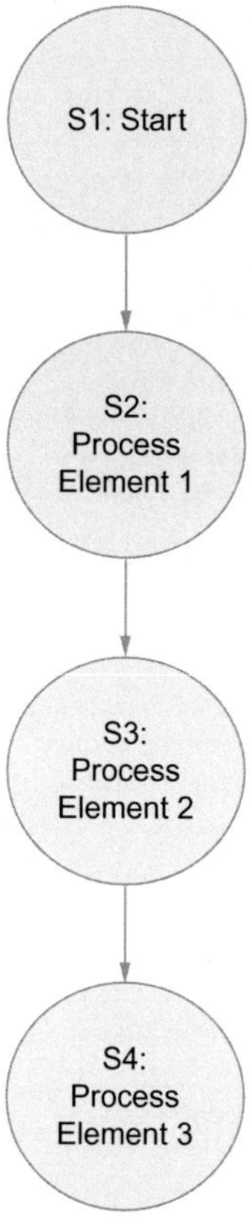

FIG. 7.12

State machine sequence.

```
Program-X
        Begin Program-X
                Do Begin-X
                Do Function-X Until All-Data-Processed
                Do End-X
        End Program-X
```

FIG. 7.13

Pseudocode function.

here to provide the reader with general awareness of the existence of SysML. The typical Microchip user will have no need for this language. However, if working for a large software business, the user may be given responsibility for a small component of a larger SysML project. Users should be capable of recognizing the language and of articulating the right questions when seeking assistance.

```
    If Task A-Active
         Then
              Do Compute and Print Task A
         Else
              Skip
(A) End-If

    Case Type
         When Type = 'A'
              Do Type-A-Processing
         When Type = 'B'
              Do Type-B-Processing
         When Type = 'C'
              Do Type-C-Processing
         Else
              Do Error Processing
(B) End-Case

    Case Type                       ; Type is a 2 bit value
         Do Type-A-Processing       ; Type is 00 for Type-A
         Do Type-B-Processing       ; Type is 01 for Type-B
         Do Type-C-Processing       ; Type is 10 for Type-C
(C) End-Case
```

FIG.7.14

Pseudocode alternation. (A) Pseudocode single alternation If-Then-Else. (B) Pseudocode multiple alternation If-Then-Else. (C) Pseudocode multiple alternation—Jump tables.

```
    Do Until All-Data-Processed
         Do Get-Next-Transaction
         Do Calculate Value
         Do Put-Record
(A) End-Do

    Do for each Transaction
         Do Get-Next-Transaction
         Do Lookup Value
         If Value > Range
         Then
              Do Report-Error
         Else
              Do Process-Transaction
         End-If
(B) End-Do

    Do for each Transaction
         Do For I=1, 10
         Move A to B (I)
         Compute A = C + B (I) + 2
(C) End-Do

    Move A to B
    Compute A = C + B + 2
(D) Print
```

FIG. 7.15

Pseudocode repetition and pseudocode sequence. (A) Pseudocode repetition type-1 Do Until. (B) Pseudocode repetition type-2 If-Then. (C) Pseudocode repetition type-3 For-Next. (D) Pseudocode sequence.

Tools and support for SysML are woefully inadequate for all but the largest projects with the largest budgets. That said, there are several classic tools, SmartDraw and Visio, that can aid the novice user. SysML is a subsystem of UML so searching for UML support tools is generally the heading under which basic tools are found. At this writing, there is an open source package, Papyrus, Eclipse Papyrus 2.0.2

Neon Release that is considered still active, as of about 1 month ago. This open source package supports SysML 1.1 and 1.4 and UML 2.5.0. Supporters of Papyrus note that "Papyrus is an open source, SysML modeling tool that shows potential but is currently undergoing a re-build and is not yet ready for serious Modeling-Based Systems Engineering (MBSE) usage." Visio is touted to support UML 2.2 and SysML 1.x, with stencils and templates. SmartDraw groups all SysML support under the heading of UML 2.0 with slightly different wording and diagrams. See the table following for a cross-comparison of SmartDraw naming conventions and SysML standards (see Fig. 7.16). Note that requirements and parametrics are not available without customizing diagrams within SmartDraw.

SysML Language Overview

According to Object Management Group (OMG) SysML, there are four pillars of SysML: (1) Structure, (2) Behavior, (3) Requirements, and (4) Parametrics. In two of the pillars, Structure and Behavior, there are four possible entities each, known as "SysML diagram frames." Each entity is added as applicable. Within Behavior, there are Activity, Sequence, State Machine, and Use Case diagrams. See Fig. 7.17

SmartDraw MBSE UML Diagram Sets	SysML
Activity	Activity Subset of Behavior Diagrams
Class	Class
Communication	N/A
Component	Composite subset of Structure Diagrams
Deployment	N/A
Object	N/A
Package	Package Subset of Structure Diagrams
Sequence	Sequence Subset of Behavior Diagrams
State Machine	State Machine Subset of Behavior Diagrams
Use Case	Use Case Subset of Behavior Diagrams

FIG. 7.16

SmartDraw cross-comparison.

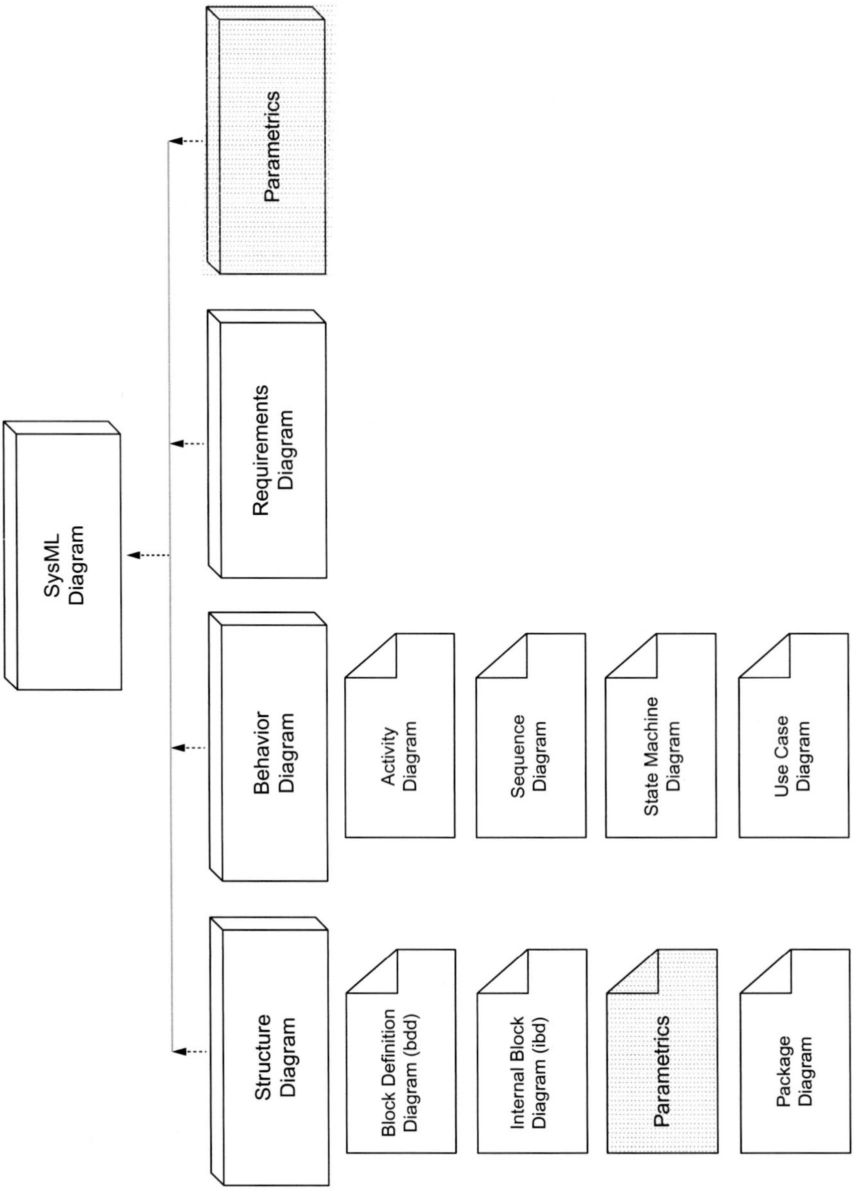

FIG. 7.17
SysML language overview and flowchart.

for a visual overview. Within Structure, there are Block Definition, Internal Block, Parametric, and Package. Requirement Diagram stands alone. Parametric Diagram may stand alone or may alternately be included under Structure. Note that State Machines are used in conjunction with pillar of (2) Behavior, as they define event-driven behaviors.

SysML diagram frames represent a model element. Diagram context is indicated in the header in the left corner of the model element. The Header is comprised of: **diagramKind [modelElementType] modelElementName [diagramName].** The Diagram Description includes Version, Description, Completion status, and Reference (User-defined fields.) The **diagramKind** is act, bdd, ibd, sd, etc. The **modelElementType** is package, block, activity, and so on. The **modelElementName** is a program-specific name, derived from the context. The **diagramName** is also a user-defined diagram name.

SysML Example

SysML is intended to help specify systems and their components. Using the front loading clothes washer example, SysML is illustrated herein with various diagrams applicable to the clothes washer example. The Package diagram, Fig. 7.18, is a summary document constructed at the end, after the diagrams are in place. Diagram kind and diagram name of part labels (A)–(L) are as follows: (A) Requirement diagram and Washer System Requirements. (B) Block definition diagram and Washer Domain. (C) Use Case diagram and Operate Washer. (D) Sequence diagram and Wash cycle. (E) Activity diagram and Control and Power Subsystem. (F) State Machine diagram and Operational States. (G) Internal block diagram and Washer Domain. (H) Block definition diagram and Analysis Context. (I) Parametric diagram and Drum Acceleration Analysis. (J) Timing diagram and Wash Performance Timeline. (K) Block definition diagram and Motor Specification. (L) Package diagram and Model Organization.

This section, and the SysML language in general, is modeled after "A practical guide to SysML: The Systems Modeling Language" written by Sanford Friedenthal, Alan Moore, and Rick Steiner. This practical, in depth, and comprehensive book on SysML is well regarded.

Each diagram type is summarized here as it applies to the Washer Example.

1. Stand-alone diagram:
   Requirement diagram; text-based requirements and design elements.
2. Behavior diagrams include:
   Activity diagram; order of actions transforming inputs to outputs.
   Sequence diagram; messages communicated between parts.
   State machine diagram; event-based transitions between states.
   Use Case diagram; functionality of system as used by external entities.

3. Structure diagrams include:
   Block definition diagram; represents block structures and their content.
   Internal block diagram; internal representation of interconnection between blocks.
   Parametric diagram; formulas and mathematical truths as implemented.
   Package diagram; summary document of diagram relationships.
4. Depending on specific implementations, the Parametric diagram may be pulled from the Structure diagrams and, instead, be separated as a stand-alone diagram.

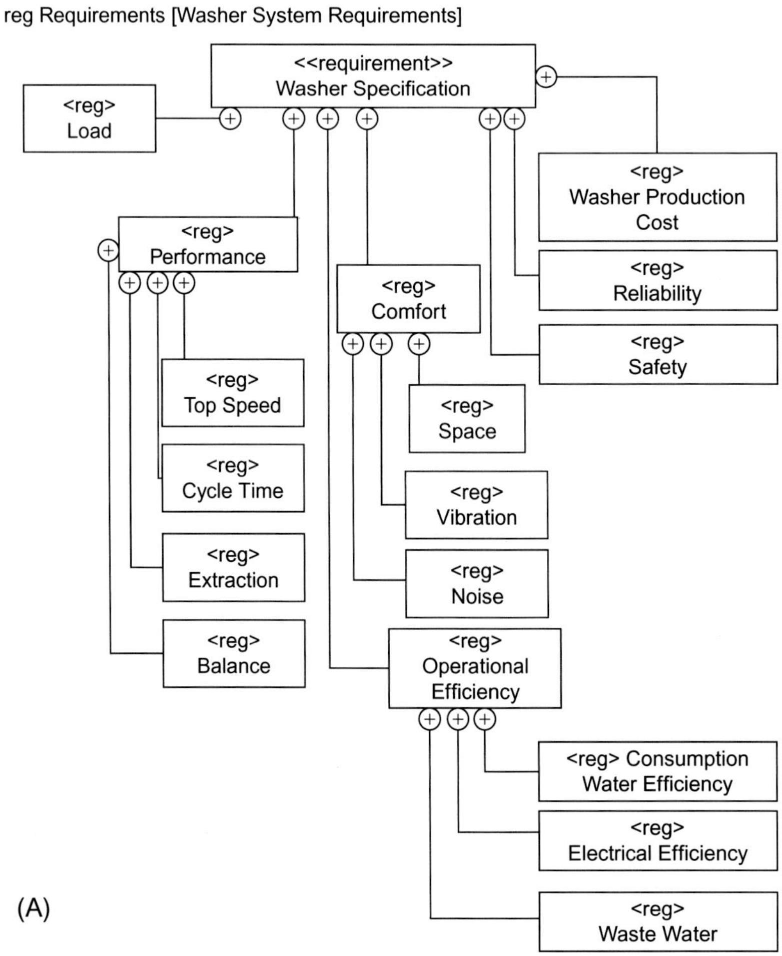

FIG. 7.18

Washer example diagrams. (A) Requirement diagram and Washer System Requirements.

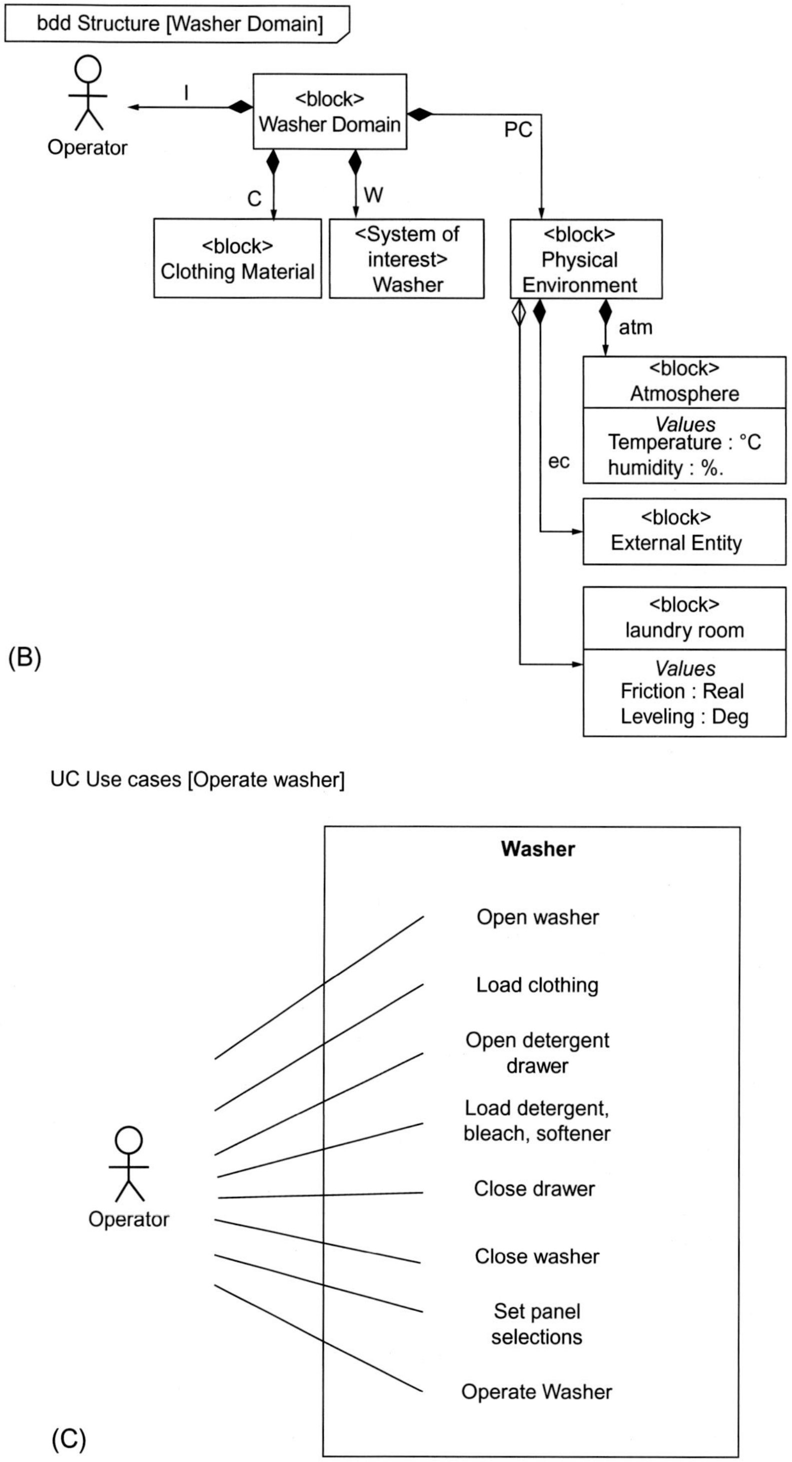

FIG. 7.18

(B) Block definition diagram and Washer Domain. (C) Use Case diagram and Operate Washer.

(Continued)

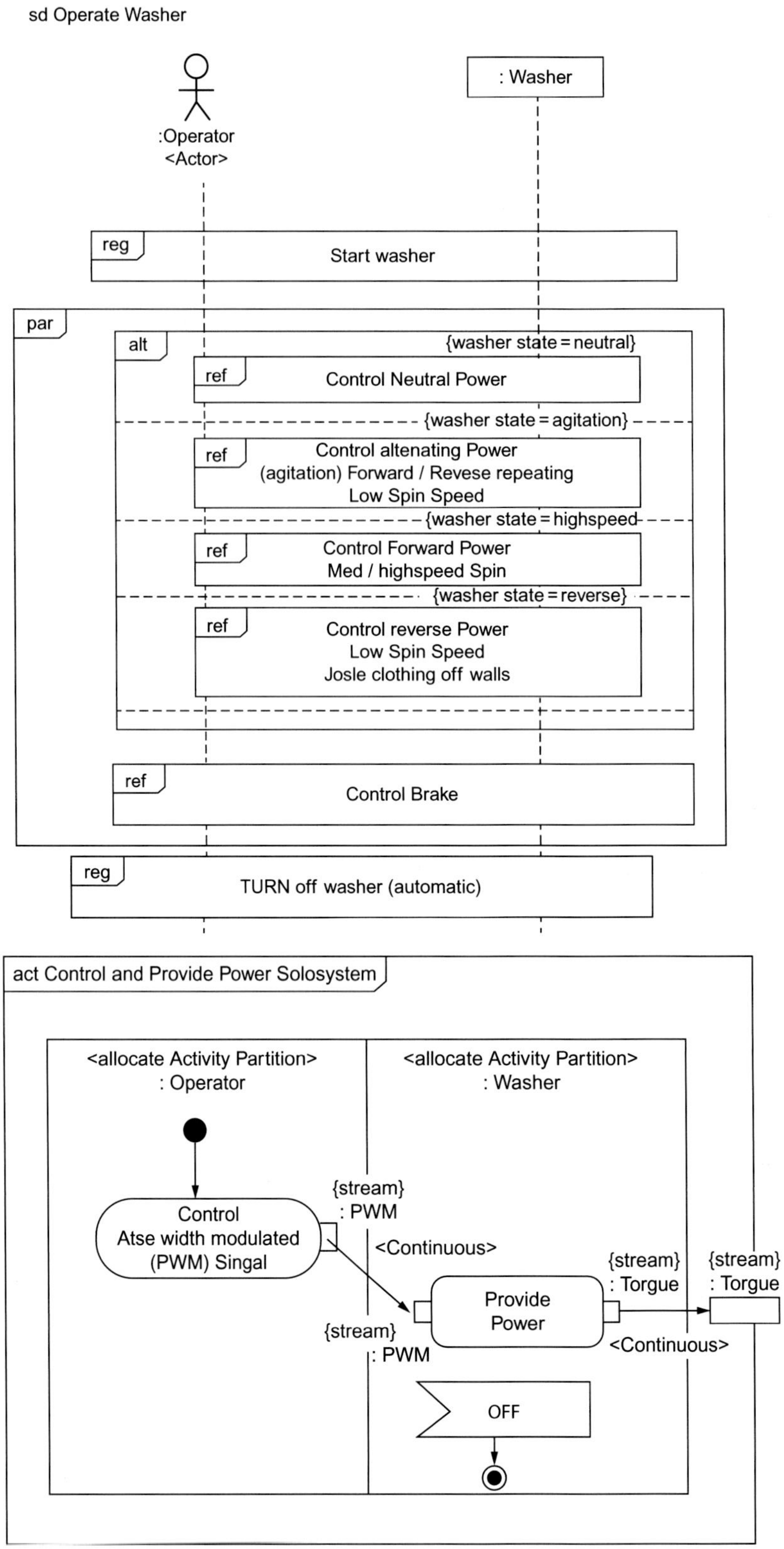

(D)

(E)

FIG. 7.18

(D) Sequence diagram and Wash cycle. (E) Activity diagram and Control and Power Subsystem.

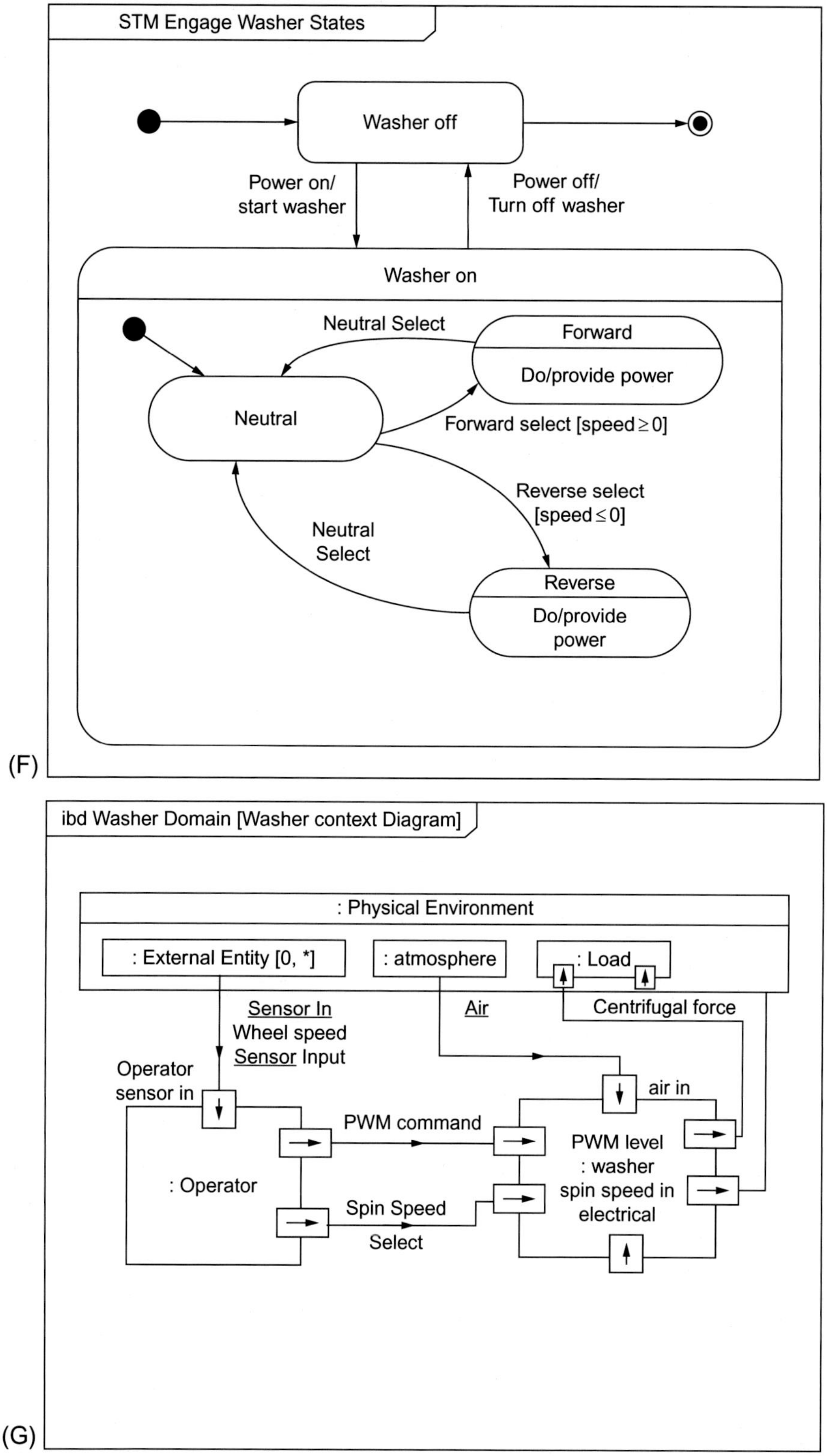

FIG. 7.18

(F) State Machine diagram and Operational States. (G) Internal block diagram and Washer Domain.

(Continued)

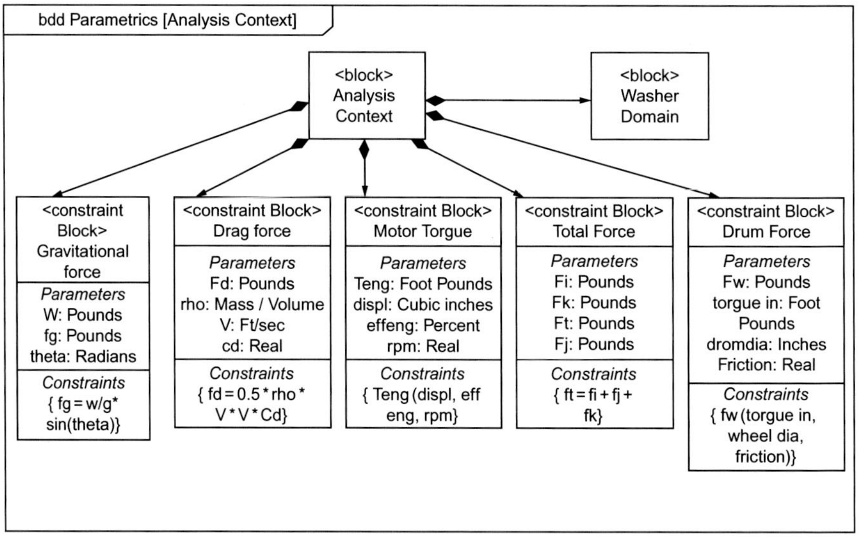

(H)

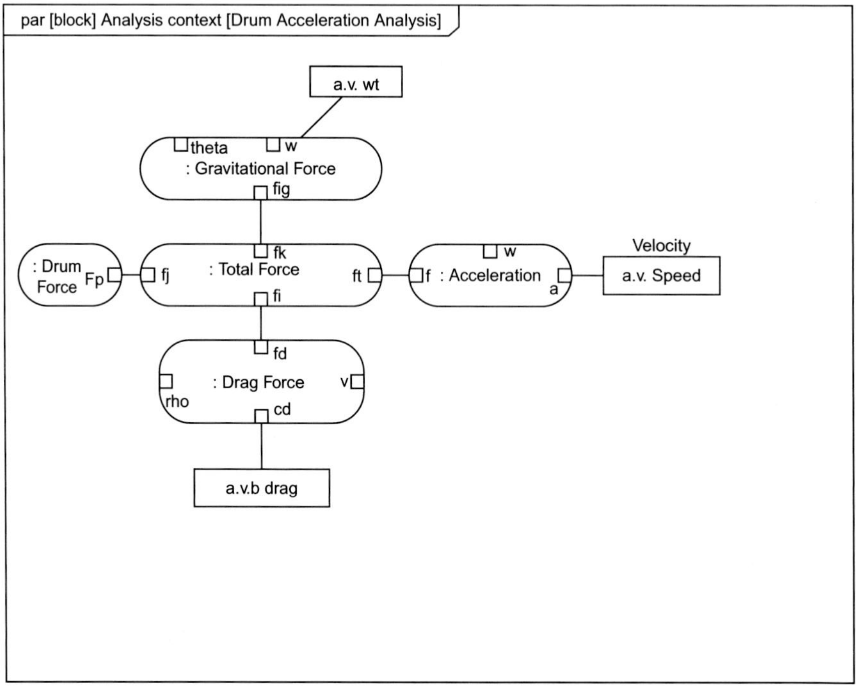

(I)

FIG. 7.18

(H) Block definition diagram and Analysis Context. (I) Parametric diagram and Drum Acceleration Analysis.

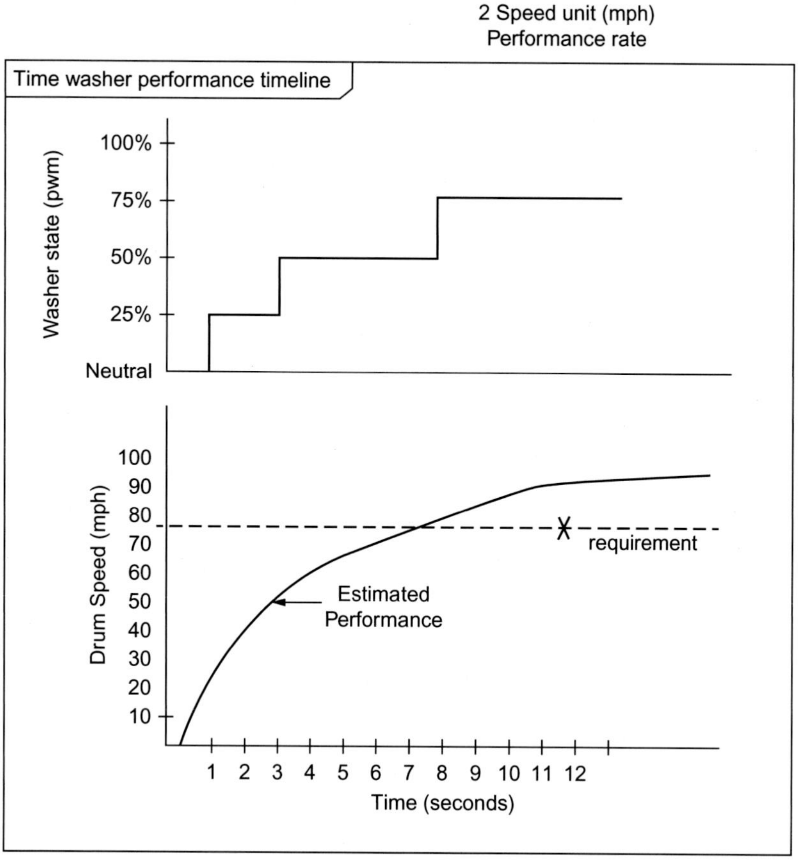

(J)

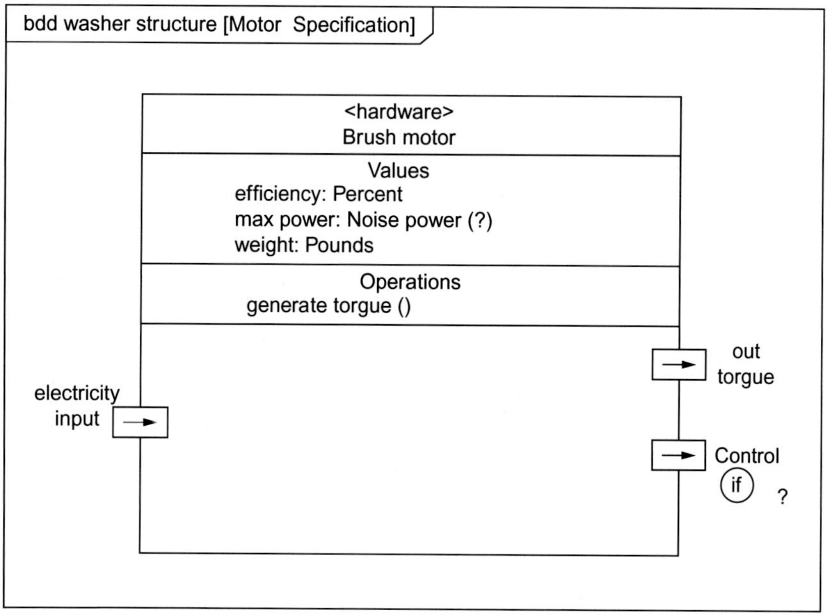

(K)

FIG. 7.18

(J) Timing diagram and Wash Performance Timeline. (K) Block definition diagram and Motor Specification.

(Continued)

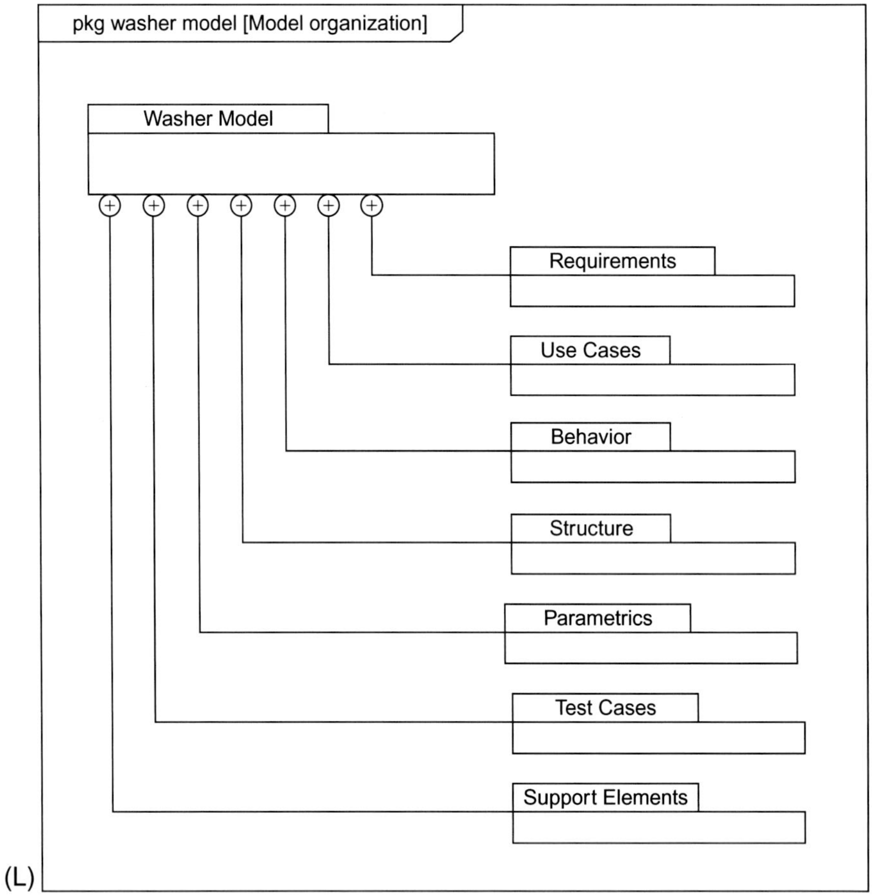

FIG. 7.18

(L) Package diagram and Model Organization.

WORD SEARCH: PROGRAM DESIGN AND DEVELOPMENT

C	E	C	N	E	U	Q	E	S	A	F	R	E	C	N
S	A	S	E	T	A	T	S	B	L	D	O	O	F	R
U	E	L	Y	N	O	I	T	I	T	E	P	E	R	V
V	W	H	C	S	M	F	M	A	E	R	A	B	Z	L
Y	R	U	C	U	T	G	J	G	R	E	Y	T	D	D
F	J	R	P	N	L	E	K	W	N	D	P	X	E	F
I	L	P	O	I	A	A	M	S	A	R	A	V	D	U
N	Z	O	S	R	E	R	T	S	T	O	E	Q	O	N
P	Q	T	W	E	E	U	B	I	I	L	V	F	C	C
U	S	K	P	C	P	I	Y	I	O	D	G	W	O	T
T	S	E	E	T	H	S	N	P	N	N	S	P	D	I
S	U	E	U	Q	D	A	M	R	G	O	S	S	U	O
J	R	O	V	L	C	E	R	T	A	Y	L	K	E	N
M	A	C	H	I	N	E	C	T	K	W	W	A	S	U
C	W	D	S	T	R	U	C	T	U	R	E	D	P	O

ALTERNATION
BRANCHES
CALCULATIONS
DEVELOPMENT
FLOWCHART
FUNCTION
INPUTS
LISTS
MACHINE
REPETITION
OUTPUTS
PSEUDOCODE
STATE
SEQUENCE
SSD
STRUCTURED
SYSTEMS
WARNIERORR

PUZZLE: SPIRAL

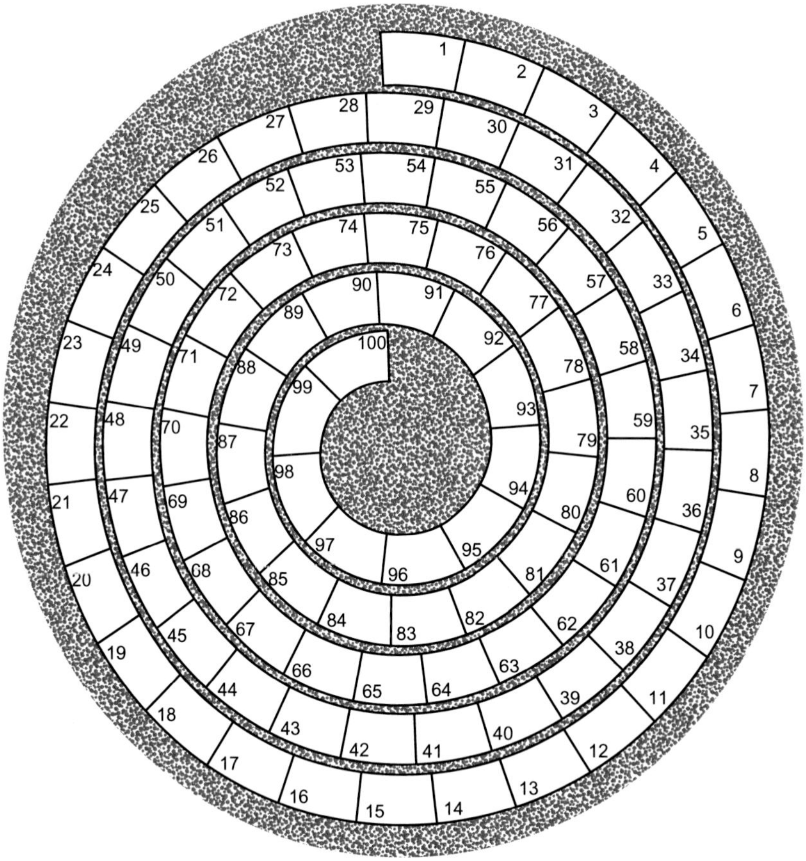

INWARD letter range	Description	Words
1-7	Software is another name for ____	
8-11	One ___ is one instance of code	
12-18	Use ____ to switch to sub-routines in the lower ½ of a page	
19-21	Integrated Design Environment	
22-28	Assembler produces ____ language	
29-34	Repeatable blocks of code	
35-38	RP1 & RP0 select the____	
39-42	____ the subroutine	
43-46	Unconditional program counter change	
47-50	Page 1 indicates the 2nd ____	

INWARD letter range	Description	Words
51-61	AN stands for ____ Note	
62-68	Program ____ points to your current location.	
69-79	Long ____ branches	
80-88	____ is one method of program design and documentation.	
89-97	One ____ is one loop around.	
98-100	Single	

Additional Hints

INWARD letter range	Description	Words
1-3	Short for professional	
22-24	Short for an early Apple Computer	
40-42	Everyone	
45-46	Go____	
47-50	Place to store money	
51-53	Computer program for cell phones	
56-58	Raining ____s and dogs	
76-77	Don't leave the stove ____	
86-88	Creative expression	
97-100	Another name for zero	

In the other direction….

OUTWARD letter range	Description	Words
4-3	__to	
22-20	Short for medicine	
29-28	____, myself, and I	
61-60	A two-year-old's favorite word	
83-80	Red riding hood's foe	

How to solve Spiral

Solve this Spiral by working in two directions. Follow the INWARD clues and fill in the correct answers by starting at Box 1 and working toward the center. Or, use the OUTWARD clues, starting at box 100, and work your way out toward the rim. Both sets of clues will help you over the difficult spots. There are columns, on the right of each table of clues, to provide work space for you. In addition, there is a section of "Additional Hints," which provides supplementary INWARD clues for different letter ranges.

FURTHER READING

A paper by Gary Forbach, "Structured System design: Analysis, design, and construction," details the use of Warnier-Orr diagrams, in a succinct paper. He provides a cohesive explanation of three structures: Sequence, Repetition, and Alternation, and how to apply Warnier-Orr diagrams to design a software program. His explanation provides a better base of understanding and practical application than either "Structured Systems Development," by Ken Orr, or "Logical Construction of Programs," by J.D. Warnier. Both Orr's and Warnier's books are, however, a fascinating walk back in history to learn about how Warnier-Orr diagrams were developed.

"Structured system design: Analysis, design, and construction," by Gary B. Forbach, Washburn University, Topeka, Kansas 66621. Published in *Behavior Research* Methods & Instrumentation, 1981, Vol. 13(2), 163–171.

Bohm, C., & Jacopini, G. Flow diagrams, turing machines and languages with only two formation rules. *Communications of ECM9*, 1966, 5, 366–371. This article is the basis for and proof that there are only four states of function, repetition, sequence, and alternation, that must be accommodated in any software design.

Orr, K. T. Structured systems development. New York: Yourdon Press, 1977.

Warnier, J. D. [Logical construction of programs] (3rd ed.) (B.M. Flanagan, translated). New York: Van Nostrand Reinhold, 1976.

4 Pillars of SysML—ABS Example, Object Management Group (OMG). 2008.

Systems Modeling Language. OMG Systems Modeling Language (OMG SysML™) Version 1.4 OMG Document Number: formal/2015-06-03 Normative Reference: http://www.omg.org/spec/SysML/1.4/ Machine consumable files: http://www.omg.org/spec/SysML/20150709.

"A practical guide to SysML: The Systems Modeling Language" written by Sanford Friedenthal, Alan Moore, and Rick Steiner.

CHAPTER

Data and control structures

8

CHAPTER OUTLINE

ASSEMBLY LANGUAGE CONCEPTS

CONFIG WORD

"Config Words" are also known as "Configuration Words," "Configuration Bits," or "Fuse Bits." "All Microchip microcontrollers contain a series of configurable settings that are arranged outside of your normal program.

The Art of Assembly Language Programming Using PIC® Technology. https://doi.org/10.1016/B978-0-12-812617-2.00008-0

"These are special directives that tell the microcontroller what to do on the most basic level, i.e., at a level lower than your program" [1]. An example is shown to illustrate how the config bits generally appear. Notes that each set are specific to your type of processor.

Follows is an example of a set of Configuration Bits for a PIC24 Series processor [1].

```
code language="cpp"]
// Microcontroller config words (fuses) – get these wrong and nothing will work
correctly, if at all!!
_CONFIG 1(JTAGEN_OFF & GCP_OFF & GWRP_OFF & ICS_PGx1 &
FWDTEN_OFF & WINDIS_OFF & FWPSA_PR32 & WDTPS_PS8192)
_CONFIG 2(IESO_OFF & FNOSC_FRCPLL & OSCIOFNC_OFF &
POSCMOD_NONE & PLL96MHZ_ON & PLLDIV_DIV2 & FCKSM_
CSECME & IOL1WAY_OFF)
_CONFIG 3(WPFP_WPFP0 & SOSCSEL_IO & WUTSEL_FST & WPDIS_
WPDIS & WPCFG_WPCFGDIS & WPEND_WPENDMEM)
_CONFIG 4(DSWDTPS_DSWDTPS3 & DSWDTOSC_LPRC & RTCOSC_
LPRC & DSBOREN_OFF & DSWDTEN_OFF)
[/code]
[1]
```

Information on your particular Configuration Bits for your processor, look in the Data Sheet under the section of "Special Features" [1]. In the PIC10F data sheet, we find "Special Features" that address the Configuration Bits on **page 35**, of Appendix E. Microchip notes that "The … microcontrollers have a host of [special circuits] intended to maximize system reliability, minimize cost through elimination of external components, provide power saving operating modes and offer code protection." Follow along on page 35 as we note the feature set. These features are Reset: Power-on Reset (POR), Device Reset Timer (DRT), Watchdog Timer (WDT), Wake-up from Sleep on pin change, Wake-up from Sleep on comparator change. Also included are: Sleep, Code Protection, ID Locations, In-Circuit Serial Programming™ and Clock Out. Note that these are achieved with a small subset of the Configuration Word. The Config Word, for this processor, uses only bits from bit 2 through bit 4. All other bits are unimplemented and read as zero. Page 35 has full details on how to configure each bit.

To set the Config Word, look in the header file for your processor. "You'll find a series of macros that are used to set the configuration bits. Below the macro declarations there is the list of options which correspond to the options listed in the datasheets" [1].

Refer to "PIC10F200/202/204/206 Memory Programming Specifications (Document DS41228) to determine how to access the Configuration Word. [It] is not user addressable during device operation" [1]. This is set prior to program operation.

USING DEFINE

Using #DEFINE, we take a look at "two examples that both … produce the same code" [2]:

```
#define D0_ON b'00010000'

D0_ON    EQU    10h
```

But there's a technical difference:

"#define" defines a MACRO, which causes a text substitution when you use it. (This is sort-of copied from the C language. Most assemblers (including MPASM) have another type of macro common to assemblers that works differently.)

"equ" defines a label-style SYMBOL, which gets assigned a value, and you can use it where you would have used that value.

So since D0_ON is just a number, [the] examples behave the same. "**A major reason to use the #define form is that the same set of definitions can be used for C programs as well as assembler. A major reason to use the "equ" form is that the symbol will show up in cross reference listings and debuggers"** [2].

PROGRAM STRUCTURES

SUBPROGRAMS

"*Subprograms* are the basic units of modularity" [3]. These allow the programmer to form logical blocks of code and to program in chunks that are more easily digestible.

SUBROUTINES

Subroutines are subprograms that perform a specific task and may be called as needed by other blocks of code, either in main or within another subprogram. Subroutines are blocks of code that may be repeatedly called by the main program to serve a given function. Subroutines are executed during program run time. See Chapter 4 Beginning Code for an example.

MACROS

Macros are similar to Subroutines in that they may be used to execute blocks of code multiple times. Macros differ significantly from subroutines as they are used by the assembler, expanding out the code at each instance of the macro. See Chapter 4 Beginning Code for an example. The physical sequence of code is repeated, not called. So, if you use a macro 25 times, you will have 25 copies of the macro code.

PROCEDURES

Procedures are subprograms that allow parameter passing to the subprogram but do not equate to a parameter value. Parameters include all *variables* used within the procedure. There are *global variables* that are used within the program and within called subroutines. There are also *local variables* that are restricted to use within the given procedure or function. Values may be passed from the main routine to a subroutine in the calling header, which defines these values.

FUNCTIONS

Functions are subprograms that (may or may not use global or local parameters) return a value in the name of the function, and treat that function name as a variable.

RECURSIVE PROCEDURES

Recursive procedures can call themselves, either *directly*, by calling itself within its own body, or can *indirectly* call itself from within other subprograms that eventually turn to call the original procedures. Following is an example of a recursive procedure, presented in pseudocode.

Label	Opcode	Operands	Comments
	PROCEDURE	trap (n:integer)	
	BEGIN		
	IF	n=0	
	THEN	put_line ("escape ! !");	
	ELSE	**trap(n);**	
	ENDIF;		
	END	trap;	

TERMINATION

Observe, in the former code, that the value n never changes, so n never reaches 0 and this code will run indefinitely. If we change the value of n to "n-1," the value of n will reach 0. We refer to this affect as *termination.*

Label	Opcode	Operands	Comments
	PROCEDURE	trap (n:integer)	
	BEGIN		
	IF	n=0	
	THEN	put_line ("escape ! !");	
	ELSE	**trap(n-1);**	
	END IF;		
	END	trap;	

DATA STRUCTURES

We are going to switch gears for a bit and look at data structures that manipulate alphanumeric characters, instead of just numbers. More detail on numbers follows in the coming chapters. Numbering systems are covered in Chapter 9; with detailed arithmetic functions in Chapter 10.

HANDLING DATA IN ASCII

"The American Standard Code for Information Interchange," or ASCII code, was created in 1963 by the "American Standards Association" Committee or "ASA", the agency changed its name in 1969 [to] "American National Standards Institute" or "ANSI" as it [has been] known since.

"At first, [ASCII] only included capital letters and numbers. In 1967, lowercase letters and some control characters, forming what is known as US-ASCII, [in other words] the characters 0 through 127" [3].

The American Standard Code for Information Interchange (ASCII) is the current standard alphanumeric table of data, shown in Fig. 8.1. There are only 128 possible characters. This means that only 7 bits are required to access characters in ASCII code. The Most Significant Bit (MSB) of an ASCII character is zero [in the original 128 character table.]

"In 1981, IBM developed an extension of 8-bit ASCII code [and] in this version were replaced some obsolete control characters for graphic characters. Also 128 characters were added, with new symbols, signs, graphics and latin letters, all punctuation signs and characters needed to write texts in other languages, such as Spanish. For this purpose, there were 128 characters added to the ASCII characters ranging from 128 to 255" [3].

A second standard, Extended Binary Coded Decimal Interchange Code (EBCDIC), does bear mentioning. However, it has been relegated to a really distant second position to ASCII.

CHARACTER CODED DATA

In handling ASCII data, one must abide by some basic principles that follow.

1. "The codes for the numbers and letters form ordered subsequences. **Since the codes for the upper-case letters (41h through 5Ah) are ordered alphabetically, you can alphabetize strings by sorting them according to their numerical values.**
2. Many ASCII devices do not use the entire character set.
3. ASCII control characters often have widely varying interpretations.
4. Some widely used ASCII characters are:
   0Ah = line feed (LF)
   0Dh = carriage return (CR)
   20h = space
   3Fh = question mark (?)
   7Fh = delete character (DEL)" [4]
5. **For practical use, just memorize the number 65d as it is the start of the capitalized alphabet.**

<table>
<tr><th>ASCII #</th><th>Character</th><th>ASCII #</th><th>Character</th><th>ASCII #</th><th>Character</th><th>ASCII #</th><th>Character</th><th>ASCII #</th><th>Character</th><th>ASCII #</th><th>Character</th><th>ASCII #</th><th>Character</th><th>ASCII #</th><th>Character</th></tr>
<tr><td>0</td><td>nul</td><td>16</td><td>dle</td><td>32</td><td>sp</td><td>48</td><td>0</td><td>64</td><td>@</td><td>80</td><td>P</td><td>96</td><td>`</td><td>112</td><td>p</td></tr>
<tr><td>1</td><td>soh</td><td>17</td><td>dc1</td><td>33</td><td>!</td><td>49</td><td>1</td><td>65</td><td>A</td><td>81</td><td>Q</td><td>97</td><td>a</td><td>113</td><td>q</td></tr>
<tr><td>2</td><td>stx</td><td>18</td><td>dc2</td><td>34</td><td></td><td>50</td><td>2</td><td>66</td><td>B</td><td>82</td><td>R</td><td>98</td><td>b</td><td>114</td><td>r</td></tr>
<tr><td>3</td><td>etx</td><td>19</td><td>dc3</td><td>35</td><td>#</td><td>51</td><td>3</td><td>67</td><td>C</td><td>83</td><td>S</td><td>99</td><td>c</td><td>115</td><td>s</td></tr>
<tr><td>4</td><td>eot</td><td>20</td><td>dc4</td><td>36</td><td>$</td><td>52</td><td>4</td><td>68</td><td>D</td><td>84</td><td>T</td><td>100</td><td>d</td><td>116</td><td>t</td></tr>
<tr><td>5</td><td>enq</td><td>21</td><td>nak</td><td>37</td><td>%</td><td>53</td><td>5</td><td>69</td><td>E</td><td>85</td><td>U</td><td>101</td><td>e</td><td>117</td><td>u</td></tr>
<tr><td>6</td><td>ack</td><td>22</td><td>syn</td><td>38</td><td>&</td><td>54</td><td>6</td><td>70</td><td>F</td><td>86</td><td>V</td><td>102</td><td>f</td><td>118</td><td>v</td></tr>
<tr><td>7</td><td>bel</td><td>23</td><td>etb</td><td>39</td><td>'</td><td>55</td><td>7</td><td>71</td><td>G</td><td>87</td><td>W</td><td>103</td><td>g</td><td>119</td><td>w</td></tr>
<tr><td>8</td><td>bs</td><td>24</td><td>can</td><td>40</td><td>(</td><td>56</td><td>8</td><td>72</td><td>H</td><td>88</td><td>X</td><td>104</td><td>h</td><td>120</td><td>x</td></tr>
<tr><td>9</td><td>ht</td><td>25</td><td>em</td><td>41</td><td>)</td><td>57</td><td>9</td><td>73</td><td>I</td><td>89</td><td>Y</td><td>105</td><td>i</td><td>121</td><td>y</td></tr>
<tr><td>10</td><td>nl</td><td>26</td><td>sub</td><td>42</td><td>*</td><td>58</td><td>:</td><td>74</td><td>J</td><td>90</td><td>Z</td><td>106</td><td>j</td><td>122</td><td>z</td></tr>
<tr><td>11</td><td>vt</td><td>27</td><td>esc</td><td>43</td><td>+</td><td>59</td><td>;</td><td>75</td><td>K</td><td>91</td><td>[</td><td>107</td><td>k</td><td>123</td><td>{</td></tr>
<tr><td>12</td><td>np</td><td>28</td><td>fs</td><td>44</td><td>,</td><td>60</td><td><</td><td>76</td><td>L</td><td>92</td><td>\</td><td>108</td><td>l</td><td>124</td><td>|</td></tr>
<tr><td>13</td><td>cr</td><td>29</td><td>gs</td><td>45</td><td>-</td><td>61</td><td>=</td><td>77</td><td>M</td><td>93</td><td>]</td><td>109</td><td>m</td><td>125</td><td>}</td></tr>
<tr><td>14</td><td>so</td><td>30</td><td>rs</td><td>46</td><td>.</td><td>62</td><td>></td><td>78</td><td>N</td><td>94</td><td>^</td><td>110</td><td>n</td><td>126</td><td>~</td></tr>
<tr><td>15</td><td>si</td><td>31</td><td>us</td><td>47</td><td>/</td><td>63</td><td>?</td><td>79</td><td>O</td><td>95</td><td>_</td><td>111</td><td>o</td><td>127</td><td>del</td></tr>
</table>

FIG. 8.1

ASCII Table (7-bit).

CHARACTER OPERATIONS

Example: Length of a String of Characters

PURPOSE:

Determine the length of a string of characters. The string starts in memory location 0014h; the end of the string is marked by an ASCII carriage return character ("CR", 0Dh). Place the length of the string (excluding the carriage return) into memory location 0013h [5]

TEST CASE:

A.	(0014h) = 0Dh	
RESULT:	(0013h) = 00h	

TEST CASE:

B.	(0014h) = 52h	'R'
	(0015h) = 41h	'A'
	(0016h) = 54h	'T'
	(0017h) = 48h	'H'
	(0018h) = 45h	'E'
	(0019h) = 52h	'R'
	(001Ah) = 0Dh	CR
RESULT:	(0013h) = 06h	

SOLUTION:

Label	Opcode	Operand	Comment
LENGTH	EQU	0X13	
RES_VECT CODE		0X0000	;PROCESSOR RESET ;VECTOR
	GOTO	START	;GO TO BEGINNING OF ;PROGRAM
			;MAIN_PROG CODE
	START		
	MOVLW	52h	;ARRAY ELEMENTS PLACED
	MOVWF	14h	
	MOVLW41H		
	MOVWF	15h	
	MOVLW	54h	
	MOVWF	16h	
	MOVLW	48h	
	MOVWF	17h	
	MOVLW	45h	
	MOVWF	18h	

Continued

Label	Opcode	Operand	Comment
	MOVLW	52h	
	MOVWF	19h	
	MOVLW	0Dh	
	MOVWF	1Ah	
	MOVLW	14h	;ARRAY START POSITION
	MOVWF	FSR	;(POINTER)
	MOVLW	00h	;CLEAR LENGTH VALUE
	MOVWF	LENGTH	
LOOP	MOVLW	0Dh	;PLACE 0DH VALUE IN W FOR COMPARISON
	SUBWF	INDF, 0	;SUBTRACT 0DH FROM CONTENTS OF INDF
	BTFSC	STATUS,Z	;OF ARRAY, ZERO IF IT IS 0DH ;SKIP TO DONE
	GOTO	DONE	
	INCF	LENGTH, 1	
	INCF	FSR, 1	
	GOTO	LOOP	
DONE			
	END		

CHARACTER OPERATIONS

Example: Pattern Match

PURPOSE: Compare two strings of ASCII characters to see which is larger (that is, which follows the other in alphabetical ordering). The length of the strings is in memory location 0X13; one string starts in memory location 0X14 and the other in memory location 0X18. Strings are compared digit by digit. If the strings match, clear memory location 0X17; otherwise, set memory location 0X17 to all ones (FFh)

TEST CASE:

A.	(0014h) = 53h	'C'
	(0015h) = 41h	'A'
	(0016h) = 43h	'T'
	(0018h) = 53h	'C'
	(0019h) = 41h	'A'
	(001Ah) = 43h	'T'
RESULT:	(0X17) = 00h	

TEST CASE:

B.	(0014h) = 52h	'B'
	(0015h) = 41h	'A'
	(0016h) = 43h	'T'
	(0018h) = 53h	'C'
	(0019h) = 41h	'A'
	(001Ah) = 43h	'T'
RESULT:	(0X17) = FFh	

SOLUTION:

Label	Opcode	Operand	Comment
TEST	EQU	0X11	
OUTPUT	EQU	0X12	
LENGTH	EQU	0X13	
STRING1	EQU	0X14	
STRING2	EQU	0X18	
MATCH1	EQU	0X17	
RES_VECT CODE		0X0000	;PROCESSOR RESET VECTOR
	GOTO	START	;GO TO BEGINNING OF PROGRAM
			;MAIN_PROG CODE
START			;ARRAY ELEMENTS PLACED
	MOVLW	43h	;C
	MOVWF	14h	
	MOVLW	41h	;A
	MOVWF	15h	
	MOVLW	54h	;T
	MOVWF	16h	
	MOVLW	43h	;C
	MOVWF	18h	
	MOVLW	41h	;A
	MOVWF	19h	
	MOVLW	54h	;T
	MOVWF	1Ah	
	MOVLW	0X03	
	MOVWF	LENGTH	
	MOVLW	0X14	
	MOVWF	FSR	

Continued

Label	Opcode	Operand	Comment
LOOP	MOVF	INDF,0	
	MOVWF	TEST	
	MOVLW	0X04	
	ADDWF	FSR,1	
	MOVF	INDF,0	
	SUBWF	TEST,0	
	BTFSS	STATUS, Z	
	GOTO	NOMATCH	
	GOTO	MATCH	
NOMATCH	MOVLW	0XFF	
	MOVWF	MATCH1	
	GOTO	DONE	
MATCH	CLRF	MATCH1	
	MOVLW	0X04	
	SUBWF	FSR, 1	
	INCF	FSR, 1	
	DECF	LENGTH, 1	
	BTFSS	STATUS, Z	
	GOTO	LOOP	
DONE	GOTO $		;endless loop
	END		

CONTROL STRUCTURES

As you'll recall from Chapter 7, there are four basic processes in programming any device: function, sequence, alternation (or multiple selections), and repetition (or iteration). Shown below, there are examples of the implementation of each structure.

FUNCTION

A *Function* differs from a Procedure in that it returns a value when the calling routine is complete. The CALL places the PC+1 on the stack. The RETLW pops the last entry to return, via the Program Counter, to the originating CALL.

Label	Opcode	Operand	Comment
	CALL	S	
S			
	RETLW	Value	

SEQUENCE

The *Sequence* control structure implements two or more processes in order. In Sequence, the processes are completed one after another.

Label	Opcode	Operand	Comment
S1			;
S2			;

ALTERNATION (OR MULTIPLE SELECTIONS)

Alternation is a control structure that uses a variable or literal value to specify one of several options. The alternation process is basically a series of if-then-else-endif statements. Pseudocode follows. This can also be implemented with a "Case" structure in most high level languages.

Label	Opcode	Operand	Comment
	IF	condition 1	
	THEN		
	GOTO	A	
	ELSE IF	condition 2	
	THEN		
	GOTO	B	
	ELSE IF	condition 3	
	THEN		
	GOTO	C	
	ELSE IF	condition 4	
	THEN		
	GOTO	D	
	ENDIF		
	ENDIF		
	ENDIF		

This structure would be implemented in assembly with BTFSS or BTFSC comparison, followed by a GOTO.

Label	Opcode	Operand	Comment
	BTFSC	condition 1	
	GOTO	A	
	BTFSC	condition 2	
	GOTO	B	
	BTFSC	condition 3	
	GOTO	C	
	BTFSC	condition 4	
	GOTO	D	

Decision Tables

Decision Tables, as follows excerpt from Chapter 4, mark the conditions under which the tests should direct program flow. These tables are useful in identifying each condition in a list. The table is an example of an alternation condition table.

Descriptor	Soil Level
Whites	N
Colors/Normal	N
Stain Wash	H
Easy Care	N
Active Wear	H
Delicates	N
Hand Wash	N
Speed Wash	X
Drain and Spin	Off
Rinse and Spin	Off

REPETITION (OR ITERATION)

Repetition, also known as *Iteration*, establishes a loop to repeat portions of the test value that must be met and directs the code to exit the loop under an established condition, using either BTFSS or BTFSC. Two processes are identified as S1 and S2.

Label	Opcode	Operand	Comment
LOOP	S1		;Process one
	S2		;Process two
	BTFSS	exit condition1	
	GOTO	LOOP	
	NEXT	Process	

WORD SEARCH: DATA STRUCTURES

P	Y	M	U	V	E	U	E	I	D	V	I	L	E	N
C	P	Y	B	B	M	C	N	M	E	Z	W	R	C	O
E	W	E	C	L	N	P	G	P	F	I	O	A	A	I
Z	F	D	V	E	S	R	C	L	I	G	H	L	S	T
B	I	P	U	A	O	Q	X	E	N	H	P	M	A	C
C	I	Q	N	E	L	M	D	M	E	H	N	G	S	N
X	E	D	V	O	V	D	E	E	A	B	K	W	W	U
S	N	T	N	O	I	T	A	N	R	E	T	L	A	F
X	N	P	R	X	Q	T	U	T	C	I	I	C	S	A
B	Z	Z	M	O	X	M	I	A	M	O	Q	M	E	W
C	P	A	T	T	E	R	N	T	F	S	N	F	K	M
M	M	D	Y	R	L	M	M	I	E	E	C	F	S	K
S	V	V	I	R	N	V	X	O	K	P	O	F	I	L
T	T	C	I	I	U	B	I	N	W	W	E	Y	X	G
S	V	X	O	F	Z	M	U	U	X	J	S	R	W	P

ALPHANUMERIC
ALTERNATION
ASCII
CONFIG
DEFINE
EBCDIC
FUNCTION
IMPLEMENTATION
PATTERN
REPETITION
SEQUENCE

PUZZLE: JIGSAW SUDOKU 66H

JIGSAW SUDOKU 66h

7h thru Fh

9	A			B	E			8
					D	F		
B					8		E	
	8		F	7		D		
	7					8		
					B			
	C	E			7			
				9				

How to solve Jigsaw Sudoku

To complete the following puzzle, fill in the squares so that each digit is a hexadecimal digit between 7h and Fh. Each Jigsaw piece will sum to the hexadecimal value of 66h. Each digit will appear exactly once in each row, in each column, and in each enclosed nine-block Jigsaw piece.

FURTHER READING

AN544 from Embedded Control Handbook.

REFERENCES

1 Anon, http://www.ModronicsAustralia.com, as Download on 10/01/2017, n.d. (158 Verbatim Words Copied).
2 Anon, EEVblog Electronics Community Forum With Contributors EENG and westfw, n.d.
3 Anon, http://www.theasciicode.com.ar from Argentina, n.d.
4 L.A. Leventhal, 56809 Assembly Language Programming, n.d.
5 J. Hext, Programming Structures, Volume 1 Machines and Programs, n.d., p. 243.

CHAPTER

Logic and numbering systems

9

CHAPTER OUTLINE

BOOLEAN ALGEBRA

Boolean algebra defines a set of operations on the values **true** and **false**. True is represented as a binary value of 1. False is represented as a binary value of 0. Boolean algebra was invented by George Boole, a mathematician who designed the first computer named the *Analytical Engine*. It is through Boolean algebra that we physically construct *hardware gates* that take in inputs and generate outputs. All of these may also be constructed out of *software using the logical operators*.

The Art of Assembly Language Programming Using PIC® Technology. https://doi.org/10.1016/B978-0-12-812617-2.00009-2

LOGICAL OPERATORS

Logical Operators are assembly language commands that act on two input values and give one output value. Let's take a look at a generic black box with the input values of A and B and the output value of F. To understand how the Logical Operators work, we engage the use of *truth tables*. Truth tables show, in tabular form, the inputs on the left side columns, A and B, and the output, F, on the right side column of the table (Fig. 9.1).

There are three basic functions: AND, OR (IOR), XOR. We will take a look at the AND function, for ANDWF and ANDLW; the inclusive OR function, for IORWF and IORLW; and the XOR exclusive OR function for XORWF and XORLW (see Fig. 9.2).

AND/NAND gates

The AND gate is represented by a D shaped object. The AND gate requires that both A and B must be true (1) before the output F can be true (1). All other input combinations result in a false (0) output F (Fig. 9.3).

The NAND, or not AND, gate, is represented by an AND gate with a bubble on the output indicating the NOT of the output. All input combinations result in a true

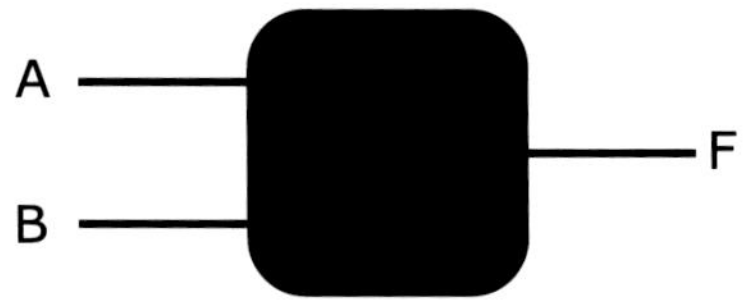

FIG. 9.1

Boolean black box.

Opcode	Operands	Description	Status Affected
ANDWF	f,d	AND W with f	Z
ANDLW	k	AND literal k with W	Z
IORWF	f,d	Inclusive OR W with f	Z
IORLW	k	Inclusive OR literal k with W	Z
XORWF	f,d	Exclusive OR W with f	Z
XORLW	k	Exclusive OR literal k with W	Z

FIG. 9.2

Logical operators: ANDWF, IORWF, XORWF, ANDLW, IORLW, XORLW.

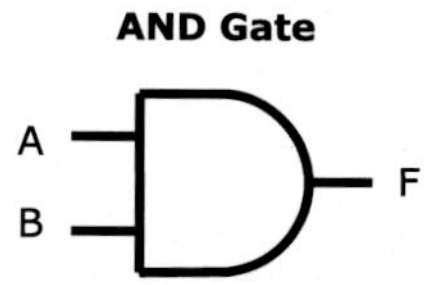

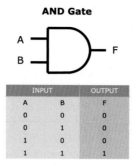

INPUT		OUTPUT
A	B	F
0	0	0
0	1	0
1	0	0
1	1	1

FIG. 9.3

AND gate and truth table.

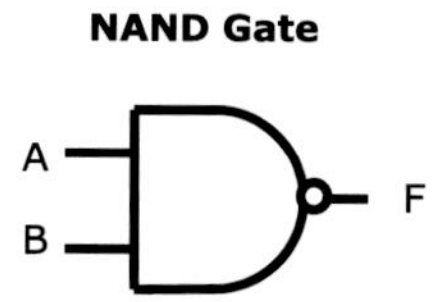

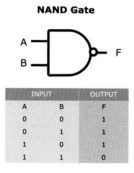

INPUT		OUTPUT
A	B	F
0	0	1
0	1	1
1	0	1
1	1	0

FIG. 9.4

NAND gate and truth table.

(1) except for the one state where both A and B are true; this state is a false (0) (see Fig. 9.4). The NAND is a very special function in that multiples of it can be used to create any of the other gates. We will look at this special relationship at the end of this chapter.

The AND Microchip assembly language logical operators are ANDWF and ANDLW (see Fig. 9.5). The logical operators are bit wise functions. The first operator, ANDWF, ands the working register, W, and the value contained in f. The destination, d, will be 0 if the solution is placed in the W register. The destination, d, will be 1 if the

Opcode	Operands	Description	Status Affected
ANDWF	f,d	AND W with f	Z
ANDLW	k	AND literal/constant with W	Z

FIG. 9.5

Logical operators ANDWF and ANDLW.

solution is placed in the memory location f. The second operator, ANDLW, ands the literal k with the working register W. The solution is placed in the W register.

IOR/NOR gates

The OR gate is represented by a gate with a concave side where the inputs are located (Figs. 9.6 and 9.7). The basic OR gate may also be called the IOR gate, for *inclusive* OR. This means that the output will be true if both A and B are true; also, inclusive of the states where only one input is true.

As with the NAND function, NOR negates the output values for the gate. This negation is indicated by the bubble on the output of the gate. If any of the inputs are true, the output will be false. The output will be true when both inputs are false.

The inclusive OR, IOR, Microchip assembly language logical operators are IORWF and IORLW (see Fig. 9.8). The logical operators are bit wise functions. The first operator, IORWF, ors the working register, W, and the value contained in f. The destination, d, will be 0 if the solution is to be placed in the W register. The destination, d, will be 1 if the solution is to be placed in the memory location f. The second operator, IORLW, inclusive ors the literal k with the working register, W. The solution is placed in the W register.

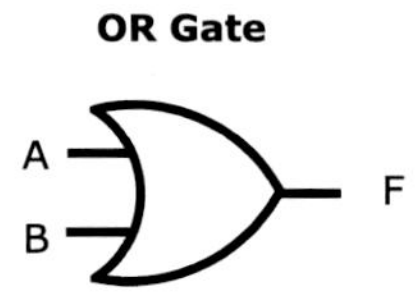

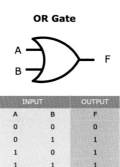

INPUT		OUTPUT
A	B	F
0	0	0
0	1	1
1	0	1
1	1	1

FIG. 9.6

IOR gate and truth table.

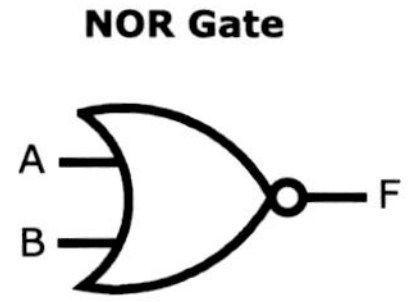

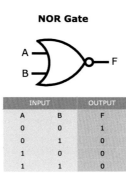

INPUT		OUTPUT
A	B	F
0	0	1
0	1	0
1	0	0
1	1	0

FIG. 9.7

NOR gate and truth table.

Opcode	Operands	Description	Status Affected
IORWF	f,d	Inclusive OR W with f	Z
IORLW	k	Inclusive OR literal/constant k with W	Z

FIG. 9.8

Logical operators IORWF and IORLW.

XOR/XNOR gates

The exclusive or, XOR, is illustrated similar to the OR, or IOR, gate with a concave input side (Figs. 9.9 and 9.10). The exclusive part of XOR is illustrated with a double concave shape on the input side. The XOR gate gives a true, 1, output if one or the other of the inputs are true, 1. When both inputs, A and B, are true, the XOR gate yields a false (0).

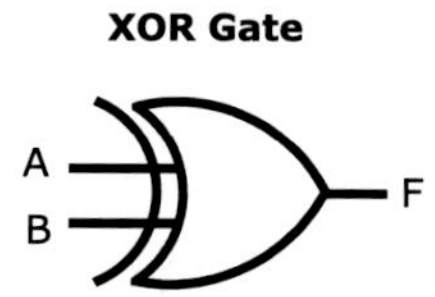

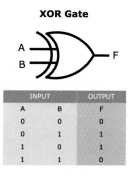

INPUT		OUTPUT
A	B	F
0	0	0
0	1	1
1	0	1
1	1	0

FIG. 9.9

XOR gate and truth table.

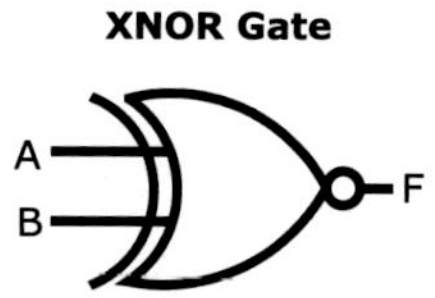

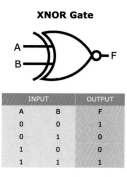

INPUT		OUTPUT
A	B	F
0	0	1
0	1	0
1	0	0
1	1	1

FIG. 9.10

XNOR gate and truth table.

Opcode	Operands	Description	Status Affected
XORWF	f,d	Exclusive OR W with f	Z
XORLW	k	Exclusive OR literal k with W	Z

FIG. 9.11

Logical Operators XORWF and XORLW.

The XNOR yields a NOT of the XOR gate and truth table. When both inputs are false, 0, or when both inputs are true, 1, then the XNOR is true (1). When one of the inputs is true and the other input is false, the XNOR yields a false (0).

The Exclusive OR, XOR, Microchip assembly language logical operators are XORWF and XORLW (Fig. 9.11). The logical operators are bit wise functions. The first operator, XORWF, exclusive ors the working register, W, and the value contained in f. The destination, d, will be 0 if the solution is to be placed in the W register. The destination, d, will be 1 if the solution is to be placed in the memory location f. The second operator, XORLW, exclusive ors the literal k with the working register W. The solution is placed in the W register.

Gray code

Note that the truth tables, shown here, are set in an order that places the inputs in a binary order: 00 (for zero), 01 (for one), 10 (for two), and 11 (for three).

There is a better alternative known as Gray code. This sequence provides a specific order that permits only one bit to change at a time: 00 (for zero), 01 (for one), 11 (for three), and then 10 (for two). Interestingly, this technique has advantages in making more robust code. If implemented in hardware, this approach has other distinct advantages in preventing spurious output from electromechanical switches and in minimizing the effect of error in conversion of analog signals to digital. This technique is referred to as *reflective binary code (RBC)* or, simply, *Gray code,* after Frank Gray who discovered and patented this technique.

NAND, the universal operation

The NAND is a very special function; multiples of it can be used to create any of the other gates. We can implement various functions by using one or more NAND operators.

NAND Gate

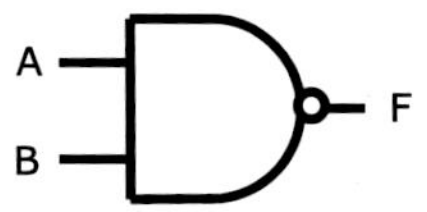

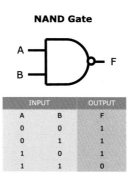

INPUT		OUTPUT
A	B	F
0	0	1
0	1	1
1	0	1
1	1	0

If we tie both inputs together, the output will be a simple NOT function.

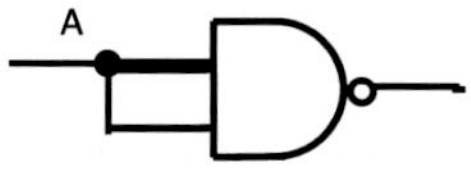

POLYNOMIAL SYSTEMS

DECIMAL SYSTEM

The decimal system is an example of a polynomial number system. The reason for the name is that the value of any numeral d is given by a polynomial. For example, the value of 2437 is given by the expression:

$$2\times 10^3 + 4\times 10^2 + 3\times 10^1 + 7\times 10^0$$

Or more simply

$$(2\times 1000)+(4\times 100)+(3\times 10)+(7\times 1)$$

In the example, the polynomial in 10 is represented, whose coefficients are the digits 2437. In general, any integer can be used as the *base* or *radix*. Some important bases, used in assembly language programming, are as follows:

- Base 2—binary number system;
- Base 8—octal number system;
- Base 10—decimal number system;
- Base 16—hexadecimal number system.

The context usually makes it clear which base is being used, the standard default being 10. If working with multiple *bases*, or *radices*, the base is subscripted to the right of the value, as 2437_{10}, or indicated, as with PIC®, by the alphabetic character d: 2437d, for decimal number system.

BINARY, OCTAL, HEXADECIMAL, DECIMAL

This chapter begins with the familiar standard numbering system of decimal. There are observations one can make on the decimal numbering system. These observations allow us to then make parallel comparisons with other mathematical systems. In this context, one can, by comparison, introduce additional systems of binary, octal, and hexadecimal. Unless otherwise noted, numbers presented in this textbook shall by default be decimal numbers.

Let's take a look at decimal numbers. Here presented is a random 8-digit number: 53,756,598. In the decimal numbering system, one works with base 10 numbers, with numerals 0, 1, 2, 3, 4, 5, 6, 7, 8, 9. So, herein, we use this random number as a base 10 number example: 53,756,598.

10,000,000	**1,000,000**	**100,000**	**10,000**	**1000**	**100**	**10**	**1**
10×10×10× 10×10×10×10	10×10×10× 10×10×10	10×10×10× 10×10	10×10× 10×10	10×10× 10	10×10	10	1
10^7	10^6	10^5	10^4	10^3	10^2	10^1	10^0
5	**3**	**7**	**5**	**6**	**5**	**9**	**8**

Note that, in decimal, base 10, the right-most first digit is the ones, as it is base 10 to the zeroth power. The second digit is base 10, 10 to the first power, or 10. The third digit is the hundreds digit, 10 to the second power, or 100. The fourth digit is 10 to the third power, or 1000. The fifth digit is 10 to the fourth power, or 10,000. And so on, until the eighth digit as 10 to the seventh power.

BINARY SYSTEM

Binary is the name of the numbering system that is base 2. This system, therefore, has numerals 0 and 1. That means that numerals 2, 3, 4, 5, 6, 7, 8, and 9 do not exist in this numbering system. Row 1 is the decimal (base 10) equivalent value.

As an example, use the decimal example value: 214. (This is a totally arbitrary value.) Look at the highest decimal value represented here: 128. Divide the value that we wish to represent, 214, by the value 128: 1.671875. *Truncate* (or "chop off") the decimal section 0.671875. The 1 indicates that 1×128 is the highest value that can be placed in the 2^7 position.

128	**64**	**32**	**16**	**8**	**4**	**2**	**1**
2x2x2x2x 2x2x2	2x2x2x2x2x2	2x2x2x2x2	2x2x2x2	2x2x2	2x2	2	1
2^7	2^6	2^5	2^4	2^3	2^2	2^1	2^0
1							

Now we subtract (1×128) from the original value, 214, to reveal what can be represented by the rest of the value. So subtracting 128 from 214, this leaves 86 to be represented by the rest of the binary number.

To find the 2^6 value, divide 86 by (2^6 or) 64: 1.34375. *Truncate* the decimal value to 1. The remainder is $86-(1 \times 64)$: 22.

128	**64**	**32**	**16**	**8**	**4**	**2**	**1**
2x2x2x2x 2x2x2	2x2x2x2x2x2	2x2x2x2x2	2x2x2x2	2x2x2	2x2	2	1
2^7	2^6	2^5	2^4	2^3	2^2	2^1	2^0
1	**1**						

Continue to calculate the next value. How many 32s are in 22? Zero. Place the 0 in the 32 column.

128	**64**	**32**	**16**	**8**	**4**	**2**	**1**
2x2x2x2x 2x2x2	2x2x2x2x2x2	2x2x2x2x2	2x2x2x2	2x2x2	2x2	2	1
2^7	2^6	2^5	2^4	2^3	2^2	2^1	2^0
1	**1**	**0**					

Continuing with the next value, 16. How many 16s are there in 22? One. Place a 1 in the 16 column. Subtracting 16 from 22 gives the remainder of 6.

128	64	32	16	8	4	2	1
2×2×2×2× 2×2×2	2×2×2×2×2×2	2×2×2×2×2	2×2×2×2	2×2×2	2×2	2	1
2^7	2^6	2^5	2^4	2^3	2^2	2^1	2^0
1	1	0	1				

How many 8s are there in 6? Zero. Place a 0 in column 8. Since we are to subtract (0 * 8), or 0, the value we seek is still 6.

128	64	32	16	8	4	2	1
2×2×2×2× 2×2×2	2×2×2×2×2×2	2×2×2×2×2	2×2×2×2	2×2×2	2×2	2	1
2^7	2^6	2^5	2^4	2^3	2^2	2^1	2^0
1	1	0	1	0			

How many 4s are there in 6? One. Place a one in the 4s column. Subtract (1 * 4) or 4 from 6: 2.

128	64	32	16	8	4	2	1
2×2×2×2× 2×2×2	2×2×2×2×2×2	2×2×2×2×2	2×2×2×2	2×2×2	2×2	2	1
2^7	2^6	2^5	2^4	2^3	2^2	2^1	2^0
1	1	0	1	0	1		

How many 2s are in 2? One. Subtract 2 from 2 for zero remaining ones. Place a 1 in the 2s column. Place a 0 in the 1s column.

128	64	32	16	8	4	2	1
2×2×2×2× 2×2×2	2×2×2×2×2×2	2×2×2×2×2	2×2×2×2	2×2×2	2×2	2	1
2^7	2^6	2^5	2^4	2^3	2^2	2^1	2^0
1	1	0	1	0	1	1	0

That's it! Representing the value 214 is our binary number 11010110b.

OCTAL SYSTEM

Octal is the name of the numbering system that is base 8. This system, therefore, has numerals 0, 1, 2, 3, 4, 5, 6, and 7. That means that numerals 8 and 9 do not exist in this numbering system. Row 1 is, again, the decimal (base 10) equivalent value.

2,097,152	262,144	32,768	4096	512	64	8	1
8×8×8×8×8×8×8	8×8×8×8×8×8	8×8×8×8×8	8×8×8×8	8×8×8	8×8	8	1
8^7	8^6	8^5	8^4	8^3	8^2	8^1	8^0

HEXADECIMAL SYSTEM

Hexadecimal is the name of the numbering system that is base 16. This system, therefore, has numerals 0, 1, 2, 3, 4, 5, 6, 7, 8, 9, 10, 11, 12, 13, 14, and 15. That means that two-digit decimal numbers 10, 11, 12, 13, 14, and 15 must be represented by a single numeral to exist in this numbering system. To address the two-digit decimal values, the alphabetic characters A, B, C, D, E, and F are used to represent these values in hexadecimal and are treated as valid numerals. Row 1 is, again, the decimal (base 10) equivalent value.

268,435,456	16,777,216	1,048,576	65,536	4096	256	16	1
16×16×16×16×16×16×16	16×16×16×16×16×16	16×16×16×16×16	16×16×16×16	16×16×16	16×16	16	1
16^7	16^6	16^5	16^4	16^3	16^2	16^1	16^0
0	**0**	**0**	**0**	**0**			

Using an example value of 538, what is the base 16 equivalent value. First place zeros in all columns that are known to be too big. 256 is the start of values below 538, so put a 0 in all columns to the left of 256.

268,435,456	16,777,216	1,048,576	65,536	4096	256	16	1
16×16×16×16×16×16×16	16×16×16×16×16×16	16×16×16×16×16	16×16×16×16	16×16×16	16×16	16	1
16^7	16^6	16^5	16^4	16^3	16^2	16^1	16^0
0	**0**	**0**	**0**	**0**	**2**		

$538-(2\times 256)=26.$

268,435,456	16,777,216	1,048,576	65,536	4096	256	16	1
16×16×16×16×16×16×16	16×16×16×16×16×16	16×16×16×16×16	16×16×16×16	16×16×16	16×16	16	1
16^7	16^6	16^5	16^4	16^3	16^2	16^1	16^0
0	**0**	**0**	**0**	**0**	**2**		

$26-(1\times 16)=10$

268,435,456	16,777,216	1,048,576	65,536	4096	256	16	1
16×16×16× 16×16× 16×16	16×16×16× 16×16×16	16×16×16× 16×16	16×16× 16×16	16×16×16	16×16	16	1
16^7	16^6	16^5	16^4	16^3	16^2	16^1	16^0
0	**0**	**0**	**0**	**0**	**2**	**1**	

10 − (10 * 1) = 0. Recall, 10 decimal is represented by Ah hexadecimal. The complete value is then 21Ah

268,435,456	16,777,216	1,048,576	65,536	4096	256	16	1
16×16×16× 16×16× 16×16	16×16×16× 16×16×16	16×16×16× 16×16	16×16× 16×16	16×16×16	16×16	16	1
16^7	16^6	16^5	16^4	16^3	16^2	16^1	16^0
0	**0**	**0**	**0**	**0**	**2**	**1**	**A**

This will come to you with practice and memorization of typical values. Rarely do we work with numbers greater than 4096 in hexadecimal.

MIXED RADIX SYSTEMS

Mixed Radix Systems are rarely used in assembly language programming. Two examples, however, are of note: time and distance measurements.

Time is usually expressed as hours-minutes-seconds, with an example of hour 12, minutes 23, and seconds 47. The numeral would be 122,347 with the radices 10, 10, 6, 10, 6, and 10. The first digit, 1, is base 10, and second numeral is 2, base 10; with the 2 digit value, 12, in base 6. The third and fourth digits being 23. This is, again, first base 10 (numeral 2), then base 10 (numeral 3), then base 6 (numerals 23). The number of seconds is then base 10 (numeral 4), and base 10 (numeral 7). The total number of seconds may be obtained from the following calculation:

$$(1\times 36{,}000)+(2\times 3600)+(2\times 600)+(3\times 60)+(4\times 10)+(6\times 1)=44{,}626$$

This number is then the total number of seconds since midnight. To return the value back to hours-minutes-seconds format, complete successive divisions, in the **reverse** order, by 10, 6, 10, 6, 10, and 10.

Measuring length in yards, feet, and inches may also be treated similarly, however falsely, as a mixed radix system. The last numerals, in number of inches, fail to mathematically work.

There are three applications in computing: representation of *n*-dimensional arrays, the ordering of blocks on a disk, and the factorial number system. These examples are left as an exercise for the interested reader.

MULTIBYTE INTEGERS

HEXADECIMAL TO BINARY CONVERSIONS

In the last example, we obtained the hexadecimal value of 021Ah. Note that each numeric value is a two-byte number, with a half byte for each value. The half byte is 4 bits in binary.

Hexadecimal Value	Binary Equivalent in 4-bit *nybbles*
0	0000
2	0010
1	0001
A	1010

The hexadecimal value is 021Ah and the binary equivalent is 0000 0010 0001 1010, as shown in the above table.

LITTLE-ENDIAN VS. BIG-ENDIAN

When you go to store this two-byte value in memory, you arrive at a conundrum: do you store the little-end of the number (least significant byte or "little-endian") or the big-end of the number (most significant byte or big endian?) There is no "right" or "best" way to store the bytes in multibyte quantities. **The most important take-away is that you should make your selection wisely and be completely consistent through your code.**

For a historical note on these terms:

Endian refers to the order in which bytes are stored. The term is taken from a story in Gulliver's Travels by Jonathan Swift about wars fought between those who thought eggs should be cracked on the Big End and those who insisted on the Little End. With chips, as with eggs, it doesn't really matter as long as you know which end is up.

The processor hardware dictates whether little-endian or big-endian standards are followed. Endian-related operations native to the architecture are dictated by the architecture. For microchip processors, you can use either standard. There are a few situations when little-endian would be preferred, thus some favor using little-endian throughout.

SIGNED AND UNSIGNED INTEGERS

SIGNED AND UNSIGNED

There are three computing methods used to represent signed and unsigned integers. They are as follows:

- Sign and magnitude
- One's complement
- Two's complement

Each method has its own caveats. Once each method is understood, one should weigh the options, select carefully, and, upon selecting one method, be consistent throughout the code.

As we assess each method, we use $x+(-x)=0$ as our test case.

Sign and magnitude

The sign magnitude approach dedicates an entire bit to just the sign. As shown in the table, the highest order bit is the sign and the remaining bits are the magnitude. In the example, four bits limit the range of values from 111 ----> 000, or +7 in decimal, for positive numbers and from 000 ---> 111, or −7 in decimal, for negative numbers. The primary caveat in this method is the presence of two zeros, +0 and −0.

n	+*n*	−*n*
0	0000	1000
1	0001	1001
2	0010	1010
3	0011	1011
4	0100	1100
5	0101	1101
6	0110	1110
7	0111	1111
8	–	–

One's complement

The one's complement system negates the number by inverting all the bits. Observe that a bit can be inverted by subtracting it from 1.

n	+*n*	−*n*
0	0000	1111
1	0001	1110
2	0010	1101
3	0011	1100
4	0100	1011
5	0101	1010
6	0110	1001
7	0111	1000
8	Invalid	Invalid

Preferred over sign-magnitude, one's complement can be given as a simple sum-of-weights. For the nybble, di has the same positive weight as negative, except for d0 whose weight is $-(2^{(n-1)}-1)$.

d0	d1	d2	d3
−7	4	2	1

This method still produces two values for 0, as +0 (0000) and −0 (1111). The caveat here is that the logical conditions of $x < 0$ and $x <= 0$ are neither equivalent to $d0 = 1$. So, arithmetic tests must be distinguished from logical tests.

Two's complement

The two's complement system uses the original weight of d0 as $2^{(n-1)}$ and the negative two's complement value changes it to $-2^{(n-1)}$. This is one less value than in one's complement but is a more logical and consistent choice that maintains the basic power of two approach.

n	+*n*	−*n*
0	0000	0000
1	0001	1111
2	0010	1110
3	0011	1101
4	0100	1100
5	0101	1011
6	0110	1010
7	0111	1001
8	1000 overflow	1000 overflow

Note that 1001 has the decimal value −7 and 1111 is −1. Negating 0000 yields 0000, however, then there exists a "spare" bit pattern of 1000…0, as the number $-2^{(n-1)}$.

d0	d1	d2	d3
−8	4	2	1

Negating the value sets the overflow flag, as maximum $+n$ is $2^{(n-1)} - 1$, which does not happen in sign/magnitude or one's complement. However, the difficulties experienced with two's complement are not as difficult to remedy as the problem of two zeros. Two's complement also provides logical equivalents. See table below.

Arithmetic Test	Logical Equivalent
x>=0	d0=0
x<0	d0=1
x=0	d=000...0
x=−1	d=111...1

FLOATING POINT

In floating point, each value is represented by the set of significant digits, known as the mantissa, and the exponent. Representation of the value in this form requires that the programmer "pack" the value, into this format, prior to storage into memory. To perform arithmetic operations, one must convert, or "unpack" the values prior to the operation, and "pack" the values, after the arithmetic operation. In some representations, the most significant bit, or MSB, is used as the negative sign as in sign/magnitude representation. For example, −625 is shown in the following table diagram. This representation is called scientific notation, where the numbers are normalized so the leading digit is always the first nonzero value. To maintain a favorable number of significant digits, up to approximately 15 decimal digits, this format permits the extension of the mantissa into a second word.

Sign (Not Present=0, Present=1)	Exponent (Multiply by 10^e, Where e Is Given)	Mantissa (Normalized to One Significant Digit to the Left of the Decimal Point)
1	2	6.25

An alternate representation, without the sign bit, is shown following. Herein, the sign information may be represented by one's complement or two's complement in the mantissa section of the number.

Mantissa	Exponent
6.25	2

BCD AS A TYPE

BINARY CODED DECIMAL

Binary Coded Decimal, BCD, provides an alternative means of representing decimal numbers. The premise is that you can use a hexadecimal encoding for the decimal digits by simply using 0–9 and ignoring numbers A through F. This may, at first blush, appear attractive as it is simple and easy to understand, as compared with one's complement and two's complement. However, in exchange, the programmer must monitor all arithmetic functions carefully so that overflow does not occur, which would not be inherently caught, as the hardware is not "aware" of the truncation of the hexadecimal system at A.

For example, in BCD, we may represent a decimal value of 403 as three bytes of "4" then "0," and "3." A second value of 528 would be represented as decimal value of "5" then "2" and "8." If the arithmetic function Add were used, "3" would be added to "8" and create a value of "11" in decimal. The computer will represent this in the lowest byte as numeral "B," which must then be converted to "11" and then correctly allocate the digits "1" and "1," with the first "1" correctly added to "0" and "2" for a decimal "tens" position of "3," and so on.

The approach to this issue is to simply run a conversion routine on either side of the arithmetic function. The programmer must create the "BCD to hex" routine on all values prior to the arithmetic operation, such as Add. Then, after the computation, the programmer must reverse the conversion, creating a "hex to BCD" function, on all values. The ultimate use of this method must, in some way, be of more value than simply holding the value for a display, as the arithmetic functions do not work in your favor.

ASCII—SINGLE-BYTE CHARACTERS

ASCII

Single-byte character data, such as ASCII, is not affected by Endianness. If you store any ASCII character string in memory, it always looks the same, no matter what the Endianness of the hardware. Each character is one byte long and the start character of the string is always stored at the lowest memory location. The single-byte character order resembles Big-Endian type order. The left-most byte leads first. It also reads correctly left-to-right in the memory dump. For the string "ABCD", the "A" is stored "first."

Four-Character String: "ABCD."

Byte Value	ASCII Character
68	D
67	C
66	B
65	A

If you dump character data from memory, it reads correctly left to right in the dump, too:

```
----- MEMORY BYTES ----------   --- ASCII CHARACTERS ---
00 65 66 67 68 00 00 00 00 00....ABCD....
```

The most convenient method of using alphabetical characters is to simply memorize the first character of the alphabet, in lowercase and uppercase values. Uppercase "A" is 65. Lowercase "a" is 97.

UNICODE—MULTIBYTE CHARACTERS

UNICODE

None of the Endian-independence holds for multibyte characters, e.g., Unicode, where each character takes more than one byte to represent. To correctly read multibyte characters you need to know the Endianness used to store them.

WORD SEARCH

B	G	J	I	K	E	H	I	O	Q	W	W	H	O	A
Z	K	A	K	V	D	E	U	C	E	I	I	V	U	T
O	N	E	S	C	O	M	P	L	E	M	E	N	T	W
U	D	G	K	L	C	S	D	C	Y	L	B	H	L	O
N	X	X	A	J	I	I	M	R	O	C	U	A	L	S
I	F	K	S	N	N	N	A	E	D	D	I	C	A	C
V	W	N	G	Y	U	N	L	T	T	M	E	B	M	O
E	D	A	F	Q	I	I	F	Q	O	S	O	D	I	M
R	N	U	M	B	E	R	I	N	G	O	Y	L	C	P
S	P	L	N	W	U	T	Y	R	L	P	H	S	E	L
A	M	O	D	N	R	L	B	E	F	W	B	V	D	E
L	K	G	G	K	O	L	A	M	I	C	E	D	A	M
T	N	I	O	P	G	N	I	T	A	O	L	F	X	E
A	S	C	I	I	K	Z	U	I	O	B	G	C	E	N
B	O	X	V	Y	T	D	F	Y	S	P	Z	F	H	T

ASCII
BCD
BINARY
BOOLEAN
CODED
DECIMAL
FLOATINGPOINT
HEXADECIMAL
LOGIC
NUMBERING
ONESCOMPLEMENT
POLYNOMIAL
SYSTEMS
TWOSCOMPLEMENT

PUZZLE: DIAMOND 22H

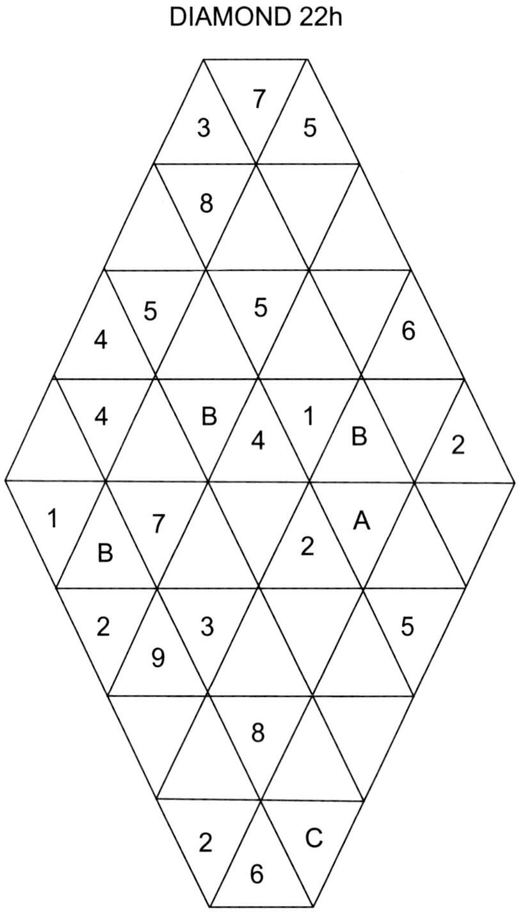

How to solve Diamond

The diamond-shaped grid contains 16 overlapping, six-sided hexagons. Each hexagon is divided into six triangles. The sum of the numbers of the six triangles is always equal to the number in the title, the hexadecimal value 22h. To complete the puzzle, place a number between 1h and Ch into each empty triangle.

FURTHER READING

From an online source, coverage of Little-Endian and Big-Endian: http://teaching.idallen.com/cst8281/10w/notes/110_byte_order_endian.html.

J. Hext, Programming Structures, Volume 1 Machines and Programs, n.d., p. 243.

Assembly Language for Intel-Based Computers, 5th edition, by Kip R. Irvine.

Smart Draw Templates for gates and truth tables.

CHAPTER

Mathematical operations 10

CHAPTER OUTLINE

#INCLUDE MATHEMATICAL ROUTINES

There are application notes (AN) that cover a large set of math routines, including square root, pseudorandom number generation, Gaussian distributed random number generation, numerical differentiation, and numerical integration. These math routines utilize the primary basic arithmetic routines by calling these as subroutines. Herein is coverage limited to addition, subtraction, multiplication, and division.

APPLICATION NOTES IN STANDARD AND HIGH PERFORMANCE MICROPROCESSORS

The primary mathematical application notes are AN526e, AN544d, AN575, AN617, and AN660. These application notes may be written specifically for one processor

The Art of Assembly Language Programming Using PIC® Technology. https://doi.org/10.1016/B978-0-12-812617-2.00010-9

or one processor family. Note that most of the routines, which may have been written for one specific processor, may be easily changed to accommodate a slightly different processor. If the processor is, for example, for a PIC® PIC16C5x and you are writing for an *enhanced* PIC® PIC17Cxx, the routine should work as-is. However, if you are writing for a PIC16C5x, and the routine is for an *enhanced* processor, you will need to change some of the instructions. First, trace through the specific program, noting the lines that have commands that are unique to the specific processor. Make sure that you fully understand the routines' functioning. Select the unique commands for the particular processor and substitute a set of commands that will work for the lower end processor. Test your routines thoroughly.

SINGLE AND DOUBLE PRECISION ARITHMETIC

Again, herein are the basic arithmetic routines of addition, subtraction, multiplication, and division. Routines use terminology as follows. "Single Precision" addition, for example, indicates that the routine will calculate a 16-bit double byte from the addition of two 8-bit bytes. This is written as "8×8." "Double Precision" addition, for example, indicates that the routine will calculate either 32 bits, in four bytes, or 16 bits, in two bytes, from two 16-bit words. This is written as "16×16."

SIGNED ARITHMETIC

"Unsigned" means exactly that; there is no accommodation for negative numbers. Sign handling can be easily added to this routine as the rules for signed arithmetic are simple. Where there are unsigned routines, one may simply track the sign separately from the unsigned calculation. For two negative numbers in addition, simply run the addition calculation and then add the sign. For one negative and one positive number, simply run the subtraction calculation. If the negative number is a higher number than the positive number, add the sign to the difference. If the positive number is a higher number than the positive number, simply keep the calculated number positive.

Note that the preferred numbering convention, for signed calculations, is two's complement representation. Recall, two's complement is calculated as one's complement (invert all digits) and then add one. (Please refer back to Chapter 9 for a detailed review of two's complement representation.)

SELECT OPTIMIZATION: SPEED OR MEMORY

These routines are provided for two different optimizations: speed or memory. One routine is *speed optimized* by writing straight line code with no loops. The other routine is *memory optimized* by writing looped code to reduce the code size. For example, in AN526, performance specifications are shown in a performance table in the application note copied here.

Specification	Memory/Code	Instruction Cycles/Speed
Speed efficient	35	37
Code efficient	16	71

CONDITIONAL ASSEMBLY

To switch optimization, simply set the variables at the top of the routine to cause the assembler to omit assembly of selective routines. Carefully review the comments at the start of the utility routines to learn the meaning of each variable. In double precision arithmetic routines in ARITH.ASM, two variables are used to select speed or memory optimization and designate signed or unsigned; MODE_FAST equals TRUE or FALSE, and SIGNED equals TRUE or FALSE.

HI/LO CONVENTIONS

RAM locations are also based upon these selections of MODE_FAST and SIGNED. CBLOCK is set to 0 × 18 for variable ACC with a through d multibyte values, and HI/LO designations are listed, as an example, ACCaLO and ACCaHI. Each column in the table below indicates the same memory location with its various names.

0x01FF		0x7FFF					
ACC							
ACCa (16)		ACCb (16)		ACCc (16)		ACCd (16)	
ACCaLO (8)	ACCaHI (8)	ACCbLO (8)	ACCbHI (8)	ACCcLO (8)	ACCcHI (8)	ACCdLO (8)	ACCd HI(8)

SPECIFIC MATH UTILITY ROUTINES

ADVANCED MACRO FEATURES

Following is the explanation of the utility routines, direct from the source listing for each routine. This may be found in the Microchip Embedded Control handbook or the online file, as referenced.

AN526e Utility Math Routines

This application note provides some utility math routines for Microchip's PIC16C5X and PIC16CXXX series of 8-bit microcontrollers. The following math outlines are provided:

- 8 × 8 unsigned multiply
- 16 × 16 double precision multiply
- Fixed Point Division
- 16 × 16 double precision addition
- 16 × 16 double precision subtraction
- BCD (binary-coded decimal) to binary conversion routines
- Binary to BCD conversion routines
- BCD addition
- BCD subtraction
- Square root

These are written in native assembly language and the listing files are provided. All the routines provided can be called as subroutines. Most of the routines have two different versions: one optimized for speed and the other optimized for code size. The calling sequence of each routine is explained at the beginning of each listing file.

AN544d Binary Coded Decimal and Fixed Point Division

This application note provides some utility math routines for Microchip's second generation of high performance 8-bit microcontroller, the PIC17C42. Three assembly language modules are provided, namely, ARITH.ASM, BCD.ASM, and FXP-DIV.ASM. Currently in each file the following subroutines are implemented:

ARITH.ASM

- single precision 8×8 unsigned multiply
- 16×16 double precision multiply (signed or unsigned)
- 16/16 double precision divide (signed or unsigned)
- 16×16 double precision addition
- 16×16 double precision subtraction
- double precision square root
- double precision numerical differentiation
- double precision numerical integration
- pseudorandom number generation
- Gaussian distributed random number generation

BCD.ASM

- 8-bit binary to 2 digit BCD conversion
- 16-bit binary to 5 digit BCD conversion
- 5-bit BCD to 16-bit binary conversion
- 2-digit BCD addition

FXP-DIV.ASM

AN575 IEEE 754 Compliant Floating Point Routines

This application note presents an implementation of the following floating point math routines for the PIC® microcontroller families:

- float to integer conversion
- integer to float conversion
- normalize
- add/subtract
- multiply
- divide

Routines for the PIC16/17 families are provided in a modified IEEE 754 32-bit format together with versions in 24-bit reduced format.

AN617 Fixed Point Math Routines

This application note presents an implementation of the following fixed point math routines for the PIC® microcontroller families: addition, subtraction, multiplication, and division. Addition and subtraction are illustrated herein.

Addition

See Fig. 10.1.

Subtraction

Routines for the PICmicro microcontroller families are provided in a variety of fixed point formats, including both unsigned and signed two's complement arithmetic (Fig. 10.2).

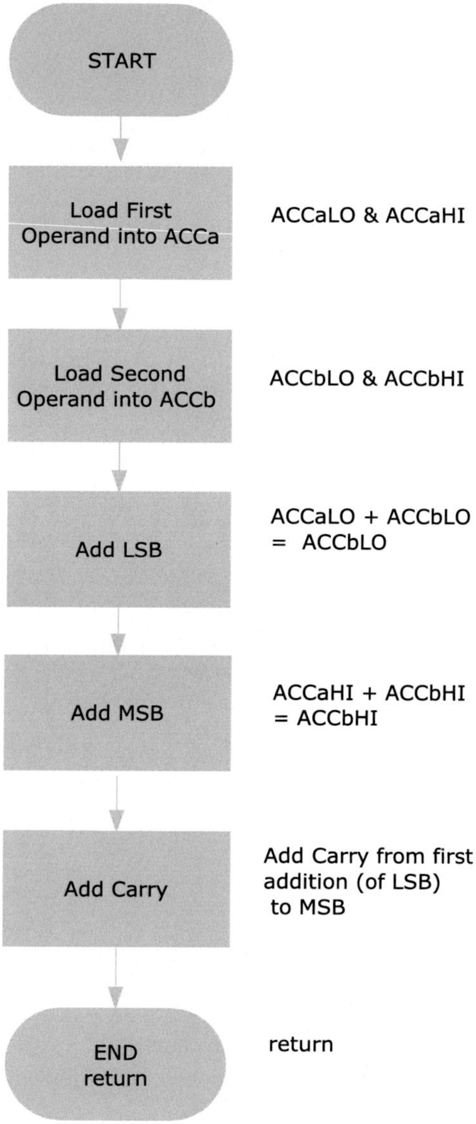

FIG. 10.1

Addition algorithm flowchart.

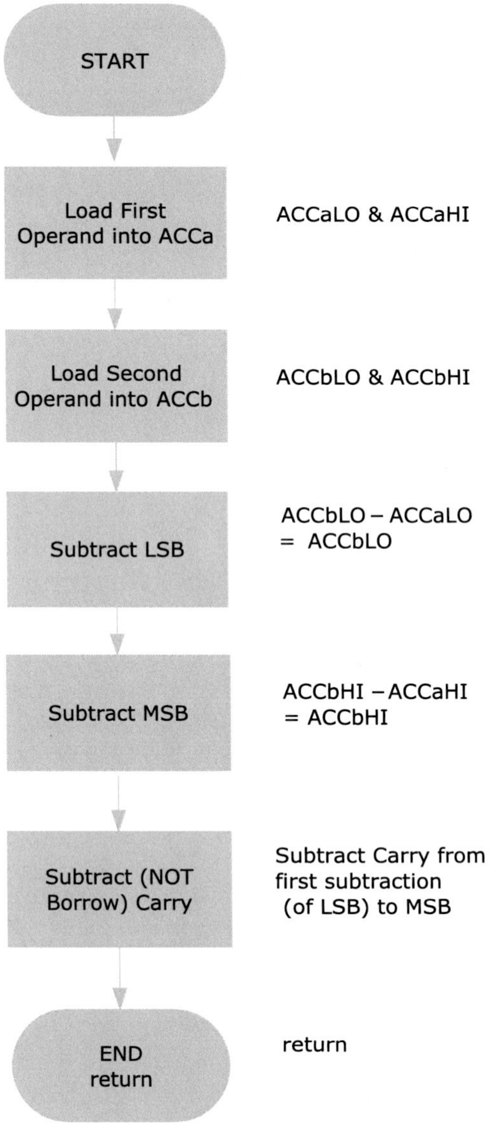

FIG. 10.2

Subtraction algorithm flowchart.

AN660 Trigonometric and Complex Math Routines

This application note presents implementations of the following math routines for the Microchip PICmicro microcontroller family:

- square root function,
- exponential function,

- base 10 exponential function,
- natural log function,
- common log function,
- trigonometric sine function,
- trigonometric cosine function,
- trigonometric sine and cosine functions,
- power function,
- floor function,
- largest integer not greater than x,
- floating point,
- logical comparison tests,
- integer random number generator.

Routines for the PIC16CXXX and PIC17CXXX families are provided in a modified IEEE 754 32-bit format together with versions in 24-bit reduced format. The techniques and methods of approximation presented here attempt to balance the usually conflicting goals of execution speed verses memory consumption, while still achieving full machine precision estimates. Although 32-bit arithmetic routines are available and constitute extended precision for the 24-bit versions, no extended precision routines are currently supported for use in the 32-bit routines, thereby requiring more sophisticated error control algorithms for full or nearly full machine precision function estimation. Differences in algorithms used for the PIC16CXXX and PIC17CXXX families are a result of performance and memory considerations and reflect the significant platform dependence in algorithm design.

FEATURED 8-BIT MCU BOARDS

The Automotive Networking Development Board is a low-cost modular development system for Microchip's 8-bit, 16-bit, and 32-bit microcontrollers targeting CAN and LIN network-related applications. The board supports devices using the 100 pin Plug-In Module (PIM) connector for easy device swapping.

See Fig. 10.3.

AUTOMOTIVE NETWORKING DEVELOPMENT BOARD

Part Number: ADM00716

The Automotive Networking Development Board is a low-cost modular development system for Microchip's 8-bit, 16-bit, and 32-bit microcontrollers targeting CAN and LIN network-related applications. The board supports devices using the 100 pin Plug-In Module (PIM) connector for easy device swapping.

CURIOSITY DEVELOPMENT BOARD

Part Number: DM16413

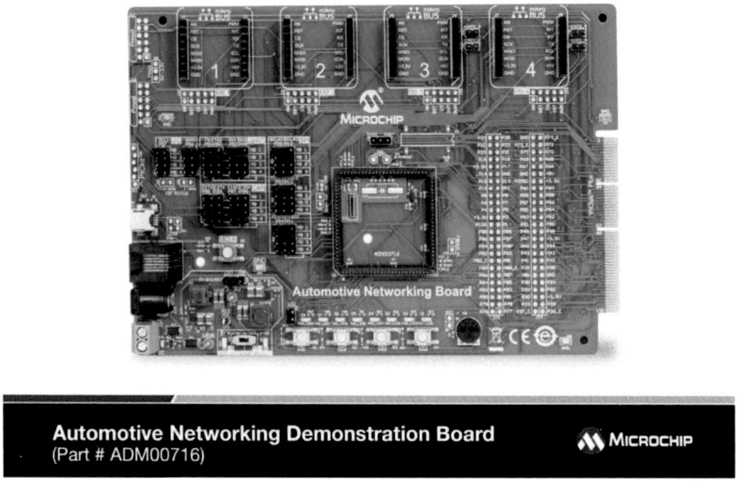

FIG. 10.3

Automotive Networking Demo Board.

Satisfy Your Curiosity, your next embedded design idea has a new home. Curiosity is a cost-effective, fully integrated 8-bit development platform targeted at first-time users, makers, and those seeking a feature-rich rapid prototyping board (Fig. 10.4).

CURIOSITY HIGH PIN COUNT (HPC) DEVELOPMENT BOARD

Part Number: DM164136

See Fig. 10.5

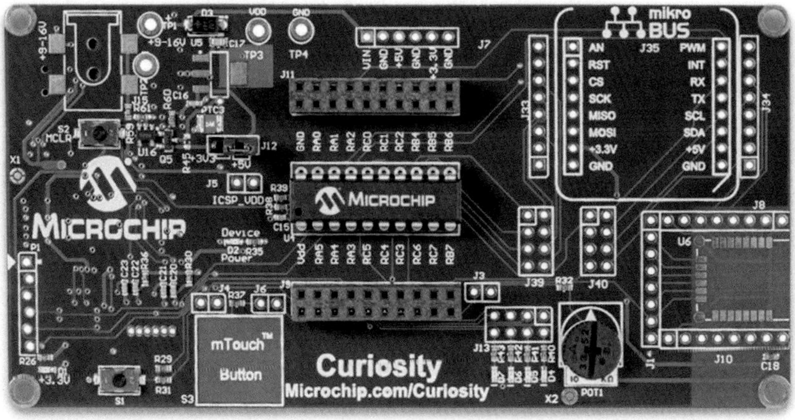

FIG. 10.4

Curiosity Development Board.

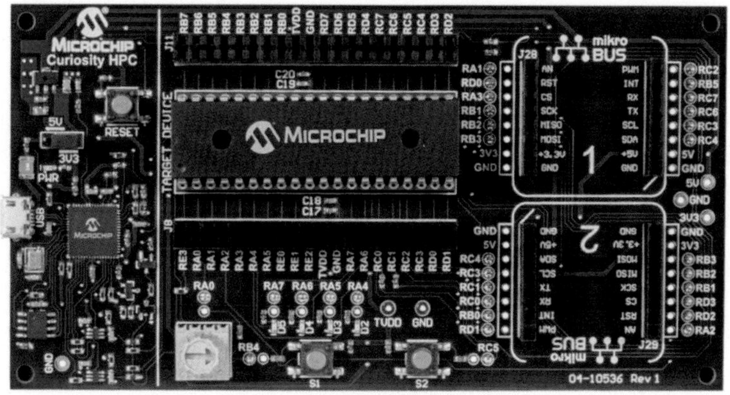

FIG. 10.5

Curiosity HPC Development Board.

EXPLORER 8 DEVELOPMENT KIT

Part Number: DM160228

See Fig. 10.6

The Explorer 8 Development Kit is a full-featured development board and platform for 8-bit PIC microcontrollers. This kit is a versatile development solution, featuring several options for external sensors, off-board communication, and human interface.

FIG. 10.6

Explorer 8.

PICDEM LAB II DEVELOPMENT PLATFORM

Part Number: DM163046

See Fig. 10.7

The PICDEM Lab II Development Board is a development and teaching platform for use with 8-bit PIC microcontrollers (MCUs). At its center, a large prototyping breadboard enables users to easily "experiment" with different values and configurations of analog components for system optimization.

Arithmetic operations on bytes
Arithmetic operations on multibyte numbers

- Using carry properly
- Detecting overflow
- How to treat signed and unsigned differently

Shifting and rotating, and how to use instructions in a multibyte shift
Division algorithms
Multiplication algorithms
Floating point
Generating custom math routines for embedded applications

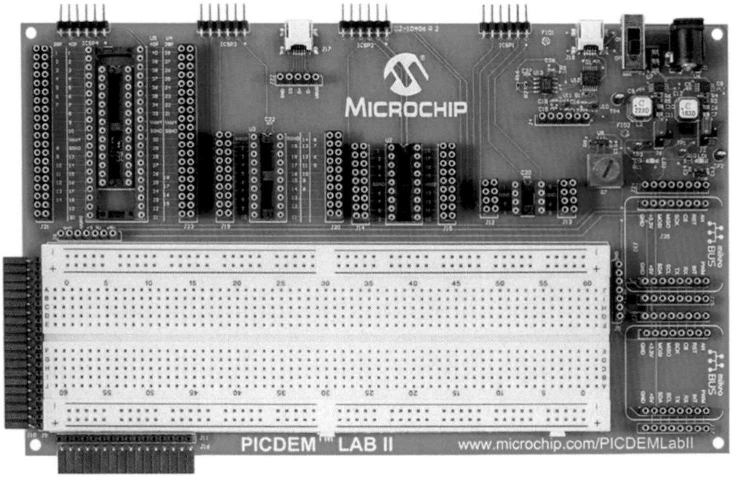

FIG. 10.7

PICDEM Lab II.

WORD SEARCH: MATHEMATICAL OPERATIONS

M	E	O	L	N	C	B	N	Q	L	C	A	C	S	A
R	U	Y	P	Z	C	O	F	I	R	S	P	A	E	H
L	M	L	E	R	I	I	M	Q	U	B	P	L	N	T
G	H	Q	T	T	I	I	T	B	L	Z	L	L	I	A
N	O	V	I	I	T	M	R	E	N	X	I	I	T	M
P	L	D	X	E	P	O	A	Q	M	E	C	N	U	V
V	D	V	D	X	U	L	V	R	L	H	A	G	O	L
A	C	S	E	T	O	N	I	M	Y	E	T	J	R	D
D	I	V	I	S	I	O	N	C	H	D	I	I	T	N
S	H	N	V	S	C	M	Q	C	A	R	O	M	R	B
F	E	G	X	I	A	G	A	X	I	T	N	N	F	A
S	W	K	D	A	S	M	J	G	M	S	I	X	D	E
N	O	I	T	C	A	R	T	B	U	S	A	O	G	C
L	U	X	V	J	B	I	D	A	P	M	O	B	N	C
L	C	Q	K	D	H	M	W	C	R	V	O	D	P	G

ADDITION
APPLICATION
ARITHMETIC
BASIC
CALLING
DIVISION
LIMITED
MATH
MULTIPLICATION
NOTES
PRIMARY
ROUTINES
SUBROUTINES
SUBTRACTION

PUZZLE: NUMBER TOWER

NUMBER TOWER

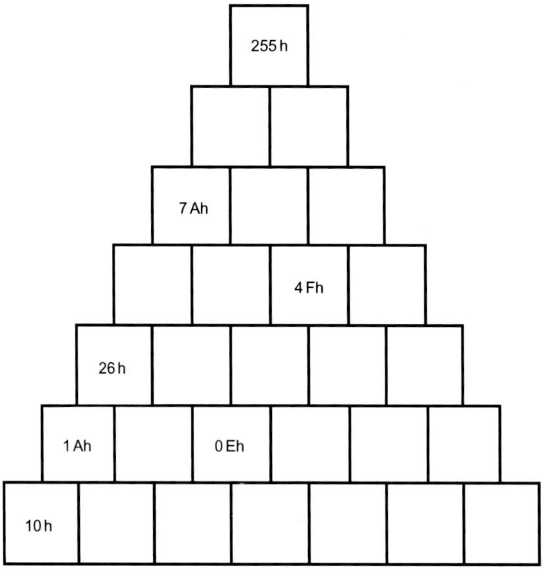

How to solve Number Tower

This is a tower of two digit hexadecimal numbers. Each number is the sum of the two squares beneath it. The tower appears as a "cousin" to the Fibonacci sequence, where each number in the sequence is the sum of the two preceding ones. Here, look where you are provided two of the three squares to start your sequence. The lower left corner would be an ideal place to begin.

Some of the above material is mine. Others are copied, in part, from one of two books noted below. Material copied from the books is below the word limit.

Pocket Size Logic Puzzles
PAPP International Inc.
177 Merizzi Street, Montreal (QC)
CANADA H4T 1Y3
www.pappintl.com

Variety Puzzle Collection Volume 21
Kappa Books Publishers, LLC.
www.kappabooks.com

Both distributed by:
Greenbrier International, Inc.
Chesapeake, VA 23320

FURTHER READING

www.microchip.com website information.
Embedded Control Handbook 1994/5.

CHAPTER

Word search solution

11

CHAPTER 1

Solution

```
+ + + + + + Y + E + D + + + +
N + M + + L + M + T A + + + +
+ O + A I + B + + N T + + + +
P + I M R E + + + E A + + + +
+ R A T D G + + D M S L + + +
+ F A D A + O O + P H O + + +
+ + E C + T C R + O E R + + +
+ D + + T O N + P L E T + + +
+ + + + D I + E + E T N + + +
+ + + U + + C + M V + O + + +
+ + E + + + + A + E + C + + +
+ S + + + + + + L D L + + + +
P A L G O R I T H M + P + + +
Y L B M E S S A C O R E M + +
+ + + + + + + + + + + + + I +
```

```
(Over,Down,Direction)
ALGORITHM(2,13,E)
ASSEMBLY(8,14,W)
CONTROL(12,11,N)
CORE(9,14,E)
DATASHEET(11,1,S)
DEVELOPMENT(10,12,N)
EMBEDDED(9,1,SW)
FAMILY(2,6,NE)
IMPLEMENTATION(14,15,NW)
PRACTICAL(1,4,SE)
PROGRAM(9,8,NW)
PSEUDOCODE(1,13,NE)
```

The Art of Assembly Language Programming Using PIC® Technology. https://doi.org/10.1016/B978-0-12-812617-2.00012-2

CHAPTER 2

Solution

```
+ S + + + + K + + + P + + E P
+ N + + + C + + + R + + R M I
+ O + S O + + + O + + U I + P
+ I + L U + + G + + T C E + E
+ T C + + B R + + C R + L + L
+ C + + + A A + E O + + C + I
+ U + + M + + T P + + + Y + N
+ R + B + + I R A + + + C + I
+ T U + + H O + + D + + + + N
+ S + + C C F L A S H + + + G
+ N S R E T S I G E R E L I F
+ I A S P E R I P H E R A L S
+ + S R O T A L L I C S O + +
+ O + + + + + + + + + + + + +
R + + + + + + + + + + + + + +
```

(Over,Down,Direction)

ARCHITECTURE (3,12,NE)
CLOCK (3,5,NE)
CYCLE (13,8,N)
DATABUS(10,9,NW)
FILEREGISTERS(15,11,W)
FLASH(7,10,E)
INSTRUCTIONS(2,12,N)
MICROPROCESSOR(14,2,SW)
OSCILLATOR(13,13,W)
PERIPHERALS(5,12,E)
PIPELINING(15,1,S)
PROGRAMBUS(11,1,SW)

CHAPTER 3

Solution

```
N O P A N D W F N W L + + F +
B + + + E + E + + O L L + C +
+ T + C W + + C X + I V A B B
+ + F L S + + O F + + T O C T
+ S D S W U R + + F + + P M F
Z N + L C L B + + R + + + O S
A + T + W G + W + L F V O M S
+ E + + + O + F F C + + + + +
R + + + + T + W S L E E P + W
+ C O M F O + V I F S R C R S
F W R O I + + O + W + I L + W
Z S F C N I R M I D + C R F A
F W R O X L F N + D R + W T P
+ + + + W S C + + A R + D + F
+ + + + B F + + + + F + T + +
```

```
(Over,Down,Direction)
ADDWF(10,14,N)
ANDLW(1,7,NE)
ANDWF(4,1,E)
BCF(14,3,N)
BSF(5,15,NE)
BTFSC(1,2,SE)
BTFSS(15,3,S)
CALL(14,4,NW)
CLRF(10,8,N)
CLRW(12,12,NE)
CLRWDT(13,10,S)
COMF(2,10,E)
DECF(6,1,SE)
DECFSZ(6,1,SW)
GOTO(6,7,S)
INCF(9,12,SW)
INCFSZ(6,12,W)
IORLW(9,10,SW)
IORWF(5,11,W)
MOVF(14,7,W)
MOVLW(14,5,NW)
MOVWF(8,12,N)
NOP(1,1,E)
OPTION(14,6,NW)
RETLW(1,9,NE)
RLF(12,10,SE)
RRF(11,13,S)
SLEEP(9,9,E)
SUBWF(5,4,SE)
SWAPF(15,10,S)
TRIS(14,13,NW)
XORLW(9,3,SW)
XORWF(5,13,W)
```

CHAPTER 4

Solution

```
S T + N + + E + + A + + + + +
E N + N O + X + + P + + P + P
S E O O + M A + + P Y + R R C
E R M I N I M U M L + + O + O
H E + T T + P O B I + G C + N
T F R U + N L M C C R S E T T
N F E L E M E N T A R Y S C E
E I S O + S S V M T + S S E N
R D U S S + + M N I + T O R T
A + L A + + I + + O + E R I S
P + T + + N + + + N C M S D +
+ L A N G U A G E S + S + N +
S C I T A M E H T A M + + I +
+ + + + + G N I S S E R D D A
S A M P L E S A C T S E T + +
```

```
(Over,Down,Direction)
ADDRESSING(15,14,W)
APPLICATIONS(10,1,S)
ASSEMBLY(4,10,NE)
COMMON(9,6,NW)
CONTENTS(15,3,S)
CONVENTIONS(11,11,NW)
DIFFERENT(2,9,N)
ELEMENTARY(3,7,E)
EXAMPLES(7,1,S)
INDIRECT(14,13,N)
LANGUAGE(2,12,E)
MATHEMATICS(11,13,W)
MINIMUM(3,4,E)
PARENTHESES(1,11,N)
PROCESSORS(13,2,S)
PROGRAMMING(15,2,SW)
RESULT(3,6,S)
SAMPLE(1,15,E)
SOLUTION(4,9,N)
SYSTEMS(12,6,S)
TESTCASE(13,15,W)
```

CHAPTER 5

Solution

```
M + + + A + + + + + + S M + +
+ A + + + P + + + + T + U + +
L + R + + + P + + R + + S + +
+ O + G + + + L U + + + A + +
+ + O + O + + C I + + + T + +
+ + + P + R T + F C + + A + +
N O T E S U P + L + A + D + +
P R O G R A M C O U N T E R B
+ + + E + + + + W + + + I A +
+ + S + + S + + C + + + N O +
+ + + + O + + + H + + K P + N
+ + + R + + + + A + + + A + +
+ + C P I H C O R C I M G + +
+ A + + + + + + T + + + E + +
M + + + + + + + S + + + + + +
```

(Over,Down,Direction)
APPLICATION(5,1,SE)
BANK(15,8,SW)
DATASUM(13,7,N)
FLOWCHARTS(9,6,S)
LOOP(1,3,SE)
MACROS(1,15,NE)
MICROCHIP(12,13,W)
NOTES(1,7,E)
PAGE(13,11,S)
PROGRAM(7,7,NW)
PROGRAMCOUNTER(1,8,E)
STRUCTURES(12,1,SW)

CHAPTER 6

Solution

```
+ + G + + + + + S + S D + S +
+ + N + + + + O + N N + + L +
+ + I D + + F + O U + + + A +
+ + R + I T + I O + + + + T +
+ + I + W A T R + + + + + N +
+ + W A + C G E M B E D D E D
+ + R + E + + R H + + + + M +
+ E + N + + + + A + + + + A +
+ + N + + + + + R M + + + D +
L O R T N O C + D T S + + N +
C + + + + + + + W + U + + U +
+ + + + + + + + A + + O + F +
+ P O W E R + + R + + + N + +
+ + + + + + + + E + + + + I +
+ + + + + + + + + + + + + + P
```

```
(Over,Down,Direction)
CONNECTIONS(1,11,NE)
CONTROL(7,10,W)
DIAGRAMS(4,3,SE)
EMBEDDED(8,6,E)
FUNDAMENTALS(14,12,N)
GROUND(7,6,NE)
HARDWARE(9,7,S)
PINOUT(15,15,NW)
POWER(2,13,E)
SOFTWARE(9,1,SW)
WIRING(3,6,N)
```

CHAPTER 7

Solution

```
C E C N E U Q E S A + + + + +
S A S E T A T S + L D + + + +
+ E L Y N O I T I T E P E R +
+ + H C S + + + + E R + + + +
+ R + C U T + + + R E + + + D
F + R + N L E + + N D + + E F
I L + O I A A M S A R + V D U
N + O S R + R T S T O E + O N
P + T W + E U B I I L + + C C
U S + + C P I + + O D + + O T
T + + + T H + N P N N S + D I
S + + U + + A M R + + S S U O
+ + O + + + E R + A + + + E N
M A C H I N E + T + W + + S +
+ + + S T R U C T U R E D P +
```

```
(Over,Down,Direction)
ALTERNATION(10,1,S)
BRANCHES(8,9,NW)
CALCULATIONS(1,1,SE)
DEVELOPMENT(15,5,SW)
FLOWCHART(1,6,SE)
FUNCTION(15,6,S)
INPUTS(1,7,S)
LISTS(6,6,SW)
MACHINE(1,14,E)
ORDERED(11,8,N)
OUTPUTS(3,13,NE)
PSEUDOCODE(14,15,N)
REPETITION(14,3,W)
SEQUENCE(9,1,W)
SSD(13,12,NW)
STATE(8,2,W)
STRUCTURED(4,15,E)
SYSTEMS(3,2,SE)
WARNIERORR(11,14,NW)
```

CHAPTER 8

Solution

```
+ + + + + E + E I D + + + + N
+ + + + B + C + M E + + + + O
+ + + C + N + + P F + + + A I
+ + D + E + + + L I + + L + T
+ I + U + + + + E N + P + + C
C + Q N + + + + M E H + + + N
+ E + + O + + + E A + + + + U
S + + N O I T A N R E T L A F
+ + + + + + T U T C I I C S A
+ + + + + + M I A + O + + + +
+ P A T T E R N T + + N + + +
+ + + + R + + + I E + + F + +
+ + + I + + + + O + P + + I +
+ + C + + + + + N + + E + + G
+ + + + + + + + + + + + R + +
```

```
(Over,Down,Direction)
ALPHANUMERIC(14,3,SW)
ALTERNATION(14,8,W)
ASCII(15,9,W)
CONFIG(10,9,SE)
DEFINE(10,1,S)
EBCDIC(6,1,SW)
FUNCTION(15,8,N)
IMPLEMENTATION(9,1,S)
PATTERN(2,11,E)
REPETITION(13,15,NW)
SEQUENCE(1,8,NE)
```

CHAPTER 9

Solution

```
O + B + Y + + + + D L + T + S
+ N + O + R D + E + A + W + Y
+ + E + O E A D + + I + O + S
L + + S C L O N + + M + S + T
+ A + I C C E + I G O + C + E
+ B M + + O + A N B N + O + M
+ A C I + + M I N + Y + M + S
L + + D C + R P + + L + P C +
A S C I I E + + L + O + L I +
+ + + + B + D + + E P + E G +
+ + + M + + + A + + M + M O +
+ + U + + + + + X + + E E L +
U N I V E R S A L E + + N + +
+ E D O C I N U + + H + T T +
+ + T N I O P G N I T A O L F
```

```
(Over,Down,Direction)
ASCII(1,9,E)
BCD(2,6,SE)
BINARY(10,6,NW)
BOOLEAN(3,1,SE)
CODED(6,5,NE)
DECIMAL(7,2,SW)
FLOATINGPOINT(15,15,W)
HEXADECIMAL(11,14,NW)
LOGIC(14,12,N)
NUMBERING(2,13,NE)
ONESCOMPLEMENT(1,1,SE)
POLYNOMIAL(11,10,N)
SYSTEMS(15,1,S)
TWOSCOMPLEMENT(13,1,S)
UNICODE(8,14,W)
UNIVERSAL(1,13,E)
```

CHAPTER 10

Solution

```
M + + + + + + N + L + A C S +
+ U + P + C O + I + S P A E H
+ + L + R I I M + U + P L N T
+ + + T T I I T B + + L L I A
+ + + I I T M R E + + I I T M
+ + D + E P O A + M + C N U +
+ D + D + U L + R + H A G O +
A + S E T O N I + Y + T + R +
D I V I S I O N C + + I I + +
+ + N + + + + + C A + O + R +
+ E + + + + + + + I T N + + A
S + + + + + + + + + S I + + +
N O I T C A R T B U S A O + +
+ + + + + + + + + + + + B N +
+ + + + + + + + + + + + + + +
```

(Over,Down,Direction)
ADDITION(1,8,NE)
APPLICATION(12,1,S)
ARITHMETIC(15,11,NW)
BASIC(13,14,NW)
CALLING(13,1,S)
DIVISION(1,9,E)
LIMITED(10,1,SW)
MATH(15,5,N)
MULTIPLICATION(1,1,SE)
NOTES(7,8,W)
PRIMARY(4,2,SE)
ROUTINES(14,8,N)
SUBROUTINES(11,2,SW)
SUBTRACTION(11,13,W)

CHAPTER 12

Puzzle Solutions

CHAPTER 1

0	0	1	0	1	0	1	1	2Bh
0	1	0	1	0	1	0	1	55h
1	0	1	0	1	1	0	0	ACh
0	1	0	1	0	0	1	1	53h
1	0	1	0	1	0	1	0	AAh
1	0	1	1	0	1	0	0	34h
0	1	0	0	1	0	1	1	CBh
0	0	1	1	0	0	1	1	33h
2Ch	52h	ADh	55h	AAh	64h	9Bh	D5h	Hex

CHAPTER 2

```
O R + + + R + + + O R O + + O
+ S O + + E O + C O + S + + S
+ + C T + T + S T + + S + + C
+ + + I I A I E C + + I + + I
+ + + + L L L + + I + L + + L
+ + + + L L L + + + L L + + L
+ + + A I I A A + + + A + + T
+ + T C + C + T C + + T T + O
+ O S + + S + + O S + O + O R
R O + + + O + + + R O R + + R
+ + R O T A L L I C S A + + +
O S C A L L A T O R + + + + +
+ + + + R O T A L L I S O + +
+ + + + + + + + + + + + + + +
+ + + + + + + + + + + + + + +
```

The Art of Assembly Language Programming Using PIC® Technology. https://doi.org/10.1016/B978-0-12-812617-2.00011-0

```
(Over,Down,Direction)
ASCILLATOR(12,11,W)
OCSILLATOR(10,1,SW)
OSCALLATOR(1,12,E)
OSCALLITOR(11,10,NW)
OSCILATOR(7,2,SE)
OSCILLATER(6,10,N)
OSCILLATOR(1,1,SE)
OSCILLETOR(2,10,NE)
OSCILLTOR(15,1,S)
OSILLATOR(13,13,W)
OSSILLATOR(12,1,S)
```

CHAPTER 3

Across	Down
5. MOVWF	1. ADDWF
7. INCFSZ	2. COMF
8. BTFSS	3. INCF
9. DECFSZ	4. TRIS
10. BTFSC	6. ANDWF
12. SUBWF	8. BSF
13. RLF	9. DECF
15. RETLW	10. BCF
17. MOVF	11. CLRWDT
18. CLRF	12. SLEEP
19. XORWF	14. SWAPF
21. IORWF	15. RRF
22. GOTO	16. NOP
25. ANDWF	18. CLRW
26. IORLW	19. XORLW
	20. MOVLW
	23. OPTION
	24. CALL

CHAPTER 4

NUMBER SQUARE
use order of operations in decimal

7	×	5	÷	1	35
×		+		−	
9	−	4	+	3	8
−		+		−	
2	+	8	×	6	50
61		17		–8	

CHAPTER 5

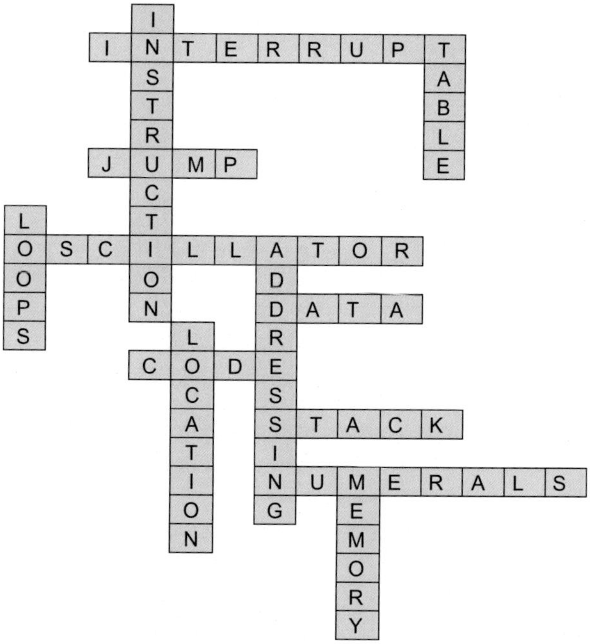

CHAPTER 6

Y	A	T	E	N
R	A	H	I	X
R	N	D	M	B
G	A	O	E	A
C	I	S	R	C

✓Hex ✓Data
✓Dec ✓Macro
✓Bin ✓or
✓Radi ✓And
✓Carry ✓Random
✓Design
✓Timer

CHAPTER 7

Spiral Puzzle

INWARD letter range	Description	Words
1-7	Software is another name for ____	Program
8-11	One ___ is one instance of code	Loop
12-18	Use ____ to switch to sub- routines in the lower ½ of a page	Vectors
19-21	Integrated Design Environment	IDE
22-28	Assembler produces ____ language	Machine
29-34	Repeatable blocks of code	Macros
35-38	RP1 & RP0 select the____	Page

INWARD letter range	Description	Words
39-42	____ the subroutine	Call
43-46	Unconditional program counter change	Goto
47-50	Page 1 indicates the 2nd ____	Bank
51-61	AN stands for ____ Note	Application
62-68	Program ____ points to your current location.	Counter
69-79	Long ____ branches	Conditional
80-88	____ is one method of program design and documentation.	Flowchart
89-97	One ____ is one loop around.	Iteration
98-100	Single	One

Additional Hints

INWARD letter range	Description	Words
1-3	Short for professional	Pro
22-24	Short for an early Apple Computer	Mac
40-42	Everyone	All
45-46	Go____	To
47-50	Place to store money	Bank
51-53	Computer program for cell phones	App
56-58	Raining ____s and dogs	Cat
76-77	Don't leave the stove ____	On
86-88	Creative expression	Art
97-100	Another name for zero	None

In the other direction....

OUTWARD letter range	Description	Words
4-3	___to	Go
22-20	Short for medicine	Med
29-28	____, myself, and I	Me
61-60	A two-year-old's favorite word	No
83-80	Red riding hood's foe	Wolf

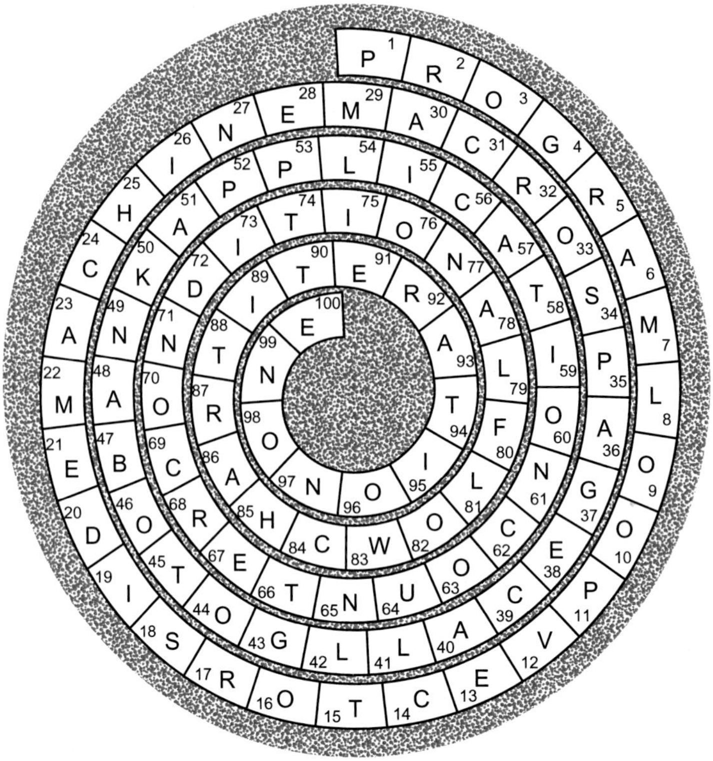
1 P
2 R
3 O
4 G
5 R
6 A
7 M
8 L
9 O
10 O
11 P
12 V
13 E
14 C
15 T
16 O
17 R
18 S
19 I
20 D
21 E
22 M
23 A
24 C
25 H
26 I
27 N
28 E
29 M
30 A
31 C
32 R
33 O
34 S
35 P
36 A
37 G
38 E
39 C
40 A
41 L
42 L
43 G
44 O
45 T
46 O
47 B
48 A
49 N
50 K
51 A
52 P
53 P
54 L
55 I
56 C
57 A
58 T
59 I
60 O
61 N
62 C
63 O
64 U
65 N
66 T
67 E
68 R
69 C
70 O
71 N
72 D
73 I
74 T
75 I
76 O
77 N
78 A
79 L
80 F
81 L
82 O
83 W
84 C
85 H
86 A
87 R
88 T
89 I
90 T
91 E
92 R
93 A
94 T
95 I
96 O
97 N
98 O
99 N
100 E

CHAPTER 8

JIGSAW SUDOKU 66 h
ANSWER KEY

7h thru Fh

9	A	D	7	B	E	C	F	8
7	E	8	C	A	D	F	9	B
B	D	F	9	C	8	A	E	7
A	8	C	F	7	9	D	B	E
C	7	B	E	D	F	8	A	9
F	9	7	8	E	A	B	C	D
D	F	9	A	8	B	E	7	C
8	C	E	B	F	7	9	D	A
E	B	A	D	9	C	7	8	F

CHAPTER 9

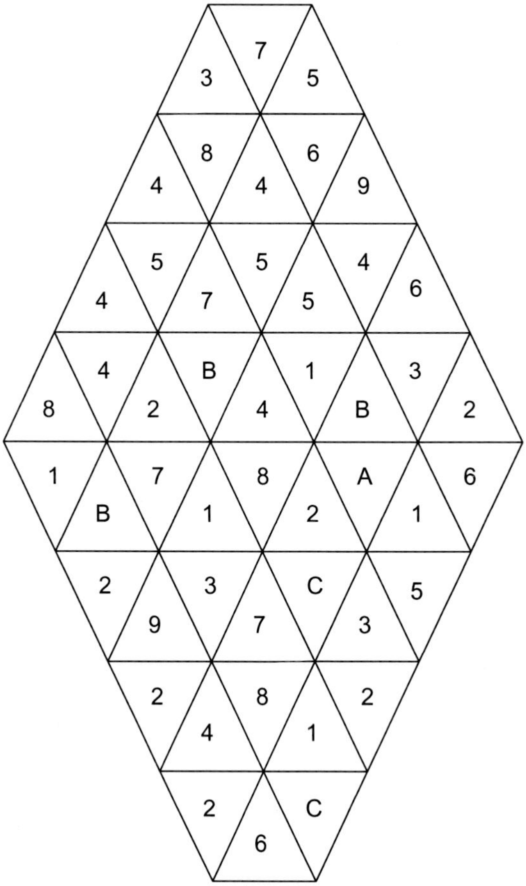

CHAPTER 10

			255h			
		103h		152h		
		7Ah	89h	C9h		
	40h	3Ah		4Fh	7Ah	
	26h	1Ah	20h	2Fh	4Bh	
1Ah	OCh	OEh		12h	1Dh	2Eh
1Oh	OAh	O2h	OCh	O6h	17h	17h

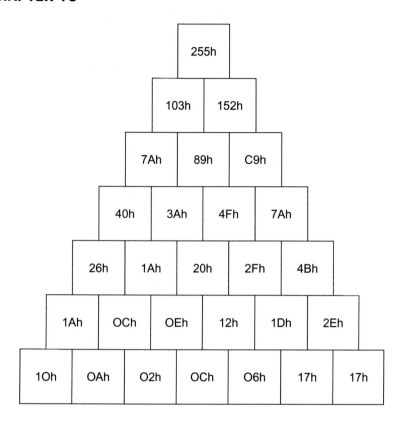

Instruction Sets

BASELINE PIC MCU INSTRUCTION SET

This applies to the PIC10F2xx, PIC12C5xx, PIC12F5xx, PIC16C5xx, and PIC16F5xx families.

Mneumonic	Operands	Description	Cycles	12-Bit Opcode (MSB...LSB)	Status Affected	Notes
Byte-oriented operations						
ADDWF	f,d	ADD W with f		0001 11df ffff	C,DC,Z	1,2,4
ANDWF	f,d	AND W with f	1	0001 01df ffff	Z	2,4
CLRF	f	Clear f	1	0000 011f ffff	Z	4
CLRW		Clear W	1	0000 0100 0000	Z	
COMF	f,d	Complement f	1	0010 01df ffff	Z	
DECF	f,d	Decrement f	1	0000 11df ffff	Z	2,4
DECFSZ	f,d	Decrement f, Skip if 0	1(2)	0010 11df ffff	None	2,4
INCF	f,d	Increment f	1	0010 10df ffff	Z	2,4
INCFSZ	f,d	Increment f, Skip if 0	1(2)	0011 11df ffff	None	2,4
IORWF	f,d	Inclusive OR W with f	1	0001 00df ffff	Z	2,4
MOVF	f,d	Move f	1	0010 00df ffff	Z	2,4
MOVWF	f	Move W to f	1	0000 001f ffff	None	1,4
NOP		No Operation	1	0000 0000 0000	None	
RLF	f,d	Rotate left f through Carry	1	0011 01df ffff	C	2,4
RRF	f,d	Rotate right f through Carry	1	0011 00df ffff	C	2,4
SUBWF	f,d	Subtract W from f	1	0000 10df ffff	C,DC,Z	1,2,4
SWAPF	f,d	Swap f	1	0011 10df ffff	None	2,4
XORWF	f,d	Exclusive OR W with f	1	0001 10df ffff	Z	2,4

Bit-oriented operations								
BCF	f,b	Bit Clear f	1	0100	bbbf	ffff	None	2,4
BSF	f,b	Bit Set f	1	0101	bbbf	ffff	None	2,4
BTFSC	f,b	Bit Test f, Skip if Clear	1(2)	0110	bbbf	ffff	None	
BTFSS	f,b	Bit Test f, Skip if Set	1(2)	0111	bbbf	ffff	None	
Literal and control operations								
ANDLW	k	AND literal with W	1	1110	kkkk	kkkk	Z	
CALL	k	Subroutine Call	2	1001	kkkk	kkkk	None	1
CLRWDT		Clear Watchdog Timer	1	0000	0000	0100	TO,PD	
GOTO	k	Unconditional Branch	2	101k	kkkk	kkkk	None	
IORLW	k	Inclusive OR literal with W	1	1101	kkkk	kkkk	Z	
MOVLW	k	Move literal to W	1	1100	kkkk	kkkk	None	
OPTION		Load OPTION register	1	0000	0000	0010	None	
RETLW	k	Return, place literal in W	2	1000	kkkk	kkkk	None	
SLEEP		Go into standby mode	1	0000	0000	0011	TO, PD	
TRIS	f	Load TRIS register	1	0000	0000	0fff	None	3
XORLW	k	Exclusive OR literal with W	1	1111	kkkk	kkkk	Z	

1. The ninth bit of the program counter will be forced to a "0" by any instruction that writes to the PC except for GOTO (see Section 3.5 "Program Counter" for more on program counter).

2. When an I/O register is modified, as a function of itself (e.g., MOVF PORTB, 1), the value used will be that value present on the pins themselves. For example, if the data latch is "1" for a pin configured as input and is driven low by an external device, the data will be written back with a "0."

3. The instruction TRIS f, where f=5, 6, or 7 causes the contents of the W register to be written to the tristate latches of PORTA, B or C, respectively.

4. A "1" forces the pin to a high impedance state and disables the output buffers.

COMPARISON TABLE

10Fxxx	12Fxxx	16Fxxx	18Fxxx		10Fxxx	12Fxxx	16Fxxx	18Fxxx
		ADDLW	ADDLW		MOVLW	MOVLW	MOVLW	MOVLW
ADDWF	ADDWF	ADDWF	ADDWF		MOVWF	MOVWF	MOVWF	MOVWF
			ADDWFC					MULLW
ANDLW	ANDLW	ANDLW	ANDLW					MULWF
ANDWF	ANDWF	ANDWF	ANDWF					NEGF
			BC		NOP	NOP	NOP	NOP
BCF	BCF	BCF	BCF		OPTION	OPTION		
			BN					POP
			BNC					PUSH
			BNN					RCALL
			BNOV					RESET
			BNZ			RETFIE	RETFIE	RETFIE
			BOV		RETLW	RETLW	RETLW	RETLW
			BRA					
BSF	BSF	BSF	BSF		RETURN		RETURN	RETURN
BTFSC	BTFSC	BTFSC	BTFSC		RLCF			RLCF
BTFSS	BTFSS	BTFSS	BTFSS		RLF	RLF	RLF	
			BTG		RRF	RRF	RRF	
			BZ					RLNCF
CALL	CALL	CALL	CALL					RRCF
CLRF	CLRF	CLRF	CLRF					RRNCF
CLRW	CLRW	CLRW						SETF
CLRWDT	CLRWDT	CLRWDT	CLRWDT		SLEEP	SLEEP	SLEEP	SLEEP
COMF	COMF	COMF	COMF					SUBFWB
			CPFSEQ				SUBLW	SUBLW

10Fxxx	12Fxxx	16Fxxx	18Fxxx		10Fxxx	12Fxxx	16Fxxx	18Fxxx
			CPFSGT		SUBWF	SUBWF	SUBWF	SUBWF
			CPFSLT					SUBWFB
			DAW		SWAPF	SWAPF	SWAPF	SWAPF
			DCFSNZ					TBLRD*
DECF	DECF	DECF	DECF					TBLRD*–
DECFSZ	DECFSZ	DECFSZ	DECFSZ					TBLRD*+
GOTO	GOTO	GOTO	GOTO					TBLRD+*
INCF	INCF	INCF	INCF					TBLWT*
INCFSZ	INCFSZ	INCFSZ	INCFSZ					TBLWT*–
			INFSNZ					TBLWT*+
IORLW	IORLW	IORLW	IORLW					TBLWT+*
IORWF	IORWF	IORWF	IORWF		TRIS			
			LFSR			TRISGPIO		
MOVF	MOVF	MOVF	MOVF					TSTFSZ
			MOVFF		XORLW	XORLW	XORLW	XORLW
			MOVLB		XORWF	XORWF	XORWF	XORWF

MICROCHIP

PIC10F200/202/204/206

6-Pin, 8-Bit Flash Microcontrollers

Devices Included In This Data Sheet:

- PIC10F200
- PIC10F202
- PIC10F204
- PIC10F206

High-Performance RISC CPU:

- Only 33 Single-Word Instructions to Learn
- All Single-Cycle Instructions except for Program Branches, which are Two-Cycle
- 12-Bit Wide Instructions
- 2-Level Deep Hardware Stack
- Direct, Indirect and Relative Addressing modes for Data and Instructions
- 8-Bit Wide Data Path
- Eight Special Function Hardware Registers
- Operating Speed:
  - 4 MHz internal clock
  - 1 µs instruction cycle

Special Microcontroller Features:

- 4 MHz Precision Internal Oscillator:
  - Factory calibrated to ±1%
- In-Circuit Serial Programming™ (ICSP™)
- In-Circuit Debugging (ICD) Support
- Power-on Reset (POR)
- Device Reset Timer (DRT)
- Watchdog Timer (WDT) with Dedicated On-Chip RC Oscillator for Reliable Operation
- Programmable Code Protection
- Multiplexed $\overline{\text{MCLR}}$ Input Pin
- Internal Weak Pull-ups on I/O Pins
- Power-Saving Sleep mode
- Wake-up from Sleep on Pin Change

Low-Power Features/CMOS Technology:

- Operating Current:
  - < 175 µA @ 2V, 4 MHz, typical
- Standby Current:
  - 100 nA @ 2V, typical
- Low-Power, High-Speed Flash Technology:
  - 100,000 Flash endurance
  - > 40 year retention
- Fully Static Design
- Wide Operating Voltage Range: 2.0V to 5.5V
- Wide Temperature Range:
  - Industrial: –40°C to +85°C
  - Extended: –40°C to +125°C

Peripheral Features (PIC10F200/202):

- Four I/O Pins:
  - Three I/O pins with individual direction control
  - One input-only pin
  - High current sink/source for direct LED drive
  - Wake-on-change
  - Weak pull-ups
- 8-Bit Real-Time Clock/Counter (TMR0) with 8-Bit Programmable Prescaler

Peripheral Features (PIC10F204/206):

- Four I/O Pins:
  - Three I/O pins with individual direction control
  - One input-only pin
  - High current sink/source for direct LED drive
  - Wake-on-change
  - Weak pull-ups
- 8-Bit Real-Time Clock/Counter (TMR0) with 8-Bit Programmable Prescaler
- One Comparator:
  - Internal absolute voltage reference
  - Both comparator inputs visible externally
  - Comparator output visible externally

TABLE 1: PIC10F20X MEMORY AND FEATURES

Device	Program Memory	Data Memory	I/O	Timers 8-bit	Comparator
	Flash (words)	SRAM (bytes)			
PIC10F200	256	16	4	1	0
PIC10F202	512	24	4	1	0
PIC10F204	256	16	4	1	1
PIC10F206	512	24	4	1	1

10.0 INSTRUCTION SET SUMMARY

The PIC16 instruction set is highly orthogonal and is comprised of three basic categories.

- **Byte-oriented** operations
- **Bit-oriented** operations
- **Literal and control** operations

Each PIC16 instruction is a 12-bit word divided into an **opcode**, which specifies the instruction type and one or more **operands** which further specify the operation of the instruction. The formats for each of the categories is presented in Figure 10-1, while the various opcode fields are summarized in Table 10-1.

For **byte-oriented** instructions, 'f' represents a file register designator and 'd' represents a destination designator. The file register designator specifies which file register is to be used by the instruction.

The destination designator specifies where the result of the operation is to be placed. If 'd' is '0', the result is placed in the W register. If 'd' is '1', the result is placed in the file register specified in the instruction.

For **bit-oriented** instructions, 'b' represents a bit field designator which selects the number of the bit affected by the operation, while 'f' represents the number of the file in which the bit is located.

For **literal and control** operations, 'k' represents an 8 or 9-bit constant or literal value.

TABLE 10-1: OPCODE FIELD DESCRIPTIONS

Field	Description
f	Register file address (0x00 to 0x7F)
W	Working register (accumulator)
b	Bit address within an 8-bit file register
k	Literal field, constant data or label
x	Don't care location (= 0 or 1) The assembler will generate code with x = 0. It is the recommended form of use for compatibility with all Microchip software tools.
d	Destination select; d = 0 (store result in W) d = 1 (store result in file register 'f') Default is d = 1
label	Label name
TOS	Top-of-Stack
PC	Program Counter
WDT	Watchdog Timer counter
$\overline{TO}$	Time-out bit
$\overline{PD}$	Power-down bit
dest	Destination, either the W register or the specified register file location
[ ]	Options
( )	Contents
→	Assigned to
< >	Register bit field
∈	In the set of
italics	User defined term (font is courier)

All instructions are executed within a single instruction cycle, unless a conditional test is true or the program counter is changed as a result of an instruction. In this case, the execution takes two instruction cycles. One instruction cycle consists of four oscillator periods. Thus, for an oscillator frequency of 4 MHz, the normal instruction execution time is 1 μs. If a conditional test is true or the program counter is changed as a result of an instruction, the instruction execution time is 2 μs.

Figure 10-1 shows the three general formats that the instructions can have. All examples in the figure use the following format to represent a hexadecimal number:

```
0xhhh
```

where 'h' signifies a hexadecimal digit.

FIGURE 10-1: GENERAL FORMAT FOR INSTRUCTIONS

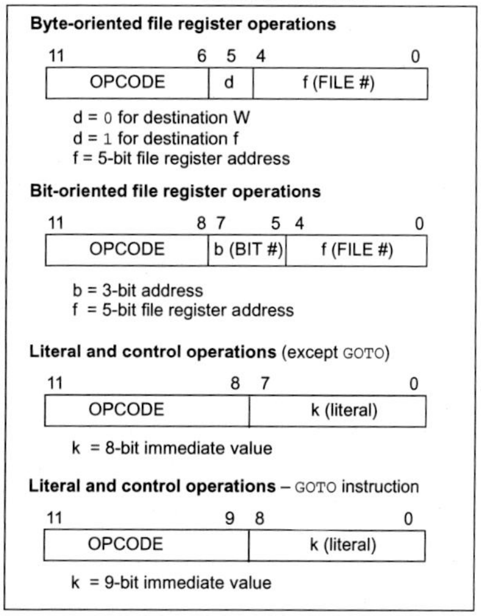

PIC10F200/202/204/206

TABLE 10-2: INSTRUCTION SET SUMMARY

Mnemonic, Operands		DescriptionCycles		12-Bit Opcode			Status Affected	Notes
				MSb		LSb		
ADDWF	f, d	Add W and f	1	0001	11df	ffff	C, DC, Z	**1, 2, 4**
ANDWF	f, d	AND W with f	1	0001	01df	ffff	Z	**2, 4**
CLRF	f	Clear f	1	0000	011f	ffff	Z	**4**
CLRW	—	Clear W	1	0000	0100	0000	Z	
COMF	f, d	Complement f	1	0010	01df	ffff	Z	
DECF	f, d	Decrement f	1	0000	11df	ffff	Z	**2, 4**
DECFSZ	f, d	Decrement f, Skip if 0	1(2)	0010	11df	ffff	None	**2, 4**
INCF	f, d	Increment f	1	0010	10df	ffff	Z	**2, 4**
INCFSZ	f, d	Increment f, Skip if 0	1(2)	0011	11df	ffff	None	**2, 4**
IORWF	f, d	Inclusive OR W with f	1	0001	00df	ffff	Z	**2, 4**
MOVF	f, d	Move f	1	0010	00df	ffff	Z	**2, 4**
MOVWF	f	Move W to f	1	0000	001f	ffff	None	**1, 4**
NOP	—	No Operation	1	0000	0000	0000	None	
RLF	f, d	Rotate left f through Carry	1	0011	01df	ffff	C	**2, 4**
RRF	f, d	Rotate right f through Carry	1	0011	00df	ffff	C	**2, 4**
SUBWF	f, d	Subtract W from f	1	0000	10df	ffff	C, DC, Z	**1, 2, 4**
SWAPF	f, d	Swap f	1	0011	10df	ffff	None	**2, 4**
XORWF	f, d	Exclusive OR W with f	1	0001	10df	ffff	Z	**2, 4**
BIT-ORIENTED FILE REGISTER OPERATIONS								
BCF	f, b	Bit Clear f	1	0100	bbbf	ffff	None	**2, 4**
BSF	f, b	Bit Set f	1	0101	bbbf	ffff	None	**2, 4**
BTFSC	f, b	Bit Test f, Skip if Clear	1(2)	0110	bbbf	ffff	None	
BTFSS	f, b	Bit Test f, Skip if Set	1(2)	0111	bbbf	ffff	None	
LITERAL AND CONTROL OPERATIONS								
ANDLW	k	AND literal with W	1	1110	kkkk	kkkk	Z	
CALL	k	Call Subroutine	2	1001	kkkk	kkkk	None	**1**
CLRWDT		Clear Watchdog Timer	1	0000	0000	0100	$\overline{TO}$, $\overline{PD}$	
GOTO	k	Unconditional branch	2	101k	kkkk	kkkk	None	
IORLW	k	Inclusive OR literal with W	1	1101	kkkk	kkkk	Z	
MOVLW	k	Move literal to W	1	1100	kkkk	kkkk	None	
OPTION	—	Load OPTION register	1	0000	0000	0010	None	
RETLW	k	Return, place Literal in W	2	1000	kkkk	kkkk	None	
SLEEP	—	Go into Standby mode	1	0000	0000	0011	$\overline{TO}$, $\overline{PD}$	
TRIS	f	Load TRIS register	1	0000	0000	0fff	None	**3**
XORLW	k	Exclusive OR literal to W	1	1111	kkkk	kkkk	Z	

Note 1: The 9th bit of the program counter will be forced to a '0' by any instruction that writes to the PC except for GOTO. See **Section 4.7 "Program Counter"**.

2: When an I/O register is modified as a function of itself (e.g. MOVF PORTB, 1), the value used will be that value present on the pins themselves. For example, if the data latch is '1' for a pin configured as input and ' is driven low by an external device, the data will be written back with a '0'.

3: The instruction TRIS f , where f = 6, causes the contents of the W register to be written to the tri-state latches of PORTB. A '1' forces the pin to a high-impedance state and disables the output buffers.

4: If this instruction is executed on the TMR0 register (and where applicable, d = 1), the prescaler will be cleared (if assigned to TMR0).

MICROCHIP

PIC12F519

8-Pin, 8-Bit Flash Microcontroller

High-Performance RISC CPU:

- Only 33 Single-Word Instructions
- All Single-Cycle Instructions except for Program Branches which are Two-Cycle
- Two-Level Deep Hardware Stack
- Direct, Indirect and Relative Addressing modes for Data and Instructions
- Operating Speed:
  - DC – 8 MHz Oscillator
  - DC – 500 ns instruction cycle
- On-chip Flash Program Memory
  - 1024 x 12
- General Purpose Registers (SRAM)
  - 41 x 8
- Flash Data Memory
  - 64 x 8

Special Microcontroller Features:

- 8 MHz Precision Internal Oscillator
  - Factory calibrated to ±1%
- In-Circuit Serial Programming™ (ICSP™)
- In-Circuit Debugging (ICD) Support
- Power-on Reset (POR)
- Device Reset Timer (DRT)
- Watchdog Timer (WDT) with Dedicated On-Chip RC Oscillator for Reliable Operation
- Programmable Code Protection
- Multiplexed $\overline{\text{MCLR}}$ Input Pin
- Internal Weak Pull-ups on I/O Pins
- Power-Saving Sleep mode
- Wake-up from Sleep on Pin Change
- Selectable Oscillator Options:
  - INTRC: 4 MHz or 8 MHz precision Internal RC oscillator
  - EXTRC: External low-cost RC oscillator
  - XT: Standard crystal/resonator
  - LP: Power-saving, low-frequency crystal

Low-Power Features/CMOS Technology:

- Standby Current:
  - 100 nA @ 2.0V, typical
- Operating Current:
  - 11 μA @ 32 kHz, 2.0V, typical
  - 175 μA @ 4 MHz, 2.0V, typical
- Watchdog Timer Current:
  - 1 μA @ 2.0V, typical
  - 7 μA @ 5.0V, typical
- High Endurance Program and Flash Data Memory Cells
  - 100,000 write Program Memory endurance
  - 1,000,000 write Flash Data Memory endurance
  - Program and Flash Data retention: >40 years
- Fully Static Design
- Wide Operating Voltage Range: 2.0V to 5.5V
  - Wide temperature range
  - Industrial: –40°C to +85°C
  - Extended: –40°C to +125°C

Peripheral Features:

- 6 I/O Pins
  - 5 I/O pins with individual direction control
  - 1 input-only pin
  - High current sink/source for direct LED drive
- 8-bit Real-Time Clock/Counter (TMR0) with 8-bit Programmable Prescaler.

PIC12F519

9.0 INSTRUCTION SET SUMMARY

The PIC12F519 instruction set is highly orthogonal and is comprised of three basic categories.

- **Byte-oriented** operations
- **Bit-oriented** operations
- **Literal and control** operations

Each PIC12F519 instruction is a 12-bit word divided into an **opcode**, which specifies the instruction type, and one or more **operands** which further specify the operation of the instruction. The formats for each of the categories is presented in Figure 9-1, while the various opcode fields are summarized in Table 9-1.

For **byte-oriented** instructions, 'f' represents a file register designator and 'd' represents a destination designator. The file register designator specifies which file register is to be used by the instruction.

The destination designator specifies where the result of the operation is to be placed. If 'd' is '0', the result is placed in the W register. If 'd' is '1', the result is placed in the file register specified in the instruction.

For **bit-oriented** instructions, 'b' represents a bit field designator which selects the number of the bit affected by the operation, while 'f' represents the number of the file in which the bit is located.

For **literal and control** operations, 'k' represents an 8 or 9-bit constant or literal value.

TABLE 9-1: OPCODE FIELD DESCRIPTIONS

Field	Description
f	Register file address (0x00 to 0x7F)
W	Working register (accumulator)
b	Bit address within an 8-bit file register
k	Literal field, constant data or label
x	Don't care location (= 0 or 1) The assembler will generate code with x = 0. It is the recommended form of use for compatibility with all Microchip software tools.
d	Destination select; d = 0 (store result in W) d = 1 (store result in file register 'f') Default is d = 1
label	Label name
TOS	Top-of-Stack
PC	Program Counter
WDT	Watchdog Timer counter
$\overline{TO}$	Time-out bit
$\overline{PD}$	Power-down bit
dest	Destination, either the W register or the specified register file location
[ ]	Options
( )	Contents
→	Assigned to
< >	Register bit field
∈	In the set of
italics	User defined term (font is courier)

All instructions are executed within a single instruction cycle, unless a conditional test is true or the program counter is changed as a result of an instruction. In this case, the execution takes two instruction cycles. One instruction cycle consists of four oscillator periods. Thus, for an oscillator frequency of 4 MHz, the normal instruction execution time is 1 μs. If a conditional test is true or the program counter is changed as a result of an instruction, the instruction execution time is 2 μs.

Figure 9-1 shows the three general formats that the instructions can have. All examples in the figure use the following format to represent a hexadecimal number:

0xhhh

where 'h' signifies a hexadecimal digit.

FIGURE 9-1: GENERAL FORMAT FOR INSTRUCTIONS

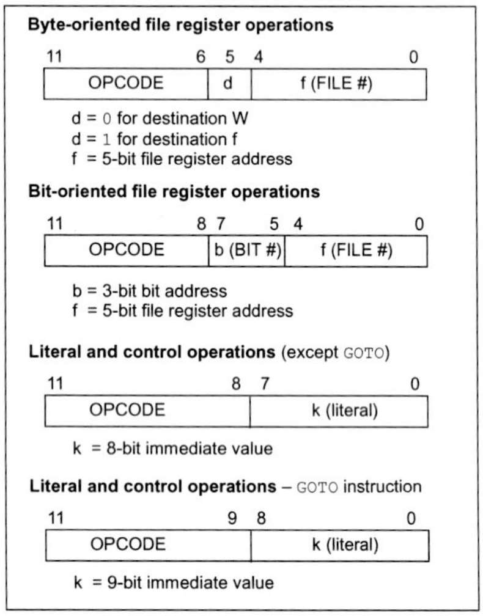

PIC12F519

TABLE 9-2: INSTRUCTION SET SUMMARY

Mnemonic, Operands		Description	Cycles	12-Bit Opcode			Status Affected	Notes
				MSb		LSb		
ADDWF	f, d	Add W and f	1	0001	11df	ffff	C, DC, Z	1, 2, 4
ANDWF	f, d	AND W with f	1	0001	01df	ffff	Z	2, 4
CLRF	f	Clear f	1	0000	011f	ffff	Z	4
CLRW	–	Clear W	1	0000	0100	0000	Z	
COMF	f, d	Complement f	1	0010	01df	ffff	Z	
DECF	f, d	Decrement f	1	0000	11df	ffff	Z	2, 4
DECFSZ	f, d	Decrement f, Skip if 0	1(2)	0010	11df	ffff	None	2, 4
INCF	f, d	Increment f	1	0010	10df	ffff	Z	2, 4
INCFSZ	f, d	Increment f, Skip if 0	1(2)	0011	11df	ffff	None	2, 4
IORWF	f, d	Inclusive OR W with f	1	0001	00df	ffff	Z	2, 4
MOVF	f, d	Move f	1	0010	00df	ffff	Z	2, 4
MOVWF	f	Move W to f	1	0000	001f	ffff	None	1, 4
NOP	–	No Operation	1	0000	0000	0000	None	
RLF	f, d	Rotate left f through Carry	1	0011	01df	ffff	C	2, 4
RRF	f, d	Rotate right f through Carry	1	0011	00df	ffff	C	2, 4
SUBWF	f, d	Subtract W from f	1	0000	10df	ffff	C, DC, Z	1, 2, 4
SWAPF	f, d	Swap f	1	0011	10df	ffff	None	2, 4
XORWF	f, d	Exclusive OR W with f	1	0001	10df	ffff	Z	2, 4
BIT-ORIENTED FILE REGISTER OPERATIONS								
BCF	f, b	Bit Clear f	1	0100	bbbf	ffff	None	2, 4
BSF	f, b	Bit Set f	1	0101	bbbf	ffff	None	2, 4
BTFSC	f, b	Bit Test f, Skip if Clear	1(2)	0110	bbbf	ffff	None	
BTFSS	f, b	Bit Test f, Skip if Set	1(2)	0111	bbbf	ffff	None	
LITERAL AND CONTROL OPERATIONS								
ANDLW	k	AND literal with W	1	1110	kkkk	kkkk	Z	
CALL	k	Call Subroutine	2	1001	kkkk	kkkk	None	1
CLRWDT	–	Clear Watchdog Timer	1	0000	0000	0100	$\overline{TO}$, $\overline{PD}$	
GOTO	k	Unconditional branch	2	101k	kkkk	kkkk	None	
IORLW	k	Inclusive OR literal with W	1	1101	kkkk	kkkk	Z	
MOVLW	k	Move literal to W	1	1100	kkkk	kkkk	None	
OPTION	–	Load OPTION register	1	0000	0000	0010	None	
RETLW	k	Return, place literal in W	2	1000	kkkk	kkkk	None	
SLEEP	–	Go into Standby mode	1	0000	0000	0011	$\overline{TO}$, $\overline{PD}$	
TRISGPIO	f	Load TRISGPIO register	1	0000	0000	0fff	None	3
XORLW	k	Exclusive OR literal to W	1	1111	kkkk	kkkk	Z	

Note 1: The 9th bit of the program counter will be forced to a '0' by any instruction that writes to the PC except for GOTO. See **Section 4.6 "Program Counter"**.

2: When an I/O register is modified as a function of itself (e.g. MOVF GPIO, 1), the value used will be that value present on the pins themselves. For example, if the data latch is '1' for a pin configured as input and is driven low by an external device, the data will be written back with a '0'.

3: The instruction TRIS f, where f = 6, causes the contents of the W register to be written to the tri-state latches of GPIO. A '1' forces the pin to a high-impedance state and disables the output buffers.

4: If this instruction is executed on the TMR0 register (and, where applicable, d = 1), the prescaler will be cleared (if assigned to TMR0).

MICROCHIP **PIC16F631/677/685/687/689/690**

20-Pin Flash-Based, 8-Bit CMOS Microcontrollers

High-Performance RISC CPU

- Only 35 Instructions to Learn:
  - All single-cycle instructions except branches
- Operating Speed:
  - DC – 20 MHz oscillator/clock input
  - DC – 200 ns instruction cycle
- Interrupt Capability
- 8-Level Deep Hardware Stack
- Direct, Indirect and Relative Addressing modes

Special Microcontroller Features

- Precision Internal Oscillator:
  - Factory calibrated to ± 1%
  - Software selectable frequency range of 8 MHz to 32 kHz
  - Software tunable
  - Two-Speed Start-up mode
  - Crystal fail detect for critical applications
  - Clock mode switching during operation for power savings
- Power-Saving Sleep mode
- Wide Operating Voltage Range (2.0V-5.5V)
- Industrial and Extended Temperature Range
- Power-on Reset (POR)
- Power-up Timer (PWRTE) and Oscillator Start-up Timer (OST)
- Brown-out Reset (BOR) with Software Control Option
- Enhanced Low-Current Watchdog Timer (WDT) with On-Chip Oscillator (Software selectable nominal 268 Seconds with Full Prescaler) with Software Enable
- Multiplexed Master Clear/Input Pin
- Programmable Code Protection
- High Endurance Flash/EEPROM Cell:
  - 100,000 write Flash endurance
  - 1,000,000 write EEPROM endurance
  - Flash/Data EEPROM retention: > 40 years
- Enhanced USART Module:
  - Supports RS-485, RS-232 and LIN 2.0
  - Auto-Baud Detect
  - Auto-wake-up on Start bit

Low-Power Features

- Standby Current:
  - 50 nA @ 2.0V, typical
- Operating Current:
  - 11 μA @ 32 kHz, 2.0V, typical
  - 220 μA @ 4 MHz, 2.0V, typical
- Watchdog Timer Current:
  - <1 μA @ 2.0V, typical

Peripheral Features

- 17 I/O Pins and 1 Input-Only Pin:
  - High current source/sink for direct LED drive
  - Interrupt-on-Change pin
  - Individually programmable weak pull-ups
  - Ultra Low-Power Wake-up (ULPWU)
- Analog Comparator Module with:
  - Two analog comparators
  - Programmable on-chip voltage reference (CVREF) module (% of VDD)
  - Comparator inputs and outputs externally accessible
  - SR Latch mode
  - Timer 1 Gate Sync Latch
  - Fixed 0.6V VREF
- A/D Converter:
  - 10-bit resolution and 12 channels
- Timer0: 8-Bit Timer/Counter with 8-Bit Programmable Prescaler
- Enhanced Timer1:
  - 16-bit timer/counter with prescaler
  - External Timer1 Gate (count enable)
  - Option to use OSC1 and OSC2 in LP mode as Timer1 oscillator if INTOSC mode selected
- Timer2: 8-Bit Timer/Counter with 8-Bit Period Register, Prescaler and Postscaler
- Enhanced Capture, Compare, PWM+ Module:
  - 16-bit Capture, max resolution 12.5 ns
  - Compare, max resolution 200 ns
  - 10-bit PWM with 1, 2 or 4 output channels, programmable "dead time", max frequency 20 kHz
  - PWM output steering control
- Synchronous Serial Port (SSP):
  - SPI mode (Master and Slave)
- I²C™ (Master/Slave modes):
  - I²C™ address mask
- In-Circuit Serial Programming™ (ICSP™) via Two Pins

15.0 INSTRUCTION SET SUMMARY

The PIC16F690 instruction set is highly orthogonal and is comprised of three basic categories:

- **Byte-oriented** operations
- **Bit-oriented** operations
- **Literal and control** operations

Each PIC16 instruction is a 14-bit word divided into an **opcode,** which specifies the instruction type and one or more **operands,** which further specify the operation of the instruction. The formats for each of the categories is presented in Figure 15-1, while the various opcode fields are summarized in Table 15-1.

Table 15-2 lists the instructions recognized by the MPASM™ assembler.

For **byte-oriented** instructions, 'f' represents a file register designator and 'd' represents a destination designator. The file register designator specifies which file register is to be used by the instruction.

The destination designator specifies where the result of the operation is to be placed. If 'd' is zero, the result is placed in the W register. If 'd' is one, the result is placed in the file register specified in the instruction.

For **bit-oriented** instructions, 'b' represents a bit field designator, which selects the bit affected by the operation, while 'f' represents the address of the file in which the bit is located.

For **literal and control** operations, 'k' represents an 8-bit or 11-bit constant, or literal value.

One instruction cycle consists of four oscillator periods; for an oscillator frequency of 4 MHz, this gives a normal instruction execution time of 1 μs. All instructions are executed within a single instruction cycle, unless a conditional test is true, or the program counter is changed as a result of an instruction. When this occurs, the execution takes two instruction cycles, with the second cycle executed as a `NOP`.

All instruction examples use the format '`0xhh`' to represent a hexadecimal number, where '`h`' signifies a hexadecimal digit.

15.1 Read-Modify-Write Operations

Any instruction that specifies a file register as part of the instruction performs a Read-Modify-Write (RMW) operation. The register is read, the data is modified, and the result is stored according to either the instruction, or the destination designator 'd'. A read operation is performed on a register even if the instruction writes to that register.

For example, a `CLRF PORTA` instruction will read PORTA, clear all the data bits, then write the result back to PORTA. This example would have the unintended consequence of clearing the condition that set the RAIF flag.

TABLE 15-1: OPCODE FIELD DESCRIPTIONS

Field	Description
f	Register file address (0x00 to 0x7F)
W	Working register (accumulator)
b	Bit address within an 8-bit file register
k	Literal field, constant data or label
x	Don't care location (= 0 or 1). The assembler will generate code with x = 0. It is the recommended form of use for compatibility with all Microchip software tools.
d	Destination select; d = 0: store result in W, d = 1: store result in file register f. Default is d = 1.
PC	Program Counter
$\overline{TO}$	Time-out bit
C	Carry bit
DC	Digit carry bit
Z	Zero bit
$\overline{PD}$	Power-down bit

FIGURE 15-1: GENERAL FORMAT FOR INSTRUCTIONS

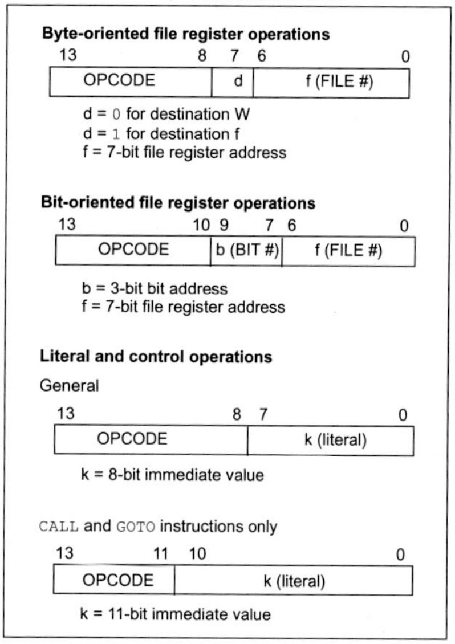

TABLE 15-2: PIC16F684 INSTRUCTION SET

Mnemonic, Operands		Description	Cycles	14-Bit Opcode MSb			LSb	Status Affected	Notes
BYTE-ORIENTED FILE REGISTER OPERATIONS									
ADDWF	f, d	Add W and f	1	00	0111	dfff	ffff	C, DC, Z	1, 2
ANDWF	f, d	AND W with f	1	00	0101	dfff	ffff	Z	1, 2
CLRF	f	Clear f	1	00	0001	lfff	ffff	Z	2
CLRW	–	Clear W	1	00	0001	0xxx	xxxx	Z	
COMF	f, d	Complement f	1	00	1001	dfff	ffff	Z	1, 2
DECF	f, d	Decrement f	1	00	0011	dfff	ffff	Z	1, 2
DECFSZ	f, d	Decrement f, Skip if 0	1(2)	00	1011	dfff	ffff		1, 2, 3
INCF	f, d	Increment f	1	00	1010	dfff	ffff	Z	1, 2
INCFSZ	f, d	Increment f, Skip if 0	1(2)	00	1111	dfff	ffff		1, 2, 3
IORWF	f, d	Inclusive OR W with f	1	00	0100	dfff	ffff	Z	1, 2
MOVF	f, d	Move f	1	00	1000	dfff	ffff	Z	1, 2
MOVWF	f	Move W to f	1	00	0000	lfff	ffff		
NOP	–	No Operation	1	00	0000	0xx0	0000		
RLF	f, d	Rotate Left f through Carry	1	00	1101	dfff	ffff	C	1, 2
RRF	f, d	Rotate Right f through Carry	1	00	1100	dfff	ffff	C	1, 2
SUBWF	f, d	Subtract W from f	1	00	0010	dfff	ffff	C, DC, Z	1, 2
SWAPF	f, d	Swap nibbles in f	1	00	1110	dfff	ffff		1, 2
XORWF	f, d	Exclusive OR W with f	1	00	0110	dfff	ffff	Z	1, 2
BIT-ORIENTED FILE REGISTER OPERATIONS									
BCF	f, b	Bit Clear f	1	01	00bb	bfff	ffff		1, 2
BSF	f, b	Bit Set f	1	01	01bb	bfff	ffff		1, 2
BTFSC	f, b	Bit Test f, Skip if Clear	1 (2)	01	10bb	bfff	ffff		3
BTFSS	f, b	Bit Test f, Skip if Set	1 (2)	01	11bb	bfff	ffff		3
LITERAL AND CONTROL OPERATIONS									
ADDLW	k	Add literal and W	1	11	111x	kkkk	kkkk	C, DC, Z	
ANDLW	k	AND literal with W	1	11	1001	kkkk	kkkk	Z	
CALL	k	Call Subroutine	2	10	0kkk	kkkk	kkkk		
CLRWDT	–	Clear Watchdog Timer	1	00	0000	0110	0100	$\overline{TO}$, $\overline{PD}$	
GOTO	k	Go to address	2	10	1kkk	kkkk	kkkk		
IORLW	k	Inclusive OR literal with W	1	11	1000	kkkk	kkkk	Z	
MOVLW	k	Move literal to W	1	11	00xx	kkkk	kkkk		
RETFIE	–	Return from interrupt	2	00	0000	0000	1001		
RETLW	k	Return with literal in W	2	11	01xx	kkkk	kkkk		
RETURN	–	Return from Subroutine	2	00	0000	0000	1000		
SLEEP	–	Go into Standby mode	1	00	0000	0110	0011	$\overline{TO}$, $\overline{PD}$	
SUBLW	k	Subtract W from literal	1	11	110x	kkkk	kkkk	C, DC, Z	
XORLW	k	Exclusive OR literal with W	1	11	1010	kkkk	kkkk	Z	

Note 1: When an I/O register is modified as a function of itself (e.g., `MOVF GPIO, 1`), the value used will be that value present on the pins themselves. For example, if the data latch is '1' for a pin configured as input and is driven low by an external device, the data will be written back with a '0'.

2: If this instruction is executed on the TMR0 register (and where applicable, d = 1), the prescaler will be cleared if assigned to the Timer0 module.

3: If the Program Counter (PC) is modified, or a conditional test is true, the instruction requires two cycles. The second cycle is executed as a `NOP`.

MICROCHIP **PIC18F2220/2320/4220/4320**

28/40/44-Pin High-Performance, Enhanced Flash MCUs with 10-Bit A/D and nanoWatt Technology

Low-Power Features:

- Power-Managed modes:
  - Run: CPU on, peripherals on
  - Idle: CPU off, peripherals on
  - Sleep: CPU off, peripherals off
- Power Consumption modes:
  - PRI_RUN: 150 μA, 1 MHz, 2V
  - PRI_IDLE: 37 μA, 1 MHz, 2V
  - SEC_RUN: 14 μA, 32 kHz, 2V
  - SEC_IDLE: 5.8 μA, 32 kHz, 2V
  - RC_RUN: 110 μA, 1 MHz, 2V
  - RC_IDLE: 52 μA, 1 MHz, 2V
  - Sleep: 0.1 μA, 1 MHz, 2V
- Timer1 Oscillator: 1.1 μA, 32 kHz, 2V
- Watchdog Timer: 2.1 μA
- Two-Speed Oscillator Start-up

Oscillators:

- Four Crystal modes:
  - LP, XT, HS: up to 25 MHz
  - HSPLL: 4-10 MHz (16-40 MHz internal)
- Two External RC modes, Up to 4 MHz
- Two External Clock modes, Up to 40 MHz
- Internal Oscillator Block:
  - 8 user-selectable frequencies: 31 kHz, 125 kHz, 250 kHz, 500 kHz, 1 MHz, 2 MHz, 4 MHz, 8 MHz
  - 125 kHz-8 MHz calibrated to 1%
  - Two modes select one or two I/O pins
  - OSCTUNE – Allows user to shift frequency
- Secondary Oscillator using Timer1 @ 32 kHz
- Fail-Safe Clock Monitor
  - Allows for safe shutdown if peripheral clock stops

Peripheral Highlights:

- High-Current Sink/Source 25 mA/25 mA
- Three External Interrupts
- Up to 2 Capture/Compare/PWM (CCP) modules:
  - Capture is 16-bit, max. resolution is 6.25 ns (TCY/16)
  - Compare is 16-bit, max. resolution is 100 ns (TCY)
  - PWM output: PWM resolution is 1 to 10-bit
- Enhanced Capture/Compare/PWM (ECCP) module:
  - One, two or four PWM outputs
  - Selectable polarity
  - Programmable dead time
  - Auto-Shutdown and Auto-Restart
- Compatible 10-Bit, Up to 13-Channel Analog-to-Digital Converter (A/D) module with Programmable Acquisition Time
- Dual Analog Comparators
- Addressable USART module:
  - RS-232 operation using internal oscillator block (no external crystal required)

Special Microcontroller Features:

- 100,000 Erase/Write Cycle Enhanced Flash Program Memory Typical
- 1,000,000 Erase/Write Cycle Data EEPROM Memory Typical
- Flash/Data EEPROM Retention: > 40 Years
- Self-Programmable under Software Control
- Priority Levels for Interrupts
- 8 x 8 Single-Cycle Hardware Multiplier
- Extended Watchdog Timer (WDT):
  - Programmable period from 41 ms to 131s
  - 2% stability over VDD and Temperature
- Single-Supply 5V In-Circuit Serial Programming™ (ICSP™) via Two Pins
- In-Circuit Debug (ICD) via Two Pins
- Wide Operating Voltage Range: 2.0V to 5.5V

Device	Program Memory		Data Memory		I/O	10-Bit A/D (ch)	CCP/ ECCP (PWM)	MSSP		USART	Comparators	Timers 8/16-bit
	Flash (bytes)	# Single Word Instructions	SRAM (bytes)	EEPROM (bytes)				SPI	Master I^2C™			
PIC18F2220	4096	2048	512	256	25	10	2/0	Y	Y	Y	2	2/3
PIC18F2320	8192	4096	512	256	25	10	2/0	Y	Y	Y	2	2/3
PIC18F4220	4096	2048	512	256	36	13	1/1	Y	Y	Y	2	2/3
PIC18F4320	8192	4096	512	256	36	13	1/1	Y	Y	Y	2	2/3

24.0 INSTRUCTION SET SUMMARY

The PIC18 instruction set adds many enhancements to the previous PIC MCU instruction sets, while maintaining an easy migration from these PIC MCU instruction sets.

Most instructions are a single program memory word (16 bits) but there are three instructions that require two program memory locations.

Each single-word instruction is a 16-bit word divided into an opcode, which specifies the instruction type and one or more operands, which further specify the operation of the instruction.

The instruction set is highly orthogonal and is grouped into four basic categories:

- **Byte-oriented** operations
- **Bit-oriented** operations
- **Literal** operations
- **Control** operations

The PIC18 instruction set summary in Table 24-2 lists **byte-oriented**, **bit-oriented**, **literal** and **control** operations. Table 24-1 shows the opcode field descriptions.

Most **byte-oriented** instructions have three operands:

1. The file register (specified by 'f')
2. The destination of the result (specified by 'd')
3. The accessed memory (specified by 'a')

The file register designator 'f' specifies which file register is to be used by the instruction.

The destination designator 'd' specifies where the result of the operation is to be placed. If 'd' is zero, the result is placed in the WREG register. If 'd' is one, the result is placed in the file register specified in the instruction.

All **bit-oriented** instructions have three operands:

1. The file register (specified by 'f')
2. The bit in the file register (specified by 'b')
3. The accessed memory (specified by 'a')

The bit field designator 'b' selects the number of the bit affected by the operation, while the file register designator 'f' represents the number of the file in which the bit is located.

The **literal** instructions may use some of the following operands:

- A literal value to be loaded into a file register (specified by 'k')
- The desired FSR register to load the literal value into (specified by 'f')
- No operand required (specified by '—')

The **control** instructions may use some of the following operands:

- A program memory address (specified by 'n')
- The mode of the `CALL` or `RETURN` instructions (specified by 's')
- The mode of the table read and table write instructions (specified by 'm')
- No operand required (specified by '—')

All instructions are a single word except for three double word instructions. These three instructions were made double word instructions so that all the required information is available in these 32 bits. In the second word, the 4 MSbs are '1's. If this second word is executed as an instruction (by itself), it will execute as a `NOP`.

All single-word instructions are executed in a single instruction cycle, unless a conditional test is true or the program counter is changed as a result of the instruction. In these cases, the execution takes two instruction cycles with the additional instruction cycle(s) executed as a `NOP`.

The double word instructions execute in two instruction cycles.

One instruction cycle consists of four oscillator periods. Thus, for an oscillator frequency of 4 MHz, the normal instruction execution time is 1 μs. If a conditional test is true, or the program counter is changed as a result of an instruction, the instruction execution time is 2 μs. Two-word branch instructions (if true) would take 3 μs.

Figure 24-1 shows the general formats that the instructions can have.

All examples use the format '`nnh`' to represent a hexadecimal number, where '`h`' signifies a hexadecimal digit.

The Instruction Set Summary, shown in Table 24-2, lists the instructions recognized by the Microchip Assembler (MPASM™). **Section 24.2 "Instruction Set"** provides a description of each instruction.

24.1 READ-MODIFY-WRITE OPERATIONS

Any instruction that specifies a file register as part of the instruction performs a Read-Modify-Write (R-M-W) operation. The register is read, the data is modified and the result is stored according to either the instruction or the destination designator 'd'. A read operation is performed on a register even if the instruction writes to that register.

For example, a "`BCF PORTB,1`" instruction will read PORTB, clear bit 1 of the data, then write the result back to PORTB. The read operation would have the unintended result that any condition that sets the RBIF flag would be cleared. The R-M-W operation may also copy the level of an input pin to its corresponding output latch.

PIC18F2220/2320/4220/4320

TABLE 24-1: OPCODE FIELD DESCRIPTIONS

Field	Description
a	RAM access bit: a = 0: RAM location in Access RAM (BSR register is ignored) a = 1: RAM bank is specified by BSR register
bbb	Bit address within an 8-bit file register (0 to 7).
BSR	Bank Select Register. Used to select the current RAM bank.
d	Destination select bit: d = 0: store result in WREG d = 1: store result in file register f
dest	Destination either the WREG register or the specified register file location.
f	8-bit register file address (0x00 to 0xFF).
fs	12-bit register file address (0x000 to 0xFFF). This is the source address.
fd	12-bit register file address (0x000 to 0xFFF). This is the destination address.
k	Literal field, constant data or label (may be either an 8-bit, 12-bit or a 20-bit value).
label	Label name.
mm	The mode of the TBLPTR register for the table read and table write instructions. Only used with table read and table write instructions:
*	No Change to register (such as TBLPTR with table reads and writes).
*+	Post-Increment register (such as TBLPTR with table reads and writes).
*-	Post-Decrement register (such as TBLPTR with table reads and writes).
+*	Pre-Increment register (such as TBLPTR with table reads and writes).
n	The relative address (2's complement number) for relative branch instructions, or the direct address for Call/Branch and Return instructions.
PRODH	Product of Multiply High Byte.
PRODL	Product of Multiply Low Byte.
s	Fast Call/Return mode select bit: s = 0: do not update into/from shadow registers s = 1: certain registers loaded into/from shadow registers (Fast mode)
u	Unused or Unchanged.
WREG	Working register (accumulator).
x	Don't care ('0' or '1'). The assembler will generate code with x = 0. It is the recommended form of use for compatibility with all Microchip software tools.
TBLPTR	21-bit Table Pointer (points to a Program Memory location).
TABLAT	8-bit Table Latch.
TOS	Top-of-Stack.
PC	Program Counter.
PCL	Program Counter Low Byte.
PCH	Program Counter High Byte.
PCLATH	Program Counter High Byte Latch.
PCLATU	Program Counter Upper Byte Latch.
GIE	Global Interrupt Enable bit.
WDT	Watchdog Timer.
$\overline{TO}$	Time-out bit.
$\overline{PD}$	Power-down bit.
C, DC, Z, OV, N	ALU Status bits: Carry, Digit Carry, Zero, Overflow, Negative.
[ ]	Optional.
( )	Contents.
→	Assigned to.
< >	Register bit field.
∈	In the set of.
italics	User-defined term (font is Courier New).

FIGURE 24-1: GENERAL FORMAT FOR INSTRUCTIONS

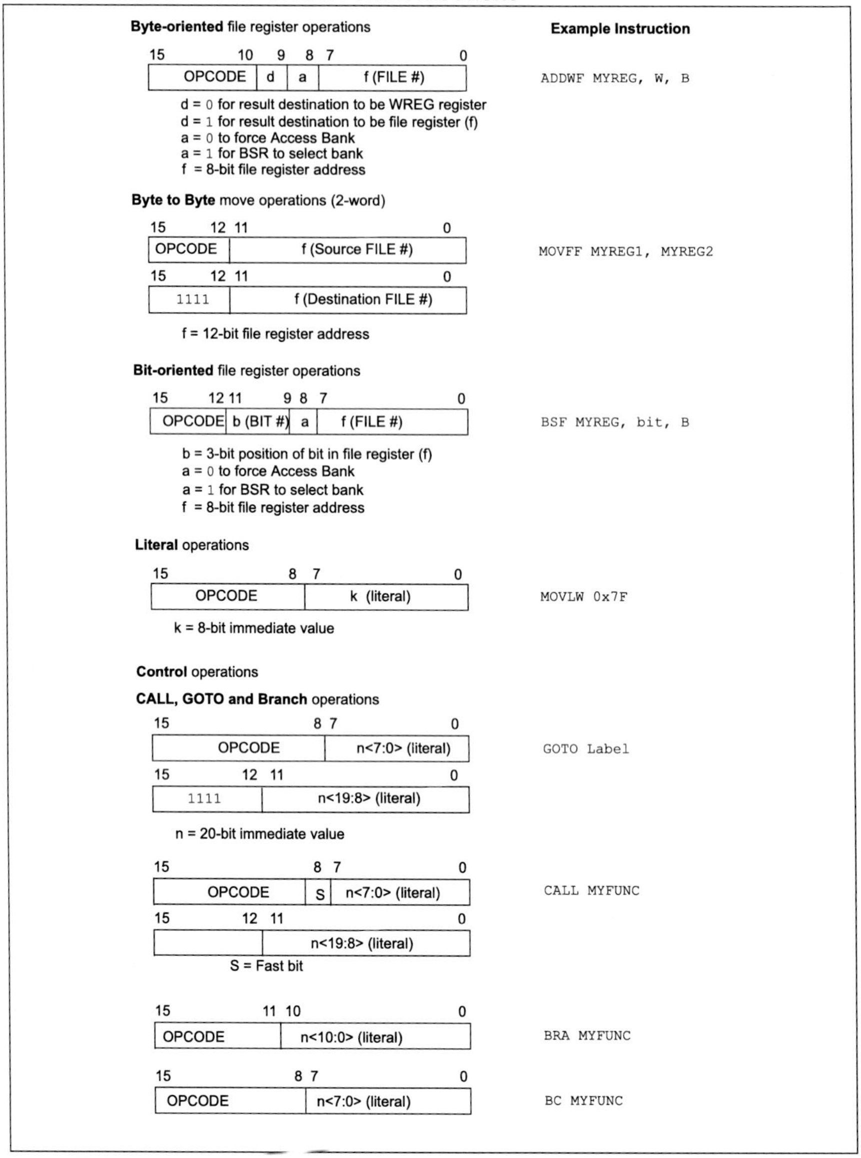
Byte-oriented file register operations
Example Instruction
15 10 9 8 7 0
OPCODE | d | a | f (FILE #)
ADDWF MYREG, W, B
d = 0 for result destination to be WREG register
d = 1 for result destination to be file register (f)
a = 0 to force Access Bank
a = 1 for BSR to select bank
f = 8-bit file register address
Byte to Byte move operations (2-word)
15 12 11 0
OPCODE | f (Source FILE #)
MOVFF MYREG1, MYREG2
15 12 11 0
1111 | f (Destination FILE #)
f = 12-bit file register address
Bit-oriented file register operations
15 12 11 9 8 7 0
OPCODE | b (BIT #) | a | f (FILE #)
BSF MYREG, bit, B
b = 3-bit position of bit in file register (f)
a = 0 to force Access Bank
a = 1 for BSR to select bank
f = 8-bit file register address
Literal operations
15 8 7 0
OPCODE | k (literal)
MOVLW 0x7F
k = 8-bit immediate value
Control operations
CALL, GOTO and Branch operations
15 8 7 0
OPCODE | n<7:0> (literal)
GOTO Label
15 12 11 0
1111 | n<19:8> (literal)
n = 20-bit immediate value
15 8 7 0
OPCODE | S | n<7:0> (literal)
CALL MYFUNC
15 12 11 0
n<19:8> (literal)
S = Fast bit
15 11 10 0
OPCODE | n<10:0> (literal)
BRA MYFUNC
15 8 7 0
OPCODE | n<7:0> (literal)
BC MYFUNC

PIC18F2220/2320/4220/4320

TABLE 24-2: PIC18FXXX INSTRUCTION SET

Mnemonic, Operands		Description	Cycles	16-Bit Instruction Word				Status Affected	Notes
				MSb			LSb		
BYTE-ORIENTED FILE REGISTER OPERATIONS									
ADDWF	f, d, a	Add WREG and f	1	0010	01da	ffff	ffff	C, DC, Z, OV, N	1, 2
ADDWFC	f, d, a	Add WREG and Carry bit to f	1	0010	00da	ffff	ffff	C, DC, Z, OV, N	1, 2
ANDWF	f, d, a	AND WREG with f	1	0001	01da	ffff	ffff	Z, N	1,2
CLRF	f, a	Clear f	1	0110	101a	ffff	ffff	Z	2
COMF	f, d, a	Complement f	1	0001	11da	ffff	ffff	Z, N	1, 2
CPFSEQ	f, a	Compare f with WREG, Skip =	1 (2 or 3)	0110	001a	ffff	ffff	None	4
CPFSGT	f, a	Compare f with WREG, Skip >	1 (2 or 3)	0110	010a	ffff	ffff	None	4
CPFSLT	f, a	Compare f with WREG, Skip <	1 (2 or 3)	0110	000a	ffff	ffff	None	1, 2
DECF	f, d, a	Decrement f	1	0000	01da	ffff	ffff	C, DC, Z, OV, N	1, 2, 3, 4
DECFSZ	f, d, a	Decrement f, Skip if 0	1 (2 or 3)	0010	11da	ffff	ffff	None	1, 2, 3, 4
DCFSNZ	f, d, a	Decrement f, Skip if Not 0	1 (2 or 3)	0100	11da	ffff	ffff	None	1, 2
INCF	f, d, a	Increment f	1	0010	10da	ffff	ffff	C, DC, Z, OV, N	1, 2, 3, 4
INCFSZ	f, d, a	Increment f, Skip if 0	1 (2 or 3)	0011	11da	ffff	ffff	None	4
INFSNZ	f, d, a	Increment f, Skip if Not 0	1 (2 or 3)	0100	10da	ffff	ffff	None	1, 2
IORWF	f, d, a	Inclusive OR WREG with f	1	0001	00da	ffff	ffff	Z, N	1, 2
MOVF	f, d, a	Move f	1	0101	00da	ffff	ffff	Z, N	1
MOVFF	f_s, f_d	Move f_s (source) to 1st word f_d (destination) 2nd word	2	1100 1111	ffff ffff	ffff ffff	ffff ffff	None	
MOVWF	f, a	Move WREG to f	1	0110	111a	ffff	ffff	None	
MULWF	f, a	Multiply WREG with f	1	0000	001a	ffff	ffff	None	
NEGF	f, a	Negate f	1	0110	110a	ffff	ffff	C, DC, Z, OV, N	1, 2
RLCF	f, d, a	Rotate Left f through Carry	1	0011	01da	ffff	ffff	C, Z, N	
RLNCF	f, d, a	Rotate Left f (No Carry)	1	0100	01da	ffff	ffff	Z, N	1, 2
RRCF	f, d, a	Rotate Right f through Carry	1	0011	00da	ffff	ffff	C, Z, N	
RRNCF	f, d, a	Rotate Right f (No Carry)	1	0100	00da	ffff	ffff	Z, N	
SETF	f, a	Set f	1	0110	100a	ffff	ffff	None	
SUBFWB	f, d, a	Subtract f from WREG with Borrow	1	0101	01da	ffff	ffff	C, DC, Z, OV, N	1, 2
SUBWF	f, d, a	Subtract WREG from f	1	0101	11da	ffff	ffff	C, DC, Z, OV, N	
SUBWFB	f, d, a	Subtract WREG from f with Borrow	1	0101	10da	ffff	ffff	C, DC, Z, OV, N	1, 2
SWAPF	f, d, a	Swap Nibbles in f	1	0011	10da	ffff	ffff	None	4
TSTFSZ	f, a	Test f, Skip if 0	1 (2 or 3)	0110	011a	ffff	ffff	None	1, 2
XORWF	f, d, a	Exclusive OR WREG with f	1	0001	10da	ffff	ffff	Z, N	
BIT-ORIENTED FILE REGISTER OPERATIONS									
BCF	f, b, a	Bit Clear f	1	1001	bbba	ffff	ffff	None	1, 2
BSF	f, b, a	Bit Set f	1	1000	bbba	ffff	ffff	None	1, 2
BTFSC	f, b, a	Bit Test f, Skip if Clear	1 (2 or 3)	1011	bbba	ffff	ffff	None	3, 4
BTFSS	f, b, a	Bit Test f, Skip if Set	1 (2 or 3)	1010	bbba	ffff	ffff	None	3, 4
BTG	f, d, a	Bit Toggle f	1	0111	bbba	ffff	ffff	None	1, 2

Note 1: When a PORT register is modified as a function of itself (e.g., `MOVF PORTB, 1, 0`), the value used will be that value present on the pins themselves. For example, if the data latch is '`1`' for a pin configured as input and is driven low by an external device, the data will be written back with a '`0`'.

2: If this instruction is executed on the TMR0 register (and where applicable, d = `1`), the prescaler will be cleared if assigned.

3: If Program Counter (PC) is modified or a conditional test is true, the instruction requires two cycles. The second cycle is executed as a `NOP`.

4: Some instructions are 2-word instructions. The second word of these instructions will be executed as a `NOP` unless the first word of the instruction retrieves the information embedded in these 16 bits. This ensures that all program memory locations have a valid instruction.

5: If the table write starts the write cycle to internal memory, the write will continue until terminated.

PIC18F2220/2320/4220/4320

TABLE 24-2: PIC18FXXX INSTRUCTION SET (CONTINUED)

Mnemonic, Operands		Description	Cycles	16-Bit Instruction Word MSb			LSb	Status Affected	Notes
CONTROL OPERATIONS									
BC	n	Branch if Carry	1 (2)	1110	0010	nnnn	nnnn	None	
BN	n	Branch if Negative	1 (2)	1110	0110	nnnn	nnnn	None	
BNC	n	Branch if Not Carry	1 (2)	1110	0011	nnnn	nnnn	None	
BNN	n	Branch if Not Negative	1 (2)	1110	0111	nnnn	nnnn	None	
BNOV	n	Branch if Not Overflow	1 (2)	1110	0101	nnnn	nnnn	None	
BNZ	n	Branch if Not Zero	1 (2)	1110	0001	nnnn	nnnn	None	
BOV	n	Branch if Overflow	1 (2)	1110	0100	nnnn	nnnn	None	
BRA	n	Branch Unconditionally	2	1101	0nnn	nnnn	nnnn	None	
BZ	n	Branch if Zero	1 (2)	1110	0000	nnnn	nnnn	None	
CALL	n, s	Call Subroutine 1st word	2	1110	110s	kkkk	kkkk	None	
		2nd word		1111	kkkk	kkkk	kkkk		
CLRWDT	—	Clear Watchdog Timer	1	0000	0000	0000	0100	$\overline{TO}$, $\overline{PD}$	
DAW	—	Decimal Adjust WREG	1	0000	0000	0000	0111	C, DC	
GOTO	n	Go to Address 1st word	2	1110	1111	kkkk	kkkk	None	
		2nd word		1111	kkkk	kkkk	kkkk		
NOP	—	No Operation	1	0000	0000	0000	0000	None	
NOP	—	No Operation **(Note 4)**	1	1111	xxxx	xxxx	xxxx	None	
POP	—	Pop Top of Return Stack (TOS)	1	0000	0000	0000	0110	None	
PUSH	—	Push Top of Return Stack (TOS)	1	0000	0000	0000	0101	None	
RCALL	n	Relative Call	2	1101	1nnn	nnnn	nnnn	None	
RESET		Software Device Reset	1	0000	0000	1111	1111	All	
RETFIE	s	Return from Interrupt Enable	2	0000	0000	0001	000s	GIE/GIEH, PEIE/GIEL	
RETLW	k	Return with Literal in WREG	2	0000	1100	kkkk	kkkk	None	
RETURN	s	Return from Subroutine	2	0000	0000	0001	001s	None	
SLEEP	—	Go into Standby mode	1	0000	0000	0000	0011	$\overline{TO}$, $\overline{PD}$	

Note 1: When a PORT register is modified as a function of itself (e.g., `MOVF PORTB, 1, 0`), the value used will be that value present on the pins themselves. For example, if the data latch is '1' for a pin configured as input and is driven low by an external device, the data will be written back with a '0'.

2: If this instruction is executed on the TMR0 register (and where applicable, d = 1), the prescaler will be cleared if assigned.

3: If Program Counter (PC) is modified or a conditional test is true, the instruction requires two cycles. The second cycle is executed as a `NOP`.

4: Some instructions are 2-word instructions. The second word of these instructions will be executed as a `NOP` unless the first word of the instruction retrieves the information embedded in these 16 bits. This ensures that all program memory locations have a valid instruction.

5: If the table write starts the write cycle to internal memory, the write will continue until terminated.

PIC18F2220/2320/4220/4320

TABLE 24-2: PIC18FXXX INSTRUCTION SET (CONTINUED)

Mnemonic, Operands		Description	Cycles	16-Bit Instruction Word				Status Affected	Notes
				MSb			LSb		
LITERAL OPERATIONS									
ADDLW	k	Add Literal and WREG	1	0000	1111	kkkk	kkkk	C, DC, Z, OV, N	
ANDLW	k	AND Literal with WREG	1	0000	1011	kkkk	kkkk	Z, N	
IORLW	k	Inclusive OR Literal with WREG	1	0000	1001	kkkk	kkkk	Z, N	
LFSR	f, k	Move Literal (12-bit) 2nd word to FSRx 1st word	2	1110 1111	1110 0000	00ff kkkk	kkkk kkkk	None	
MOVLB	k	Move Literal to BSR<3:0>	1	0000	0001	0000	kkkk	None	
MOVLW	k	Move Literal to WREG	1	0000	1110	kkkk	kkkk	None	
MULLW	k	Multiply Literal with WREG	1	0000	1101	kkkk	kkkk	None	
RETLW	k	Return with Literal in WREG	2	0000	1100	kkkk	kkkk	None	
SUBLW	k	Subtract WREG from Literal	1	0000	1000	kkkk	kkkk	C, DC, Z, OV, N	
XORLW	k	Exclusive OR Literal with WREG	1	0000	1010	kkkk	kkkk	Z, N	
DATA MEMORY ↔ PROGRAM MEMORY OPERATIONS									
TBLRD*		Table Read	2	0000	0000	0000	1000	None	
TBLRD*+		Table Read with Post-Increment		0000	0000	0000	1001	None	
TBLRD*-		Table Read with Post-Decrement		0000	0000	0000	1010	None	
TBLRD+*		Table Read with Pre-Increment		0000	0000	0000	1011	None	
TBLWT*		Table Write	2 (5)	0000	0000	0000	1100	None	
TBLWT*+		Table Write with Post-Increment		0000	0000	0000	1101	None	
TBLWT*-		Table Write with Post-Decrement		0000	0000	0000	1110	None	
TBLWT+*		Table Write with Pre-Increment		0000	0000	0000	1111	None	

Note 1: When a PORT register is modified as a function of itself (e.g., `MOVF PORTB, 1, 0`), the value used will be that value present on the pins themselves. For example, if the data latch is '1' for a pin configured as input and is driven low by an external device, the data will be written back with a '0'.

2: If this instruction is executed on the TMR0 register (and where applicable, d = 1), the prescaler will be cleared if assigned.

3: If Program Counter (PC) is modified or a conditional test is true, the instruction requires two cycles. The second cycle is executed as a NOP.

4: Some instructions are 2-word instructions. The second word of these instructions will be executed as a NOP unless the first word of the instruction retrieves the information embedded in these 16 bits. This ensures that all program memory locations have a valid instruction.

5: If the table write starts the write cycle to internal memory, the write will continue until terminated.

ASCII characters

ASCII Code Number	Symbol	Descriptor
0	NULL	(Null character)
1	SOH	(Start of Header)
2	STX	(Start of Text)
3	ETX	(End of Text)
4	EOT	(End of Transmission)
5	ENQ	(Enquiry)
6	ACK	(Acknowledgement)
7	BEL	(Bell)
8	BS	(Backspace)
9	HT	(Horizontal Tab)
10	LF	(Line feed)
11	VT	(Vertical Tab)
12	FF	(Form feed)
13	CR	(Carriage return)
14	SO	(Shift Out)
15	SI	(Shift In)
16	DLE	(Data link escape)
17	DC1	(Device control 1)
18	DC2	(Device control 2)
19	DC3	(Device control 3)
20	DC4	(Device control 4)
21	NAK	(Negative acknowledgement)
22	SYN	(Synchronous idle)
23	ETB	(End of transmission block)
24	CAN	(Cancel)
25	EM	(End of medium)
26	SUB	(Substitute)
27	ESC	(Escape)
28	FS	(File separator)
29	GS	(Group separator)
30	RS	(Record separator)
31	US	(Unit separator)
32		(Space)
33	!	(Exclamation mark)
34	"	(Quotation mark; quotes)
35	#	(Number sign)

Continued

ASCII Code Number	Symbol	Descriptor
36	**$**	(Dollar sign)
37	**%**	(Percent sign)
38	**&**	(Ampersand)
39	**'**	(Apostrophe)
40	**(**	(Round brackets or parentheses)
41	**)**	(Round brackets or parentheses)
42	*	(Asterisk)
43	+	(Plus sign)
44	**,**	(Comma)
45	**-**	(Hyphen)
46	**.**	(Dot, full stop)
47	**/**	(Slash)
48	**0**	(Number zero)
49	**1**	(Number one)
50	**2**	(Number two)
51	**3**	(Number three)
52	**4**	(Number four)
53	**5**	(Number five)
54	**6**	(Number six)
55	**7**	(Number seven)
56	**8**	(Number eight)
57	**9**	(Number nine)
58	**:**	(Colon)
59	**;**	(Semicolon)
60	<	(Less-than sign)
61	=	(Equals sign)
62	>	(Greater-than sign; Inequality)
63	**?**	(Question mark)
64	**@**	(At sign)
65	**A**	(Capital A)
66	**B**	(Capital B)
67	**C**	(Capital C)
68	**D**	(Capital D)
69	**E**	(Capital E)
70	**F**	(Capital F)
71	**G**	(Capital G)
72	**H**	(Capital H)
73	**I**	(Capital I)
74	**J**	(Capital J)
75	**K**	(Capital K)
76	**L**	(Capital L)
77	**M**	(Capital M)

ASCII Code Number	Symbol	Descriptor
78	**N**	(Capital N)
79	**O**	(Capital O)
80	**P**	(Capital P)
81	**Q**	(Capital Q)
82	**R**	(Capital R)
83	**S**	(Capital S)
84	**T**	(Capital T)
85	**U**	(Capital U)
86	**V**	(Capital V)
87	**W**	(Capital W)
88	**X**	(Capital X)
89	**Y**	(Capital Y)
90	**Z**	(Capital Z)
91	**[**	(Square brackets or box brackets)
92	****	(Backslash)
93	**]**	(Square brackets or box brackets)
94	**^**	(Caret or circumflex accent)
95	**_**	(Underscore, understrike, underbar or low line)
96	**`**	(Grave accent)
97	**a**	(Lowercase a)
98	**b**	(Lowercase b)
99	**c**	(Lowercase c)
100	**d**	(Lowercase d)
101	**e**	(Lowercase e)
102	**f**	(Lowercase f)
103	**g**	(Lowercase g)
104	**h**	(Lowercase h)
105	**i**	(Lowercase i)
106	**j**	(Lowercase j)
107	**k**	(Lowercase k)
108	**l**	(Lowercase l)
109	**m**	(Lowercase m)
110	**n**	(Lowercase n)
111	**o**	(Lowercase o)
112	**p**	(Lowercase p)
113	**q**	(Lowercase q)
114	**r**	(Lowercase r)
115	**s**	(Lowercase s)
116	**t**	(Lowercase t)
117	**u**	(Lowercase u)
118	**v**	(Lowercase v)

Continued

ASCII Code Number	Symbol	Descriptor
119	**w**	(Lowercase w)
120	**x**	(Lowercase x)
121	**y**	(Lowercase y)
122	**z**	(Lowercase z)
123	**{**	(Curly brackets or braces)
124	**|**	(Vertical-bar, vbar, vertical line or vertical slash)
125	**}**	(Curly brackets or braces)
126	**~**	(Tilde; swung dash)
127	**DEL**	(Delete)
128	**Ç**	(Majuscule C-cedilla)
129	**ü**	(Letter "u" with umlaut or diaeresis; "u-umlaut")
130	**é**	(Letter "e" with acute accent or "e-acute")
131	**â**	(Letter "a" with circumflex accent or "a-circumflex")
132	**ä**	(Letter "a" with umlaut or diaeresis; "a-umlaut")
133	**à**	(Letter "a" with grave accent)
134	**å**	(Letter "a" with a ring)
135	**ç**	(Minuscule c-cedilla)
136	**ê**	(Letter "e" with circumflex accent or "e-circumflex")
137	**ë**	(Letter "e" with umlaut or diaeresis; "e-umlaut")
138	**è**	(Letter "e" with grave accent)
139	**ï**	(Letter "i" with umlaut or diaeresis; "i-umlaut")
140	**î**	(Letter "i" with circumflex accent or "i-circumflex")
141	**ì**	(Letter "i" with grave accent)
142	**Ä**	(Letter "A" with umlaut or diaeresis; "A-umlaut")
143	**Å**	(Capital letter "A" with a ring)
144	**É**	(Capital letter "E" with acute accent or "E-acute")
145	**æ**	(Latin diphthong "ae" in lowercase)
146	**Æ**	(Latin diphthong "AE" in uppercase)
147	**ô**	(Letter "o" with circumflex accent or "o-circumflex")
148	**ö**	(Letter "o" with umlaut or diaeresis; "o-umlaut")
149	**ò**	(Letter "o" with grave accent)

ASCII Code Number	Symbol	Descriptor
150	û	(Letter "u" with circumflex accent or "u-circumflex")
151	ù	(Letter "u" with grave accent)
152	ÿ	(Lowercase letter "y" with diaeresis)
153	Ö	(Letter "O" with umlaut or diaeresis; "O-umlaut")
154	Ü	(Letter "U" with umlaut or diaeresis; "U-umlaut")
155	ø	(Slashed zero or empty set)
156	£	(Pound sign; symbol for the pound sterling)
157	Ø	(Slashed zero or empty set)
158	×	(Multiplication sign)
159	ƒ	(Function sign; f with hook sign; florin sign)
160	á	(Letter "a" with acute accent or "a-acute")
161	í	(Letter "i" with acute accent or "i-acute")
162	ó	(Letter "o" with acute accent or "o-acute")
163	ú	(Letter "u" with acute accent or "u-acute")
164	ñ	(Letter "n" with tilde; enye)
165	Ñ	(Letter "N" with tilde; enye)
166	ª	(Feminine ordinal indicator)
167	º	(Masculine ordinal indicator)
168	¿	(Inverted question marks)
169	®	(Registered trademark symbol)
170	¬	(Logical negation symbol)
171	½	(One half)
172	¼	(Quarter or one fourth)
173	¡	(Inverted exclamation marks)
174	«	(Angle quotes or guillemets)
175	»	(Guillemets or angle quotes)
176	░	
177	▒	
178	▓	
179	│	(Box drawing character)
180	┤	(Box drawing character)
181	Á	(Capital letter "A" with acute accent or "A-acute")
182	Â	(Letter "A" with circumflex accent or "A-circumflex")
183	À	(Letter "A" with grave accent)

Continued

ASCII Code Number	Symbol	Descriptor
184	©	(Copyright symbol)
185	╣	(Box drawing character)
186	║	(Box drawing character)
187	╗	(Box drawing character)
188	╝	(Box drawing character)
189	¢	(Cent symbol)
190	¥	(YEN and YUAN sign)
191	┐	(Box drawing character)
192	└	(Box drawing character)
193	┴	(Box drawing character)
194	┬	(Box drawing character)
195	├	(Box drawing character)
196	─	(Box drawing character)
197	┼	(Box drawing character)
198	ã	(Lowercase letter "a" with tilde or "a-tilde")
199	Ã	(Capital letter "A" with tilde or "A-tilde")
200	╚	(Box drawing character)
201	╔	(Box drawing character)
202	╩	(Box drawing character)
203	╦	(Box drawing character)
204	╠	(Box drawing character)
205	═	(Box drawing character)
206	╬	(Box drawing character)
207	¤	(generic currency sign)
208	ð	(Lowercase letter "eth")
209	Ð	(Capital letter "Eth")
210	Ê	(Letter "E" with circumflex accent or "E-circumflex")
211	Ë	(Letter "E" with umlaut or diaeresis; "E-umlaut")
212	È	(Letter "E" with grave accent)
213	ı	(Lowercase dot less i)
214	Í	(Capital letter "I" with acute accent or "I-acute")
215	Î	(Letter "I" with circumflex accent or "I-circumflex")
216	Ï	(Letter "I" with umlaut or diaeresis, "I-umlaut")
217	┘	(Box drawing character)
218	┌	(Box drawing character)
219	█	(Block)
220	▄	(Bottom half block)
221	¦	(Vertical broken bar)

ASCII Code Number	Symbol	Descriptor
222	Ì	(Letter "I" with grave accent)
223	▀	(Top half block)
224	Ó	(Capital letter "O" with acute accent or "O-acute")
225	ß	(Letter "Eszett"; "scharfes S" or "sharp S")
226	Ô	(Letter "O" with circumflex accent or "O-circumflex")
227	Ò	(Letter "O" with grave accent)
228	õ	(Letter "o" with tilde or "o-tilde")
229	Õ	(Letter "O" with tilde or "O-tilde")
230	µ	(Lowercase letter "Mu"; micro sign or micron)
231	þ	(Lowercase letter "Thorn")
232	Þ	(Capital letter "thorn")
233	Ú	(Capital letter "U" with acute accent or "U-acute")
234	Û	(Letter "U" with circumflex accent or "U-circumflex")
235	Ù	(Letter "U" with grave accent)
236	ý	(Lowercase letter "y" with acute accent)
237	Ý	(Capital letter "Y" with acute accent)
238	¯	(Macron symbol)
239	´	(Acute accent)
240	-	(Hyphen)
241	±	(Plus-minus sign)
242	‗	(Underline or underscore)
243	¾	(Three quarters)
244	¶	(Paragraph sign or pilcrow)
245	§	(Section sign)
246	÷	(The division sign; Obelus)
247	¸	(Cedilla)
248	°	(Degree symbol)
249	¨	(Diaeresis)
250	·	(Interpunct or space dot)
251	¹	(Superscript one)
252	³	(Cube or superscript three)
253	²	(Square or superscript two)
254	■	(Black square)
255	nbsp	(Nonbreaking space or no-break space)

Decimal-binary-hexadecimal characters

Decimal	Binary	Hexadecimal
000	00000000	00
001	00000001	01
002	00000010	02
003	00000011	03
004	00000100	04
005	00000101	05
006	00000110	06
007	00000111	07
008	00001000	08
009	00001001	09
010	00001010	0A
011	00001011	0B
012	00001100	0C
013	00001101	0D
014	00001110	0E
015	00001111	0F
016	00010000	10
017	00010001	11
018	00010010	12
019	00010011	13
020	00010100	14
021	00010101	15
022	00010110	16
023	00010111	17
024	00011000	18
025	00011001	19
026	00011010	1A
027	00011011	1B
028	00011100	1C
029	00011101	1D
030	00011110	1E
031	00011111	1F
032	00100000	20

Continued

Decimal	Binary	Hexadecimal
033	00100001	21
034	00100010	22
035	00100011	23
036	00100100	24
037	00100101	25
038	00100110	26
039	00100111	27
040	00101000	28
041	00101001	29
042	00101010	2A
043	00101011	2B
044	00101100	2C
045	00101101	2D
046	00101110	2E
047	00101111	2F
048	00110000	30
049	00110001	31
050	00110010	32
051	00110011	33
052	00110100	34
053	00110101	35
054	00110110	36
055	00110111	37
056	00111000	38
057	00111001	39
058	00111010	3A
059	00111011	3B
060	00111100	3C
061	00111101	3D
062	00111110	3E
063	00111111	3F
064	01000000	40
065	01000001	41
066	01000010	42
067	01000011	43
068	01000100	44
069	01000101	45
070	01000110	46
071	01000111	47
072	01001000	48
073	01001001	49
074	01001010	4A

Decimal	Binary	Hexadecimal
075	01001011	4B
076	01001100	4C
077	01001101	4D
078	01001110	4E
079	01001111	4F
080	01010000	50
081	01010001	51
082	01010010	52
083	01010011	53
084	01010100	54
085	01010101	55
086	01010110	56
087	01010111	57
088	01011000	58
089	01011001	59
090	01011010	5A
091	01011011	5B
092	01011100	5C
093	01011101	5D
094	01011110	5E
095	01011111	5F
096	01100000	60
097	01100001	61
098	01100010	62
099	01100011	63
100	01100100	64
101	01100101	65
102	01100110	66
103	01100111	67
104	01101000	68
105	01101001	69
106	01101010	6A
107	01101011	6B
108	01101100	6C
109	01101101	6D
110	01101110	6E
111	01101111	6F
112	01110000	70
113	01110001	71
114	01110010	72
115	01110011	73
116	01110100	74

Continued

Decimal	Binary	Hexadecimal
117	01110101	75
118	01110110	76
119	01110111	77
120	01111000	78
121	01111001	79
122	01111010	7A
123	01111011	7B
124	01111100	7C
125	01111101	7D
126	01111110	7E
127	01111111	7F
128	10000000	80
129	10000001	81
130	10000010	82
131	10000011	83
132	10000100	84
133	10000101	85
134	10000110	86
135	10000111	87
136	10001000	88
137	10001001	89
138	10001010	8A
139	10001011	8B
140	10001100	8C
141	10001101	8D
142	10001110	8E
143	10001111	8F
144	10010000	90
145	10010001	91
146	10010010	92
147	10010011	93
148	10010100	94
149	10010101	95
150	10010110	96
151	10010111	97
152	10011000	98
153	10011001	99
154	10011010	9A
155	10011011	9B
156	10011100	9C
157	10011101	9D
158	10011110	9E

Decimal	Binary	Hexadecimal
159	10011111	9F
160	10100000	A0
161	10100001	A1
162	10100010	A2
163	10100011	A3
164	10100100	A4
165	10100101	A5
166	10100110	A6
167	10100111	A7
168	10101000	A8
169	10101001	A9
170	10101010	AA
171	10101011	AB
172	10101100	AC
173	10101101	AD
174	10101110	AE
175	10101111	AF
176	10110000	B0
177	10110001	B1
178	10110010	B2
179	10110011	B3
180	10110100	B4
181	10110101	B5
182	10110110	B6
183	10110111	B7
184	10111000	B8
185	10111001	B9
186	10111010	BA
187	10111011	BB
188	10111100	BC
189	10111101	BD
190	10111110	BE
191	10111111	BF
192	11000000	C0
193	11000001	C1
194	11000010	C2
195	11000011	C3
196	11000100	C4
197	11000101	C5
198	11000110	C6
199	11000111	C7
200	11001000	C8

Continued

Decimal	Binary	Hexadecimal
201	11001001	C9
202	11001010	CA
203	11001011	CB
204	11001100	CC
205	11001101	CD
206	11001110	CE
207	11001111	CF
208	11010000	D0
209	11010001	D1
210	11010010	D2
211	11010011	D3
212	11010100	D4
213	11010101	D5
214	11010110	D6
215	11010111	D7
216	11011000	D8
217	11011001	D9
218	11011010	DA
219	11011011	DB
220	11011100	DC
221	11011101	DD
222	11011110	DE
223	11011111	DF
224	11100000	E0
225	11100001	E1
226	11100010	E2
227	11100011	E3
228	11100100	E4
229	11100101	E5
230	11100110	E6
231	11100111	E7
232	11101000	E8
233	11101001	E9
234	11101010	EA
235	11101011	EB
236	11101100	EC
237	11101101	ED
238	11101110	EE
239	11101111	EF
240	11110000	F0
241	11110001	F1
242	11110010	F2

Decimal	Binary	Hexadecimal
243	11110011	F3
244	11110100	F4
245	11110101	F5
246	11110110	F6
247	11110111	F7
248	11111000	F8
249	11111001	F9
250	11111010	FA
251	11111011	FB
252	11111100	FC
253	11111101	FD
254	11111110	FE
255	11111111	FF
256	100000000	100

Best practices

C-compilers pretend to be free of errors and, yet, there is a continuous need to update the compiler. Ideally, when programming in C, the program should compile in such a way as to take advantage of the nuances of the underlying assembly language. However, as told by Keith Curtis of Microchip, a long-time consultant, "typically the errors [in C compilation] are not so much errors as they are inefficiencies." He gives the following example. "Setting a bit in a register is typically done by reading the register, ORing on the bit, and then storing the value back into the register. However, PIC16Fxxxx parts all have a bit set instruction that allows you to set individual bits." In other words, the compiler may not pick up on this more efficient new instruction characteristic of the PIC16Fxxxx series. The consequence of this may be to have three instructions to execute the work that could be achieved with one instruction. Obviously, this is a clear inefficiency. However, whether or not you have need of this level of efficiency is known only to the programmer and their specific project. This could potentially make a section of code much longer if this efficiency is not present and you compile this within a loop that magnifies the inefficiency. Ten times through a loop and your code winds up being 30 instructions, instead of 10.

Keith has generously provided a set of 10 recommendations for addressing common inefficiencies. These other inefficiencies should be reviewed to see if the compiled code handled the compilation process correctly. In addition, some recommendations are addressed *to the programmer* as more efficient means of writing code in general. These examples are of use to the assembly language programmer as well.

Other examples [from Keith, include:]

1. Using a series of IF/THEN [instructions] instead of making a code-efficient jump table.
2. Failing to take advantage of linear RAM shadowing. [The] same RAM decodes in two locations, one intermixed with peripheral registers, and the other in a linear section of memory allowing much simpler indirect address calculations.
3. There is also a problem with the [general purpose input/output pins] GPIO, if you read the port register you get the state of the pins whether they are inputs or outputs. If you do a read, modify, write operation you will change the state of the output latches for each of the inputs. If you then make a bit an output, its level may have changed and you won't know it.

4. Bank select issues [with pages] are typical problem areas. When you go to access a variable, you have to select it's bank. Hopefully, all the needed variables are in the same bank. If they are, then you won't need to do repeated bank selects. Sometimes compilers miss this and do too many or too few.
5. Locating arrays next to important peripheral control registers is also a potential problem. If the routine access[es] the array [and] goes one step too far, you can trash an important peripheral control register creating an intermittent glitch in your system.
6. Another typical inefficiency is failing to test for interrupts before leaving the interrupt service routine. It does not cause a problem because the processor will just return to the interrupt when the current interrupt clears, but you do waste time retrieving the context and then saving it off again.
7. PRINTF and other bloat ware functions are typically included in a compiler although they are of little use in embedded design due to their size. MCC is a much better option in that it builds a custom right sized function for your needs and can be linked into the compiler's [input/output] IO functions.
8. Building a replacement function for PRINTF is not a bad idea either. It doesn't have to accept the format string, just do the appropriate conversions [such as] print_char, print_int, print_str, [and] print_crlf.
9. [Best practices] I used to do as a consultant, was to run a test program that checked for most of the glitches I had found on every new revision of the compiler. That way I knew which functions could cause trouble.
10. [Best practices] I also NEVER change compiler revisions **during** a project. This is the fastest way to go from 1–2 bugs to over 100.

PIC10F200/202/204/206

6-Pin, 8-Bit Flash Microcontrollers

Devices Included In This Data Sheet:

- PIC10F200
- PIC10F202
- PIC10F204
- PIC10F206

High-Performance RISC CPU:

- Only 33 Single-Word Instructions to Learn
- All Single-Cycle Instructions except for Program Branches, which are Two-Cycle
- 12-Bit Wide Instructions
- 2-Level Deep Hardware Stack
- Direct, Indirect and Relative Addressing modes for Data and Instructions
- 8-Bit Wide Data Path
- Eight Special Function Hardware Registers
- Operating Speed:
  - 4 MHz internal clock
  - 1 μs instruction cycle

Special Microcontroller Features:

- 4 MHz Precision Internal Oscillator:
  - Factory calibrated to ±1%
- In-Circuit Serial Programming™ (ICSP™)
- In-Circuit Debugging (ICD) Support
- Power-on Reset (POR)
- Device Reset Timer (DRT)
- Watchdog Timer (WDT) with Dedicated On-Chip RC Oscillator for Reliable Operation
- Programmable Code Protection
- Multiplexed $\overline{MCLR}$ Input Pin
- Internal Weak Pull-ups on I/O Pins
- Power-Saving Sleep mode
- Wake-up from Sleep on Pin Change

Low-Power Features/CMOS Technology:

- Operating Current:
  - < 175 μA @ 2V, 4 MHz, typical
- Standby Current:
  - 100 nA @ 2V, typical
- Low-Power, High-Speed Flash Technology:
  - 100,000 Flash endurance
  - > 40 year retention
- Fully Static Design
- Wide Operating Voltage Range: 2.0V to 5.5V
- Wide Temperature Range:
  - Industrial: -40°C to +85°C
  - Extended: -40°C to +125°C

Peripheral Features (PIC10F200/202):

- Four I/O Pins:
  - Three I/O pins with individual direction control
  - One input-only pin
  - High current sink/source for direct LED drive
  - Wake-on-change
  - Weak pull-ups
- 8-Bit Real-Time Clock/Counter (TMR0) with 8-Bit Programmable Prescaler

Peripheral Features (PIC10F204/206):

- Four I/O Pins:
  - Three I/O pins with individual direction control
  - One input-only pin
  - High current sink/source for direct LED drive
  - Wake-on-change
  - Weak pull-ups
- 8-Bit Real-Time Clock/Counter (TMR0) with 8-Bit Programmable Prescaler
- One Comparator:
  - Internal absolute voltage reference
  - Both comparator inputs visible externally
  - Comparator output visible externally

TABLE 1: PIC10F20X MEMORY AND FEATURES

Device	Program Memory	Data Memory	I/O	Timers 8-bit	Comparator
	Flash (words)	SRAM (bytes)			
PIC10F200	256	16	4	1	0
PIC10F202	512	24	4	1	0
PIC10F204	256	16	4	1	1
PIC10F206	512	24	4	1	1

PIC10F200/202/204/206

Pin Diagrams

FIGURE 1: 6-PIN SOT-23

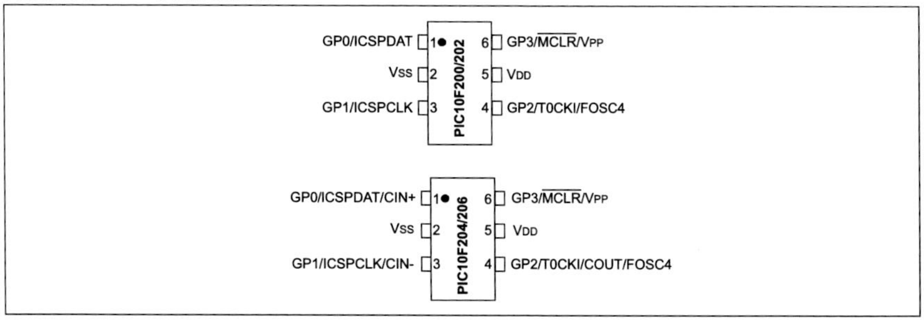

FIGURE 2: 8-PIN PDIP

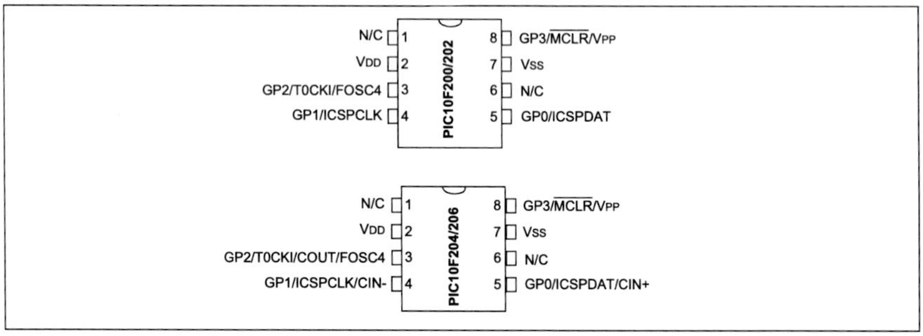

FIGURE 3: 8-PIN DFN

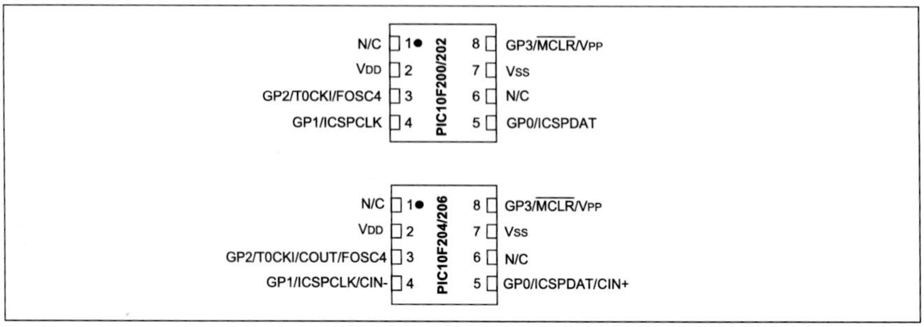

PIC10F200/202/204/206

Table of Contents

TO OUR VALUED CUSTOMERS

It is our intention to provide our valued customers with the best documentation possible to ensure successful use of your Microchip products. To this end, we will continue to improve our publications to better suit your needs. Our publications will be refined and enhanced as new volumes and updates are introduced.

If you have any questions or comments regarding this publication, please contact the Marketing Communications Department via E-mail at **docerrors@microchip.com**. We welcome your feedback.

Most Current Data Sheet

To obtain the most up-to-date version of this data sheet, please register at our Worldwide Web site at:

http://www.microchip.com

You can determine the version of a data sheet by examining its literature number found on the bottom outside corner of any page. The last character of the literature number is the version number, (e.g., DS30000000A is version A of document DS30000000).

Errata

An errata sheet, describing minor operational differences from the data sheet and recommended workarounds, may exist for current devices. As device/documentation issues become known to us, we will publish an errata sheet. The errata will specify the revision of silicon and revision of document to which it applies.

To determine if an errata sheet exists for a particular device, please check with one of the following:

- Microchip's Worldwide Web site; **http://www.microchip.com**
- Your local Microchip sales office (see last page)

When contacting a sales office, please specify which device, revision of silicon and data sheet (include literature number) you are using.

Customer Notification System

Register on our web site at **www.microchip.com** to receive the most current information on all of our products.

PIC10F200/202/204/206

1.0 GENERAL DESCRIPTION

The PIC10F200/202/204/206 devices from Microchip Technology are low-cost, high-performance, 8-bit, fully-static, Flash-based CMOS microcontrollers. They employ a RISC architecture with only 33 single-word/single-cycle instructions. All instructions are single cycle (1 μs) except for program branches, which take two cycles. The PIC10F200/202/204/206 devices deliver performance in an order of magnitude higher than their competitors in the same price category. The 12-bit wide instructions are highly symmetrical, resulting in a typical 2:1 code compression over other 8-bit microcontrollers in its class. The easy-to-use and easy to remember instruction set reduces development time significantly.

The PIC10F200/202/204/206 products are equipped with special features that reduce system cost and power requirements. The Power-on Reset (POR) and Device Reset Timer (DRT) eliminate the need for external Reset circuitry. INTRC Internal Oscillator mode is provided, thereby preserving the limited number of I/O available. Power-Saving Sleep mode, Watchdog Timer and code protection features improve system cost, power and reliability.

The PIC10F200/202/204/206 devices are available in cost-effective Flash, which is suitable for production in any volume. The customer can take full advantage of Microchip's price leadership in Flash programmable microcontrollers, while benefiting from the Flash programmable flexibility.

The PIC10F200/202/204/206 products are supported by a full-featured macro assembler, a software simulator, an in-circuit debugger, a 'C' compiler, a low-cost development programmer and a full featured programmer. All the tools are supported on IBM® PC and compatible machines.

1.1 Applications

The PIC10F200/202/204/206 devices fit in applications ranging from personal care appliances and security systems to low-power remote transmitters/receivers. The Flash technology makes customizing application programs (transmitter codes, appliance settings, receiver frequencies, etc.) extremely fast and convenient. The small footprint packages, for through hole or surface mounting, make these microcontrollers well suited for applications with space limitations. Low cost, low power, high performance, ease-of-use and I/O flexibility make the PIC10F200/202/204/206 devices very versatile even in areas where no microcontroller use has been considered before (e.g., timer functions, logic and PLDs in larger systems and coprocessor applications).

TABLE 1-1: PIC10F200/202/204/206 DEVICES

		PIC10F200	PIC10F202	PIC10F204	PIC10F206
Clock	Maximum Frequency of Operation (MHz)	4	4	4	4
Memory	Flash Program Memory	256	512	256	512
	Data Memory (bytes)	16	24	16	24
Peripherals	Timer Module(s)	TMR0	TMR0	TMR0	TMR0
	Wake-up from Sleep on Pin Change	Yes	Yes	Yes	Yes
	Comparators	0	0	1	1
Features	I/O Pins	3	3	3	3
	Input-Only Pins	1	1	1	1
	Internal Pull-ups	Yes	Yes	Yes	Yes
	In-Circuit Serial Programming™	Yes	Yes	Yes	Yes
	Number of Instructions	33	33	33	33
	Packages	6-pin SOT-23 8-pin PDIP, DFN	6-pin SOT-23 8-pin PDIP, DFN	6-pin SOT-23 8-pin PDIP, DFN	6-pin SOT-23 8-pin PDIP, DFN

The PIC10F200/202/204/206 devices have Power-on Reset, selectable Watchdog Timer, selectable code-protect, high I/O current capability and precision internal oscillator.
The PIC10F200/202/204/206 devices use serial programming with data pin GP0 and clock pin GP1.

2.0 PIC10F200/202/204/206 DEVICE VARIETIES

A variety of packaging options are available. Depending on application and production requirements, the proper device option can be selected using the information in this section. When placing orders, please use the PIC10F200/202/204/206 Product Identification System at the back of this data sheet to specify the correct part number.

2.1 Quick Turn Programming (QTP) Devices

Microchip offers a QTP programming service for factory production orders. This service is made available for users who choose not to program medium-to-high quantity units and whose code patterns have stabilized. The devices are identical to the Flash devices but with all Flash locations and fuse options already programmed by the factory. Certain code and prototype verification procedures do apply before production shipments are available. Please contact your local Microchip Technology sales office for more details.

2.2 Serialized Quick Turn Programming^{SM} (SQTP^{SM}) Devices

Microchip offers a unique programming service, where a few user-defined locations in each device are programmed with different serial numbers. The serial numbers may be random, pseudo-random or sequential.

Serial programming allows each device to have a unique number, which can serve as an entry code, password or ID number.

PIC10F200/202/204/206

3.0 ARCHITECTURAL OVERVIEW

The high performance of the PIC10F200/202/204/206 devices can be attributed to a number of architectural features commonly found in RISC microprocessors. To begin with, the PIC10F200/202/204/206 devices use a Harvard architecture in which program and data are accessed on separate buses. This improves bandwidth over traditional von Neumann architectures where program and data are fetched on the same bus. Separating program and data memory further allows instructions to be sized differently than the 8-bit wide data word. Instruction opcodes are 12 bits wide, making it possible to have all single-word instructions. A 12-bit wide program memory access bus fetches a 12-bit instruction in a single cycle. A two-stage pipeline overlaps fetch and execution of instructions. Consequently, all instructions (33) execute in a single cycle (1 μs @ 4 MHz) except for program branches.

The table below lists program memory (Flash) and data memory (RAM) for the PIC10F200/202/204/206 devices.

TABLE 3-1: PIC10F2XX MEMORY

Device	Memory	
	Program	Data
PIC10F200	256 x 12	16 x 8
PIC10F202	512 x 12	24 x 8
PIC10F204	256 x 12	16 x 8
PIC10F206	512 x 12	24 x 8

The PIC10F200/202/204/206 devices can directly or indirectly address its register files and data memory. All Special Function Registers (SFR), including the PC, are mapped in the data memory. The PIC10F200/202/204/206 devices have a highly orthogonal (symmetrical) instruction set that makes it possible to carry out any operation, on any register, using any addressing mode. This symmetrical nature and lack of "special optimal situations" make programming with the PIC10F200/202/204/206 devices simple, yet efficient. In addition, the learning curve is reduced significantly.

The PIC10F200/202/204/206 devices contain an 8-bit ALU and working register. The ALU is a general purpose arithmetic unit. It performs arithmetic and Boolean functions between data in the working register and any register file.

The ALU is 8 bits wide and capable of addition, subtraction, shift and logical operations. Unless otherwise mentioned, arithmetic operations are two's complement in nature. In two-operand instructions, one operand is typically the W (working) register. The other operand is either a file register or an immediate constant. In single operand instructions, the operand is either the W register or a file register.

The W register is an 8-bit working register used for ALU operations. It is not an addressable register.

Depending on the instruction executed, the ALU may affect the values of the Carry (C), Digit Carry (DC) and Zero (Z) bits in the STATUS register. The C and DC bits operate as a $\overline{\text{borrow}}$ and $\overline{\text{digit borrow}}$ out bit, respectively, in subtraction. See the `SUBWF` and `ADDWF` instructions for examples.

A simplified block diagram is shown in Figure 3-1 and Figure 3-2, with the corresponding device pins described in Table 3-2.

PIC10F200/202/204/206

FIGURE 3-1: PIC10F200/202 BLOCK DIAGRAM

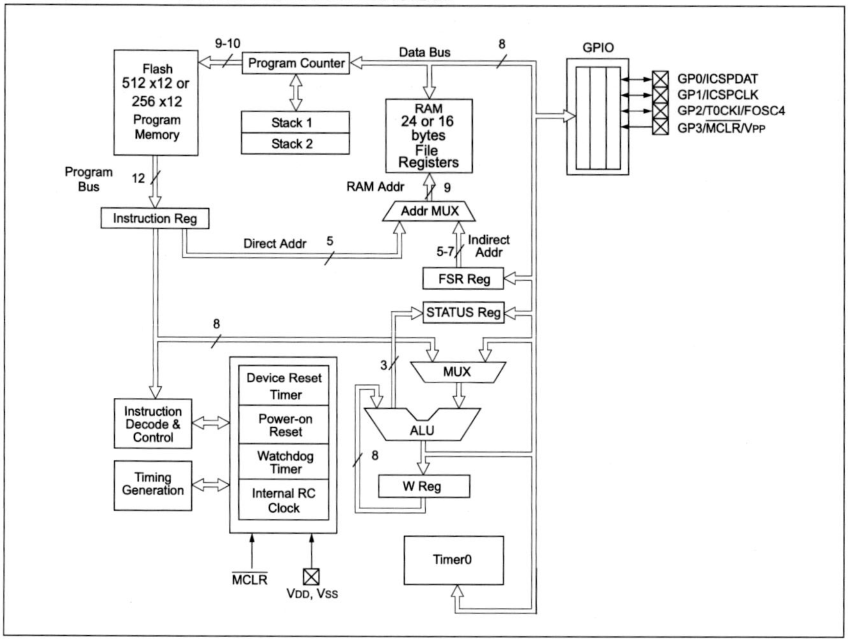

PIC10F200/202/204/206

FIGURE 3-2: PIC10F204/206 BLOCK DIAGRAM

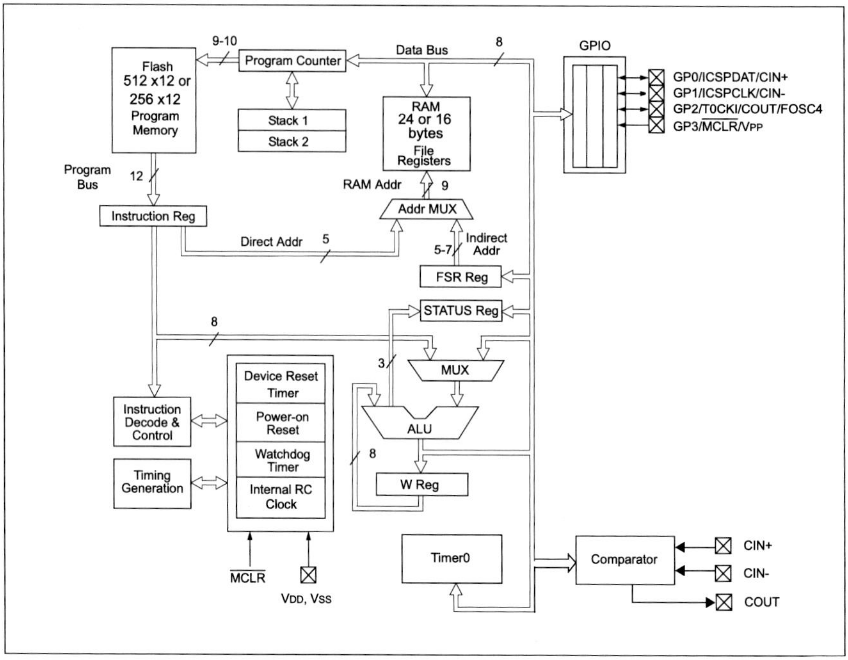

PIC10F200/202/204/206

TABLE 3-2: PIC10F200/202/204/206 PINOUT DESCRIPTION

Name	Function	Input Type	Output Type	Description
GP0/ICSPDAT/CIN+	GP0	TTL	CMOS	Bidirectional I/O pin. Can be software programmed for internal weak pull-up and wake-up from Sleep on pin change.
	ICSPDAT	ST	CMOS	In-Circuit Serial Programming™ data pin.
	CIN+	AN	—	Comparator input (PIC10F204/206 only).
GP1/ICSPCLK/CIN-	GP1	TTL	CMOS	Bidirectional I/O pin. Can be software programmed for internal weak pull-up and wake-up from Sleep on pin change.
	ICSPCLK	ST	CMOS	In-Circuit Serial Programming clock pin.
	CIN-	AN	—	Comparator input (PIC10F204/206 only).
GP2/T0CKI/COUT/FOSC4	GP2	TTL	CMOS	Bidirectional I/O pin.
	T0CKI	ST	—	Clock input to TMR0.
	COUT	—	CMOS	Comparator output (PIC10F204/206 only).
	FOSC4	—	CMOS	Oscillator/4 output.
GP3/$\overline{MCLR}$/VPP	GP3	TTL	—	Input pin. Can be software programmed for internal weak pull-up and wake-up from Sleep on pin change.
	$\overline{MCLR}$	ST	—	Master Clear (Reset). When configured as $\overline{MCLR}$, this pin is an active-low Reset to the device. Voltage on GP3/$\overline{MCLR}$/VPP must not exceed VDD during normal device operation or the device will enter Programming mode. Weak pull-up always on if configured as $\overline{MCLR}$.
	VPP	HV	—	Programming voltage input.
VDD	VDD	P	—	Positive supply for logic and I/O pins.
VSS	VSS	P	—	Ground reference for logic and I/O pins.

Legend: I = Input, O = Output, I/O = Input/Output, P = Power, — = Not used, TTL = TTL input, ST = Schmitt Trigger input, AN = Analog input

PIC10F200/202/204/206

3.1 Clocking Scheme/Instruction Cycle

The clock is internally divided by four to generate four non-overlapping quadrature clocks, namely Q1, Q2, Q3 and Q4. Internally, the PC is incremented every Q1 and the instruction is fetched from program memory and latched into the instruction register in Q4. It is decoded and executed during the following Q1 through Q4. The clocks and instruction execution flow is shown in Figure 3-3 and Example 3-1.

3.2 Instruction Flow/Pipelining

An instruction cycle consists of four Q cycles (Q1, Q2, Q3 and Q4). The instruction fetch and execute are pipelined such that fetch takes one instruction cycle, while decode and execute take another instruction cycle. However, due to the pipelining, each instruction effectively executes in one cycle. If an instruction causes the PC to change (e.g., GOTO), then two cycles are required to complete the instruction (Example 3-1).

A fetch cycle begins with the PC incrementing in Q1.

In the execution cycle, the fetched instruction is latched into the Instruction Register (IR) in cycle Q1. This instruction is then decoded and executed during the Q2, Q3 and Q4 cycles. Data memory is read during Q2 (operand read) and written during Q4 (destination write).

FIGURE 3-3: CLOCK/INSTRUCTION CYCLE

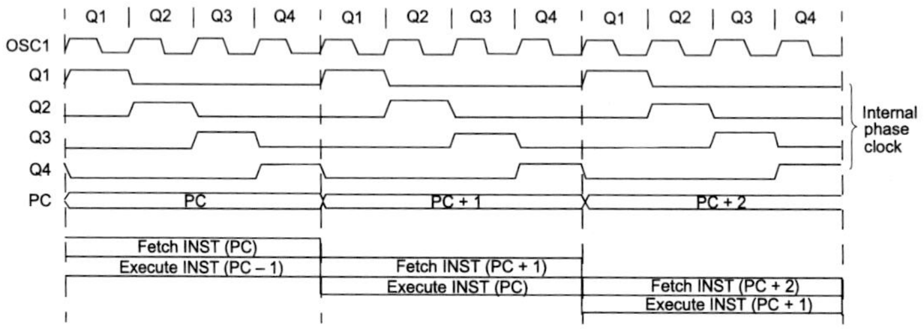

EXAMPLE 3-1: INSTRUCTION PIPELINE FLOW

```
1. MOVLW 03H
2. MOVWF GPIO
3. CALL  SUB_1
4. BSF   GPIO, BIT1
```

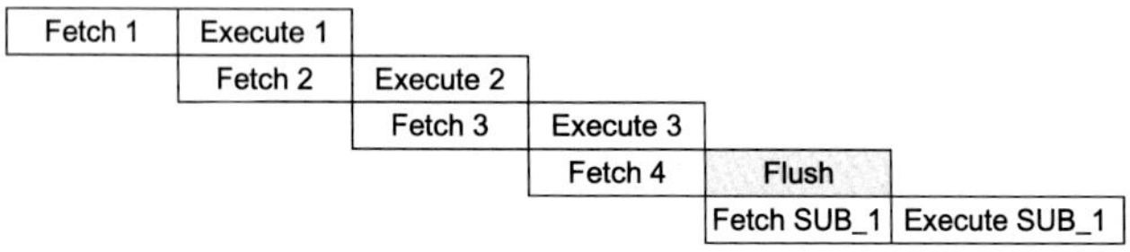

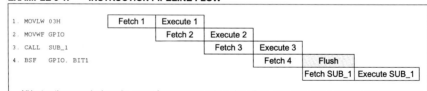

All instructions are single cycle, except for any program branches. These take two cycles, since the fetch instruction is "flushed" from the pipeline, while the new instruction is being fetched and then executed.

PIC10F200/202/204/206

4.0 MEMORY ORGANIZATION

The PIC10F200/202/204/206 memories are organized into program memory and data memory. Data memory banks are accessed using the File Select Register (FSR).

4.1 Program Memory Organization for the PIC10F200/204

The PIC10F200/204 devices have a 9-bit Program Counter (PC) capable of addressing a 512 x 12 program memory space.

Only the first 256 x 12 (0000h-00FFh) for the PIC10F200/204 are physically implemented (see Figure 4-1). Accessing a location above these boundaries will cause a wraparound within the first 256 x 12 space (PIC10F200/204). The effective Reset vector is at 0000h (see Figure 4-1). Location 00FFh (PIC10F200/204) contains the internal clock oscillator calibration value. This value should never be overwritten.

FIGURE 4-1: PROGRAM MEMORY MAP AND STACK FOR THE PIC10F200/204

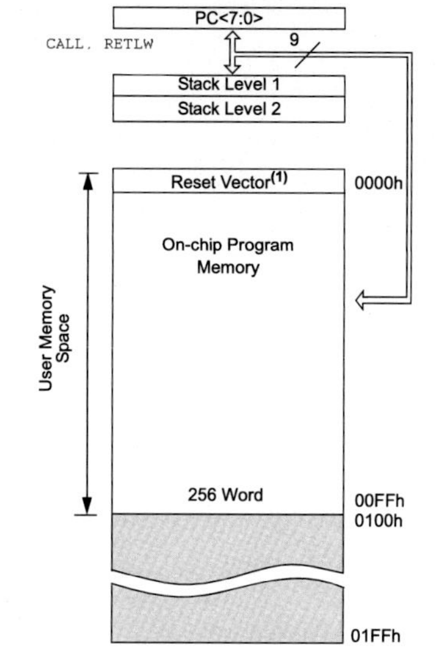

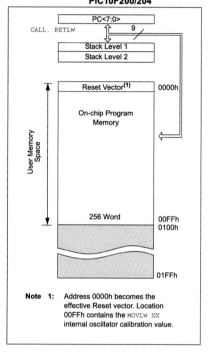

Note 1: Address 0000h becomes the effective Reset vector. Location 00FFh contains the `MOVLW XX` internal oscillator calibration value.

PIC10F200/202/204/206

4.2 Program Memory Organization for the PIC10F202/206

The PIC10F202/206 devices have a 10-bit Program Counter (PC) capable of addressing a 1024 x 12 program memory space.

Only the first 512 x 12 (0000h-01FFh) for the PIC10F202/206 are physically implemented (see Figure 4-2). Accessing a location above these boundaries will cause a wraparound within the first 512 x 12 space (PIC10F202/206). The effective Reset vector is at 0000h (see Figure 4-2). Location 01FFh (PIC10F202/206) contains the internal clock oscillator calibration value. This value should never be overwritten.

FIGURE 4-2: PROGRAM MEMORY MAP AND STACK FOR THE PIC10F202/206

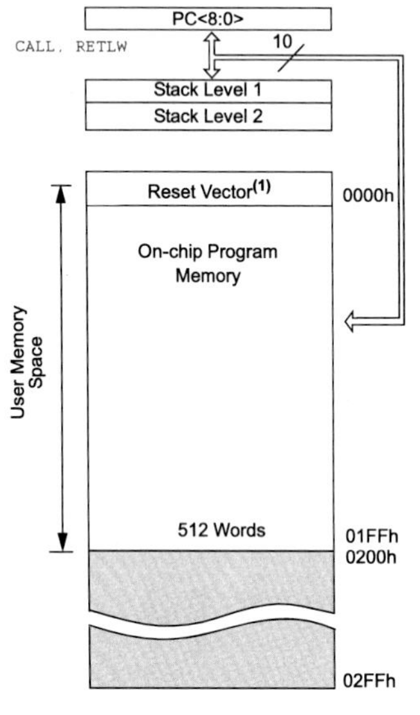

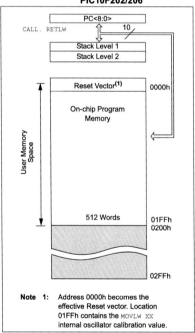

Note 1: Address 0000h becomes the effective Reset vector. Location 01FFh contains the `MOVLW XX` internal oscillator calibration value.

4.3 Data Memory Organization

Data memory is composed of registers or bytes of RAM. Therefore, data memory for a device is specified by its register file. The register file is divided into two functional groups: Special Function Registers (SFR) and General Purpose Registers (GPR).

The Special Function Registers include the TMR0 register, the Program Counter (PCL), the STATUS register, the I/O register (GPIO) and the File Select Register (FSR). In addition, Special Function Registers are used to control the I/O port configuration and prescaler options.

The General Purpose registers are used for data and control information under command of the instructions.

For the PIC10F200/204, the register file is composed of seven Special Function registers and 16 General Purpose registers (see Figure 4-3 and Figure 4-4).

For the PIC10F202/206, the register file is composed of eight Special Function registers and 24 General Purpose registers (see Figure 4-4).

4.3.1 GENERAL PURPOSE REGISTER FILE

The General Purpose Register file is accessed, either directly or indirectly, through the File Select Register (FSR). See **Section 4.9 "Indirect Data Addressing: INDF and FSR Registers"**.

PIC10F200/202/204/206

FIGURE 4-3: PIC10F200/204 REGISTER FILE MAP

File Address	
00h	INDF(1)
01h	TMR0
02h	PCL
03h	STATUS
04h	FSR
05h	OSCCAL
06h	GPIO
07h	CMCON0(2)
08h–0Fh	Unimplemented(3)
10h–1Fh	General Purpose Registers

Note 1: Not a physical register. See **Section 4.9 "Indirect Data Addressing: INDF and FSR Registers"**.

2: PIC10F204 only. Unimplemented on the PIC10F200 and reads as 00h.

3: Unimplemented, read as 00h.

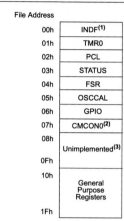

FIGURE 4-4: PIC10F202/206 REGISTER FILE MAP

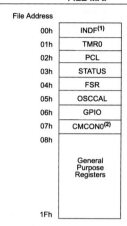

File Address	
00h	INDF(1)
01h	TMR0
02h	PCL
03h	STATUS
04h	FSR
05h	OSCCAL
06h	GPIO
07h	CMCON0(2)
08h–1Fh	General Purpose Registers

Note 1: Not a physical register. See **Section 4.9 "Indirect Data Addressing: INDF and FSR Registers"**.

2: PIC10F206 only. Unimplemented on the PIC10F202 and reads as 00h.

4.3.2 SPECIAL FUNCTION REGISTERS

The Special Function Registers (SFRs) are registers used by the CPU and peripheral functions to control the operation of the device (Table 4-1).

The Special Function Registers can be classified into two sets. The Special Function Registers associated with the "core" functions are described in this section. Those related to the operation of the peripheral features are described in the section for each peripheral feature.

TABLE 4-1: SPECIAL FUNCTION REGISTER (SFR) SUMMARY (PIC10F200/202/204/206)

Address	Name	Bit 7	Bit 6	Bit 5	Bit 4	Bit 3	Bit 2	Bit 1	Bit 0	Value on Power-On Reset(2)	Register on Page
00h	INDF	Uses Contents of FSR to Address Data Memory (not a physical register)								xxxx xxxx	19
01h	TMR0	8-bit Real-Time Clock/Counter								xxxx xxxx	23, 27
02h(1)	PCL	Low-order 8 bits of PC								1111 1111	18
03h	STATUS	GPWUF	CWUF(5)	—	$\overline{\text{TO}}$	$\overline{\text{PD}}$	Z	DC	C	00-1 1xxx (3)	15
04h	FSR	Indirect Data Memory Address Pointer								111x xxxx	19
05h	OSCCAL	CAL6	CAL5	CAL4	CAL3	CAL2	CAL1	CAL0	FOSC4	1111 1110	17
06h	GPIO	—	—	—	—	GP3	GP2	GP1	GP0	---- xxxx	20
07h(4)	CMCON0	CMPOUT	$\overline{\text{COUTEN}}$	POL	$\overline{\text{CMPT0CS}}$	CMPON	CNREF	CPREF	$\overline{\text{CWU}}$	1111 1111	28
N/A	TRISGPIO	—	—	—	—	I/O Control Register				---- 1111	31
N/A	OPTION	$\overline{\text{GPWU}}$	$\overline{\text{GPPU}}$	T0CS	T0SE	PSA	PS2	PS1	PS0	1111 1111	16

Legend: – = unimplemented, read as '0', x = unknown, u = unchanged, q = value depends on condition.

Note 1: The upper byte of the Program Counter is not directly accessible. See **Section 4.7 "Program Counter"** for an explanation of how to access these bits.

2: Other (non Power-up) Resets include external Reset through $\overline{\text{MCLR}}$, Watchdog Timer and wake-up on pin change Reset.

3: See Table 9-1 for other Reset specific values.

4: PIC10F204/206 only.

5: PIC10F204/206 only. On all other devices, this bit is reserved and should not be used.

PIC10F200/202/204/206

4.4 STATUS Register

This register contains the arithmetic status of the ALU, the Reset status and the page preselect bit.

The STATUS register can be the destination for any instruction, as with any other register. If the STATUS register is the destination for an instruction that affects the Z, DC or C bits, then the write to these three bits is disabled. These bits are set or cleared according to the device logic. Furthermore, the $\overline{TO}$ and $\overline{PD}$ bits are not writable. Therefore, the result of an instruction with the STATUS register as destination may be different than intended.

For example, `CLRF STATUS`, will clear the upper three bits and set the Z bit. This leaves the STATUS register as `000u u1uu` (where `u` = unchanged).

Therefore, it is recommended that only `BCF`, `BSF` and `MOVWF` instructions be used to alter the STATUS register. These instructions do not affect the Z, DC or C bits from the STATUS register. For other instructions which do affect Status bits, see **Section 10.0 "Instruction Set Summary"**.

REGISTER 4-1: STATUS REGISTER

R/W-0	R/W-0	U-1	R-1	R-1	R/W-x	R/W-x	R/W-x
GPWUF	CWUF[1]	—	$\overline{TO}$	$\overline{PD}$	Z	DC	C
bit 7							bit 0

Legend:			
R = Readable bit	W = Writable bit	U = Unimplemented bit, read as '0'	
-n = Value at POR	'1' = Bit is set	'0' = Bit is cleared	x = Bit is unknown

bit 7 **GPWUF:** GPIO Reset bit
1 = Reset due to wake-up from Sleep on pin change
0 = After power-up or other Reset

bit 6 **CWUF:** Comparator Wake-up on Change Flag bit[1]
1 = Reset due to wake-up from Sleep on comparator change
0 = After power-up or other Reset conditions.

bit 5 **Reserved:** Do not use. Use of this bit may affect upward compatibility with future products.

bit 4 **$\overline{TO}$:** Time-out bit
1 = After power-up, `CLRWDT` instruction or `SLEEP` instruction
0 = A WDT time-out occurred

bit 3 **$\overline{PD}$:** Power-down bit
1 = After power-up or by the `CLRWDT` instruction
0 = By execution of the `SLEEP` instruction

bit 2 **Z:** Zero bit
1 = The result of an arithmetic or logic operation is zero
0 = The result of an arithmetic or logic operation is not zero

bit 1 **DC:** Digit Carry/$\overline{Borrow}$ bit (for `ADDWF` and `SUBWF` instructions)
<u>ADDWF</u>:
1 = A carry from the 4th low-order bit of the result occurred
0 = A carry from the 4th low-order bit of the result did not occur
<u>SUBWF</u>:
1 = A borrow from the 4th low-order bit of the result did not occur
0 = A borrow from the 4th low-order bit of the result occurred

bit 0 **C:** Carry/$\overline{Borrow}$ bit (for `ADDWF`, `SUBWF` and `RRF`, `RLF` instructions)

<u>ADDWF</u>:	<u>SUBWF</u>:	<u>RRF or RLF</u>:
1 = A carry occurred	1 = A borrow did not occur	Load bit with LSb or MSb, respectively
0 = A carry did not occur	0 = A borrow occurred	

Note 1: This bit is used on the PIC10F204/206. For code compatibility do not use this bit on the PIC10F200/202.

PIC10F200/202/204/206

4.5 OPTION Register

The OPTION register is a 8-bit wide, write-only register, which contains various control bits to configure the Timer0/WDT prescaler and Timer0.

By executing the `OPTION` instruction, the contents of the W register will be transferred to the OPTION register. A Reset sets the OPTION<7:0> bits.

> **Note:** If TRIS bit is set to '0', the wake-up on change and pull-up functions are disabled for that pin (i.e., note that TRIS overrides Option control of $\overline{\text{GPPU}}$ and $\overline{\text{GPWU}}$).

> **Note:** If the T0CS bit is set to '1', it will override the TRIS function on the T0CKI pin.

REGISTER 4-2: OPTION REGISTER

W-1	W-1	W-1	W-1	W-1	W-1	W-1	W-1
$\overline{\text{GPWU}}$	$\overline{\text{GPPU}}$	T0CS	T0SE	PSA	PS2	PS1	PS0
bit 7							bit 0

Legend:			
R = Readable bit	W = Writable bit	U = Unimplemented bit, read as '0'	
-n = Value at POR	'1' = Bit is set	'0' = Bit is cleared	x = Bit is unknown

bit 7 **$\overline{\text{GPWU}}$:** Enable Wake-up on Pin Change bit (GP0, GP1, GP3)
1 = Disabled
0 = Enabled

bit 6 **$\overline{\text{GPPU}}$:** Enable Weak Pull-ups bit (GP0, GP1, GP3)
1 = Disabled
0 = Enabled

bit 5 **T0CS:** Timer0 Clock Source Select bit
1 = Transition on T0CKI pin (overrides TRIS on the T0CKI pin)
0 = Transition on internal instruction cycle clock, FOSC/4

bit 4 **T0SE:** Timer0 Source Edge Select bit
1 = Increment on high-to-low transition on the T0CKI pin
0 = Increment on low-to-high transition on the T0CKI pin

bit 3 **PSA:** Prescaler Assignment bit
1 = Prescaler assigned to the WDT
0 = Prescaler assigned to Timer0

bit 2-0 **PS<2:0>:** Prescaler Rate Select bits

Bit Value	Timer0 Rate	WDT Rate
000	1 : 2	1 : 1
001	1 : 4	1 : 2
010	1 : 8	1 : 4
011	1 : 16	1 : 8
100	1 : 32	1 : 16
101	1 : 64	1 : 32
110	1 : 128	1 : 64
111	1 : 256	1 : 128

PIC10F200/202/204/206

4.6 OSCCAL Register

The Oscillator Calibration (OSCCAL) register is used to calibrate the internal precision 4 MHz oscillator. It contains seven bits for calibration.

Note:	Erasing the device will also erase the pre-programmed internal calibration value for the internal oscillator. The calibration value must be read prior to erasing the part so it can be reprogrammed correctly later.

After you move in the calibration constant, do not change the value. See **Section 9.2.2 "Internal 4 MHz Oscillator"**.

REGISTER 4-3: OSCCAL REGISTER

R/W-1	R/W-1	R/W-1	R/W-1	R/W-1	R/W-1	R/W-1	R/W-0
CAL6	CAL5	CAL4	CAL3	CAL2	CAL1	CAL0	FOSC4
bit 7							bit 0

Legend:			
R = Readable bit	W = Writable bit	U = Unimplemented bit, read as '0'	
-n = Value at POR	'1' = Bit is set	'0' = Bit is cleared	x = Bit is unknown

bit 7-1 **CAL<6:0>:** Oscillator Calibration bits

`0111111` = Maximum frequency
•
•
•
`0000001`
`0000000` = Center frequency
`1111111`
•
•
•
`1000000` = Minimum frequency

bit 0 **FOSC4:** INTOSC/4 Output Enable bit(1)

`1` = INTOSC/4 output onto GP2
`0` = GP2/T0CKI/COUT applied to GP2

Note 1: Overrides GP2/T0CKI/COUT control registers when enabled.

PIC10F200/202/204/206

4.7 Program Counter

As a program instruction is executed, the Program Counter (PC) will contain the address of the next program instruction to be executed. The PC value is increased by one every instruction cycle, unless an instruction changes the PC.

For a `GOTO` instruction, bits 8-0 of the PC are provided by the `GOTO` instruction word. The Program Counter Low (PCL) is mapped to PC<7:0>.

For a `CALL` instruction, or any instruction where the PCL is the destination, bits 7:0 of the PC again are provided by the instruction word. However, PC<8> does not come from the instruction word, but is always cleared (Figure 4-5).

Instructions where the PCL is the destination, or modify PCL instructions, include `MOVWF PC`, `ADDWF PC` and `BSF PC,5`.

Note:	Because PC<8> is cleared in the `CALL` instruction or any modify PCL instruction, all subroutine calls or computed jumps are limited to the first 256 locations of any program memory page (512 words long).

FIGURE 4-5: LOADING OF PC BRANCH INSTRUCTIONS

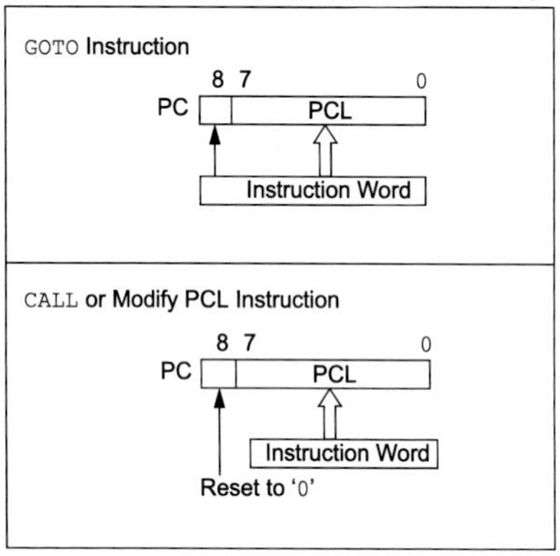

4.7.1 EFFECTS OF RESET

The PC is set upon a Reset, which means that the PC addresses the last location in program memory (i.e., the oscillator calibration instruction). After executing `MOVLW XX`, the PC will roll over to location 0000h and begin executing user code.

4.8 Stack

The PIC10F200/204 devices have a 2-deep, 8-bit wide hardware PUSH/POP stack.

The PIC10F202/206 devices have a 2-deep, 9-bit wide hardware PUSH/POP stack.

A `CALL` instruction will PUSH the current value of Stack 1 into Stack 2 and then PUSH the current PC value, incremented by one, into Stack Level 1. If more than two sequential `CALL`s are executed, only the most recent two return addresses are stored.

A `RETLW` instruction will POP the contents of Stack Level 1 into the PC and then copy Stack Level 2 contents into level 1. If more than two sequential `RETLW`s are executed, the stack will be filled with the address previously stored in Stack Level 2.

Note 1:	The W register will be loaded with the literal value specified in the instruction. This is particularly useful for the implementation of the data look-up tables within the program memory.
2:	There are no Status bits to indicate stack overflows or stack underflow conditions.
3:	There are no instruction mnemonics called PUSH or POP. These are actions that occur from the execution of the `CALL` and `RETLW` instructions.

PIC10F200/202/204/206

4.9 Indirect Data Addressing: INDF and FSR Registers

The INDF register is not a physical register. Addressing INDF actually addresses the register whose address is contained in the FSR register (FSR is a *pointer*). This is indirect addressing.

4.10 Indirect Addressing

- Register file 09 contains the value 10h
- Register file 0A contains the value 0Ah
- Load the value 09 into the FSR register
- A read of the INDF register will return the value of 10h
- Increment the value of the FSR register by one (FSR = 0A)
- A read of the INDR register now will return the value of 0Ah.

Reading INDF itself indirectly (FSR = 0) will produce 00h. Writing to the INDF register indirectly results in a no operation (although Status bits may be affected).

A simple program to clear RAM locations 10h-1Fh using indirect addressing is shown in Example 4-1.

EXAMPLE 4-1: HOW TO CLEAR RAM USING INDIRECT ADDRESSING

```
        MOVLW   0x10    ;initialize pointer
        MOVWF   FSR     ;to RAM
NEXT    CLRF    INDF    ;clear INDF
                        ;register
        INCF    FSR,F   ;inc pointer
        BTFSC   FSR,4   ;all done?
        GOTO    NEXT    ;NO, clear next
CONTINUE
                :       ;YES, continue
                :
```

The FSR is a 5-bit wide register. It is used in conjunction with the INDF register to indirectly address the data memory area.

The FSR<4:0> bits are used to select data memory addresses 00h to 1Fh.

Note:	PIC10F200/202/204/206 – Do not use banking. FSR <7:5> are unimplemented and read as '1's.

FIGURE 4-6: DIRECT/INDIRECT ADDRESSING (PIC10F200/202/204/206)

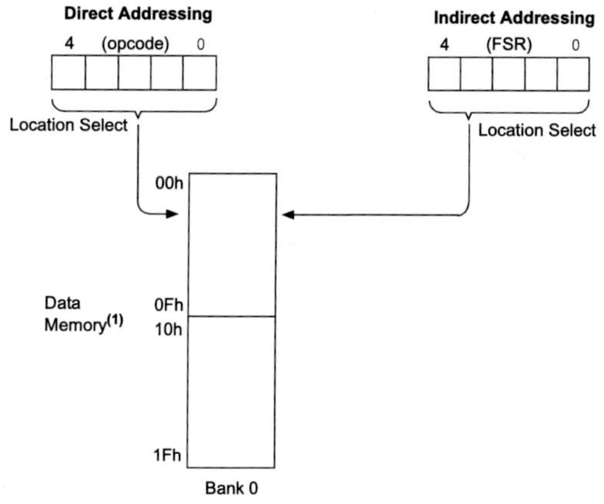

Note 1: For register map detail, see **Section 4.3 "Data Memory Organization"**.

PIC10F200/202/204/206

5.0 I/O PORT

As with any other register, the I/O register(s) can be written and read under program control. However, read instructions (e.g., `MOVF GPIO, W`) always read the I/O pins independent of the pin's Input/Output modes. On Reset, all I/O ports are defined as input (inputs are at high-impedance) since the I/O control registers are all set.

5.1 GPIO

GPIO is an 8-bit I/O register. Only the low-order 4 bits are used (GP<3:0>). Bits 7 through 4 are unimplemented and read as '0's. Please note that GP3 is an input-only pin. Pins GP0, GP1 and GP3 can be configured with weak pull-ups and also for wake-up on change. The wake-up on change and weak pull-up functions are not pin selectable. If GP3/MCLR is configured as MCLR, weak pull-up is always on and wake-up on change for this pin is not enabled.

5.2 TRIS Registers

The Output Driver Control register is loaded with the contents of the W register by executing the `TRIS f` instruction. A '1' from a TRIS register bit puts the corresponding output driver in a High-Impedance mode. A '0' puts the contents of the output data latch on the selected pins, enabling the output buffer. The exceptions are GP3, which is input-only and the GP2/T0CKI/COUT/FOSC4 pin, which may be controlled by various registers. See Table 5-1.

> **Note:** A read of the ports reads the pins, not the output data latches. That is, if an output driver on a pin is enabled and driven high, but the external system is holding it low, a read of the port will indicate that the pin is low.

The TRIS registers are "write-only" and are set (output drivers disabled) upon Reset.

TABLE 5-1: ORDER OF PRECEDENCE FOR PIN FUNCTIONS

Priority	GP0	GP1	GP2	GP3
1	CIN+	CIN-	FOSC4	I/MCLR
2	TRIS GPIO	TRIS GPIO	COUT	—
3	—	—	T0CKI	—
4	—	—	TRIS GPIO	—

5.3 I/O Interfacing

The equivalent circuit for an I/O port pin is shown in Figure 5-1. All port pins, except GP3 which is input-only, may be used for both input and output operations. For input operations, these ports are non-latching. Any input must be present until read by an input instruction (e.g., `MOVF GPIO, W`). The outputs are latched and remain unchanged until the output latch is rewritten. To use a port pin as output, the corresponding direction control bit in TRIS must be cleared (= 0). For use as an input, the corresponding TRIS bit must be set. Any I/O pin (except GP3) can be programmed individually as input or output.

FIGURE 5-1: PIC10F200/202/204/206 EQUIVALENT CIRCUIT FOR A SINGLE I/O PIN

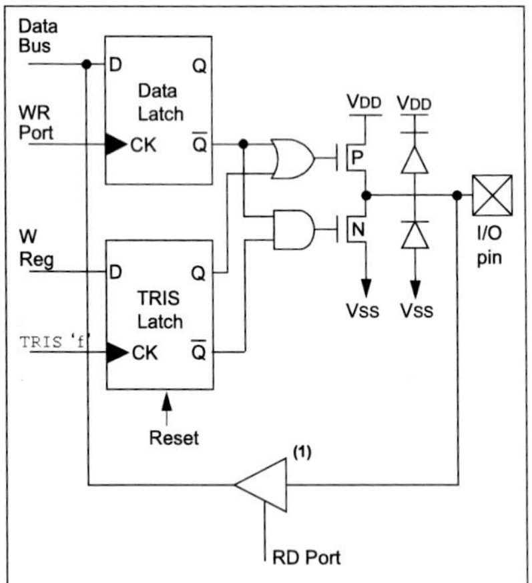

Note 1: See Table 3-2 for buffer type.

PIC10F200/202/204/206

TABLE 5-2: SUMMARY OF PORT REGISTERS

Address	Name	Bit 7	Bit 6	Bit 5	Bit 4	Bit 3	Bit 2	Bit 1	Bit 0	Value on Power-On Reset	Value on All Other Resets
N/A	TRISGPIO	—	—	—	—	I/O Control Register				---- 1111	---- 1111
N/A	OPTION	$\overline{GPWU}$	$\overline{GPPU}$	T0CS	T0SE	PSA	PS2	PS1	PS0	1111 1111	1111 1111
03h	STATUS	GPWUF	CWUF	—	$\overline{TO}$	$\overline{PD}$	Z	DC	C	00-1 1xxx	qq-q quuu(1), (2)
06h	GPIO	—	—	—	—	GP3	GP2	GP1	GP0	---- xxxx	---- uuuu

Legend: Shaded cells are not used by PORT registers, read as '0', – = unimplemented, read as '0', x = unknown, u = unchanged, q = depends on condition.

Note 1: If Reset was due to wake-up on pin change, then bit 7 = 1. All other Resets will cause bit 7 = 0.

2: If Reset was due to wake-up on comparator change, then bit 6 = 1. All other Resets will cause bit 6 = 0.

5.4 I/O Programming Considerations

5.4.1 BIDIRECTIONAL I/O PORTS

Some instructions operate internally as read followed by write operations. The BCF and BSF instructions, for example, read the entire port into the CPU, execute the bit operation and rewrite the result. Caution must be used when these instructions are applied to a port where one or more pins are used as input/outputs. For example, a BSF operation on bit 2 of GPIO will cause all eight bits of GPIO to be read into the CPU, bit 2 to be set and the GPIO value to be written to the output latches. If another bit of GPIO is used as a bidirectional I/O pin (say bit 0), and it is defined as an input at this time, the input signal present on the pin itself would be read into the CPU and rewritten to the data latch of this particular pin, overwriting the previous content. As long as the pin stays in the Input mode, no problem occurs. However, if bit 0 is switched into Output mode later on, the content of the data latch may now be unknown.

Example 5-1 shows the effect of two sequential Read-Modify-Write instructions (e.g., BCF, BSF, etc.) on an I/O port.

A pin actively outputting a high or a low should not be driven from external devices at the same time in order to change the level on this pin ("wired OR", "wired AND"). The resulting high output currents may damage the chip.

EXAMPLE 5-1: READ-MODIFY-WRITE INSTRUCTIONS ON AN I/O PORT

```
;Initial GPIO Settings
;GPIO<3:2> Inputs
;GPIO<1:0> Outputs
;
;                   GPIO latch    GPIO pins
;                   ----------    ----------
  BCF    GPIO,   1 ;---- pp01     ---- pp11
  BCF    GPIO,   0 ;---- pp10     ---- pp11
  MOVLW  007h;
  TRIS   GPIO      ;---- pp10     ---- pp11
;
```

Note 1: The user may have expected the pin values to be ---- pp00. The 2nd BCF caused GP1 to be latched as the pin value (High).

5.4.2 SUCCESSIVE OPERATIONS ON I/O PORTS

The actual write to an I/O port happens at the end of an instruction cycle, whereas for reading, the data must be valid at the beginning of the instruction cycle (Figure 5-2). Therefore, care must be exercised if a write followed by a read operation is carried out on the same I/O port. The sequence of instructions should allow the pin voltage to stabilize (load dependent) before the next instruction causes that file to be read into the CPU. Otherwise, the previous state of that pin may be read into the CPU rather than the new state. When in doubt, it is better to separate these instructions with a NOP or another instruction not accessing this I/O port.

PIC10F200/202/204/206

FIGURE 5-2: SUCCESSIVE I/O OPERATION (PIC10F200/202/204/206)

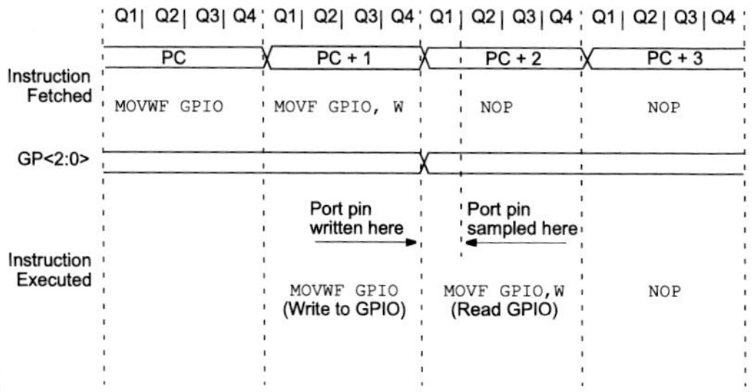

This example shows a write to GPIO followed by a read from GPIO.

Data setup time = (0.25 T_{CY} – T_{PD})

where: T_{CY} = instruction cycle

T_{PD} = propagation delay

Therefore, at higher clock frequencies, a write followed by a read may be problematic.

PIC10F200/202/204/206

6.0 TIMER0 MODULE AND TMR0 REGISTER (PIC10F200/202)

The Timer0 module has the following features:

- 8-bit timer/counter register, TMR0
- Readable and writable
- 8-bit software programmable prescaler
- Internal or external clock select:
  - Edge select for external clock

Figure 6-1 is a simplified block diagram of the Timer0 module.

Timer mode is selected by clearing the T0CS bit (OPTION<5>). In Timer mode, the Timer0 module will increment every instruction cycle (without prescaler). If TMR0 register is written, the increment is inhibited for the following two cycles (Figure 6-2 and Figure 6-3). The user can work around this by writing an adjusted value to the TMR0 register.

Counter mode is selected by setting the T0CS bit (OPTION<5>). In this mode, Timer0 will increment either on every rising or falling edge of pin T0CKI. The T0SE bit (OPTION<4>) determines the source edge. Clearing the T0SE bit selects the rising edge. Restrictions on the external clock input are discussed in detail in **Section 6.1 "Using Timer0 with an External Clock (PIC10F200/202)"**.

The prescaler may be used by either the Timer0 module or the Watchdog Timer, but not both. The prescaler assignment is controlled in software by the control bit, PSA (OPTION<3>). Clearing the PSA bit will assign the prescaler to Timer0. The prescaler is not readable or writable. When the prescaler is assigned to the Timer0 module, prescale values of 1:2, 1:4, 1:256 are selectable. **Section 6.2 "Prescaler"** details the operation of the prescaler.

A summary of registers associated with the Timer0 module is found in Table 6-1.

FIGURE 6-1: TIMER0 BLOCK DIAGRAM

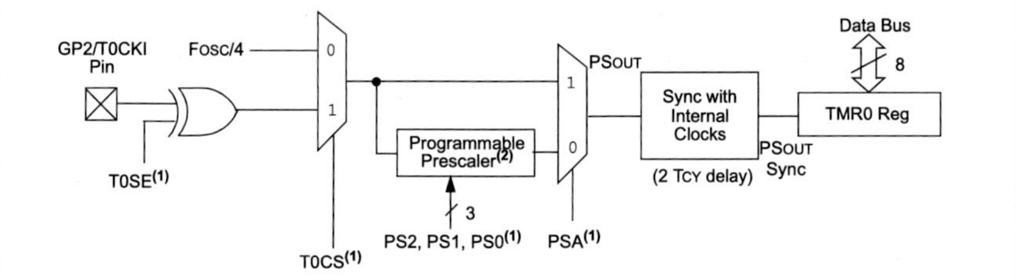

Note 1: Bits T0CS, T0SE, PSA, PS2, PS1 and PS0 are located in the OPTION register.
2: The prescaler is shared with the Watchdog Timer (Figure 6-5).

FIGURE 6-2: TIMER0 TIMING: INTERNAL CLOCK/NO PRESCALE

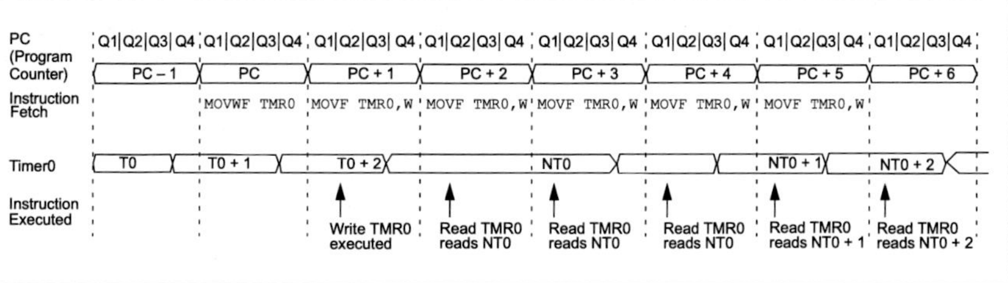

PIC10F200/202/204/206

FIGURE 6-3: TIMER0 TIMING: INTERNAL CLOCK/PRESCALE 1:2

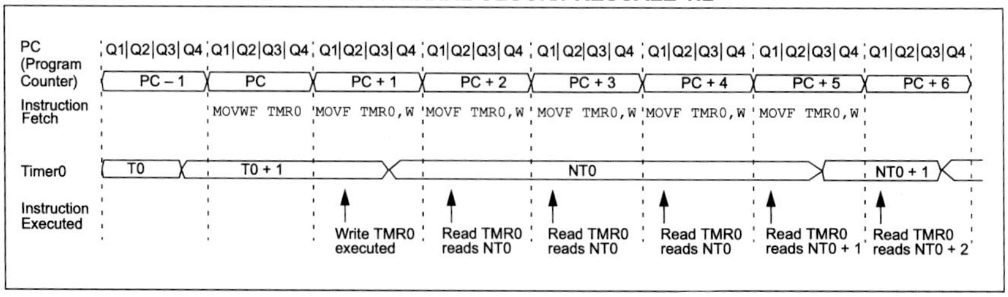

TABLE 6-1: REGISTERS ASSOCIATED WITH TIMER0

Address	Name	Bit 7	Bit 6	Bit 5	Bit 4	Bit 3	Bit 2	Bit 1	Bit 0	Value on Power-On Reset	Value on All Other Resets
01h	TMR0	Timer0 – 8-bit Real-Time Clock/Counter								xxxx xxxx	uuuu uuuu
N/A	OPTION	GPWU	GPPU	T0CS	T0SE	PSA	PS2	PS1	PS0	1111 1111	1111 1111
N/A	TRISGPIO(1)	—	—	—	—	I/O Control Register				---- 1111	---- 1111

Legend: Shaded cells not used by Timer0. – = unimplemented, x = unknown, u = unchanged.
Note 1: The TRIS of the T0CKI pin is overridden when T0CS = 1.

6.1 Using Timer0 with an External Clock (PIC10F200/202)

When an external clock input is used for Timer0, it must meet certain requirements. The external clock requirement is due to internal phase clock (TOSC) synchronization. Also, there is a delay in the actual incrementing of Timer0 after synchronization.

6.1.1 EXTERNAL CLOCK SYNCHRONIZATION

When no prescaler is used, the external clock input is the same as the prescaler output. The synchronization of T0CKI with the internal phase clocks is accomplished by sampling the prescaler output on the Q2 and Q4 cycles of the internal phase clocks (Figure 6-4). Therefore, it is necessary for T0CKI to be high for at least 2 TOSC (and a small RC delay of 2 Tt0H) and low for at least 2 TOSC (and a small RC delay of 2 Tt0H). Refer to the electrical specification of the desired device.

When a prescaler is used, the external clock input is divided by the asynchronous ripple counter-type prescaler, so that the prescaler output is symmetrical. For the external clock to meet the sampling requirement, the ripple counter must be taken into account. Therefore, it is necessary for T0CKI to have a period of at least 4 TOSC (and a small RC delay of 4 Tt0H) divided by the prescaler value. The only requirement on T0CKI high and low time is that they do not violate the minimum pulse width requirement of Tt0H. Refer to parameters 40, 41 and 42 in the electrical specification of the desired device.

6.1.2 TIMER0 INCREMENT DELAY

Since the prescaler output is synchronized with the internal clocks, there is a small delay from the time the external clock edge occurs to the time the Timer0 module is actually incremented. Figure 6-4 shows the delay from the external clock edge to the timer incrementing.

FIGURE 6-4: TIMER0 TIMING WITH EXTERNAL CLOCK

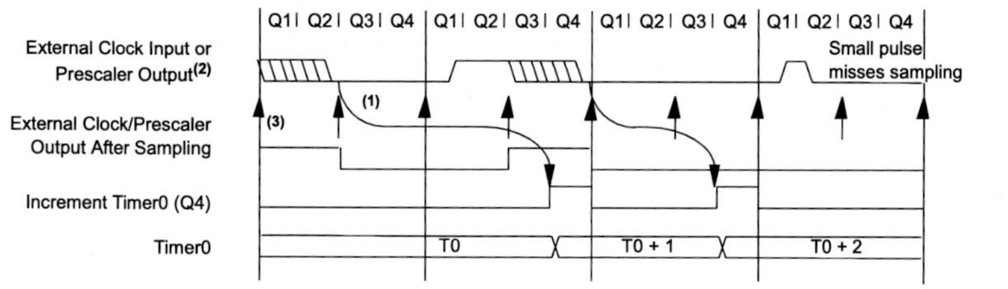

Note 1: Delay from clock input change to Timer0 increment is 3 Tosc to 7 Tosc (Duration of Q = Tosc). Therefore, the error in measuring the interval between two edges on Timer0 input = ±4 Tosc max.

2: External clock if no prescaler selected; prescaler output otherwise.

3: The arrows indicate the points in time where sampling occurs.

6.2 Prescaler

An 8-bit counter is available as a prescaler for the Timer0 module or as a postscaler for the Watchdog Timer (WDT), respectively (see **Section 9.6 "Watchdog Timer (WDT)"**). For simplicity, this counter is being referred to as "prescaler" throughout this data sheet.

> **Note:** The prescaler may be used by either the Timer0 module or the WDT, but not both. Thus, a prescaler assignment for the Timer0 module means that there is no prescaler for the WDT and vice versa.

The PSA and PS<2:0> bits (OPTION<3:0>) determine prescaler assignment and prescale ratio.

When assigned to the Timer0 module, all instructions writing to the TMR0 register (e.g., `CLRF 1`, `MOVWF 1`, `BSF 1,x`, etc.) will clear the prescaler. When assigned to WDT, a `CLRWDT` instruction will clear the prescaler along with the WDT. The prescaler is neither readable nor writable. On a Reset, the prescaler contains all '`0`'s.

6.2.1 SWITCHING PRESCALER ASSIGNMENT

The prescaler assignment is fully under software control (i.e., it can be changed "on-the-fly" during program execution). To avoid an unintended device Reset, the following instruction sequence (Example 6-1) must be executed when changing the prescaler assignment from Timer0 to the WDT.

EXAMPLE 6-1: CHANGING PRESCALER (TIMER0 → WDT)

```
CLRWDT                  ;Clear WDT
CLRF    TMR0            ;Clear TMR0 & Prescaler
MOVLW   '00xx1111'b     ;These 3 lines (5, 6, 7)
OPTION                  ;are required only if
                        ;desired
CLRWDT                  ;PS<2:0> are 000 or 001
MOVLW   '00xx1xxx'b     ;Set Postscaler to
OPTION                  ;desired WDT rate
```

PIC10F200/202/204/206

To change the prescaler from the WDT to the Timer0 module, use the sequence shown in Example 6-2. This sequence must be used even if the WDT is disabled. A `CLRWDT` instruction should be executed before switching the prescaler.

EXAMPLE 6-2: CHANGING PRESCALER (WDT→TIMER0)

```
CLRWDT                 ;Clear WDT and
                       ;prescaler
MOVLW   'xxxx0xxx'     ;Select TMR0, new
                       ;prescale value and
                       ;clock source
OPTION
```

FIGURE 6-5: BLOCK DIAGRAM OF THE TIMER0/WDT PRESCALER

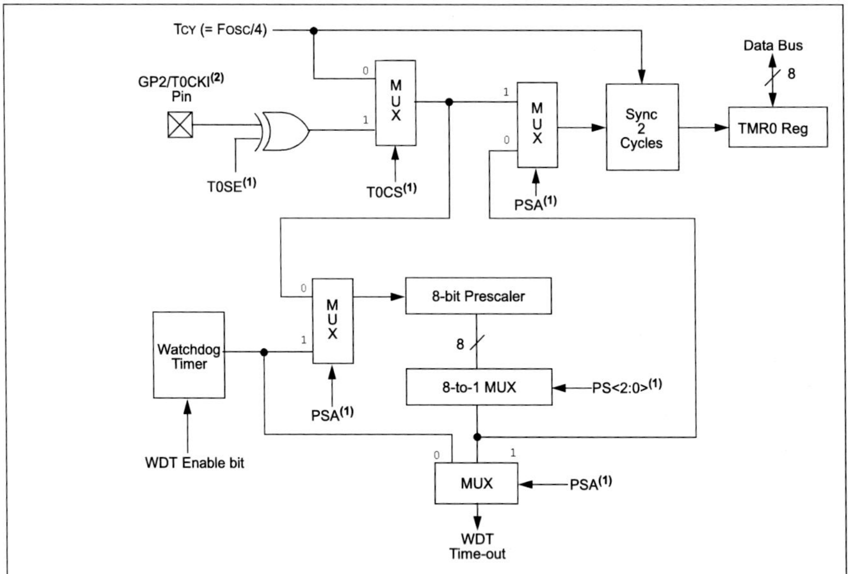

Note 1: T0CS, T0SE, PSA, PS<2:0> are bits in the OPTION register.
2: T0CKI is shared with pin GP2 on the PIC10F200/202/204/206.

7.0 TIMER0 MODULE AND TMR0 REGISTER (PIC10F204/206)

The Timer0 module has the following features:

- 8-bit timer/counter register, TMR0
- Readable and writable
- 8-bit software programmable prescaler
- Internal or external clock select:
  - Edge select for external clock
  - External clock from either the T0CKI pin or from the output of the comparator

Figure 7-1 is a simplified block diagram of the Timer0 module.

Timer mode is selected by clearing the T0CS bit (OPTION<5>). In Timer mode, the Timer0 module will increment every instruction cycle (without prescaler). If TMR0 register is written, the increment is inhibited for the following two cycles (Figure 7-2 and Figure 7-3). The user can work around this by writing an adjusted value to the TMR0 register.

There are two types of Counter mode. The first Counter mode uses the T0CKI pin to increment Timer0. It is selected by setting the T0CS bit (OPTION<5>), setting the CMPT0CS bit (CMCON0<4>) and setting the COUTEN bit (CMCON0<6>). In this mode, Timer0 will increment either on every rising or falling edge of pin T0CKI. The T0SE bit (OPTION<4>) determines the source edge. Clearing the T0SE bit selects the rising edge. Restrictions on the external clock input are discussed in detail in **Section 7.1 "Using Timer0 with an External Clock (PIC10F204/206)"**.

The second Counter mode uses the output of the comparator to increment Timer0. It can be entered in two different ways. The first way is selected by setting the T0CS bit (OPTION<5>) and clearing the CMPT0CS bit (CMCON<4>); (COUTEN [CMCON<6>]) does not affect this mode of operation. This enables an internal connection between the comparator and the Timer0.

The second way is selected by setting the T0CS bit (OPTION<5>), setting the CMPT0CS bit (CMCON0<4>) and clearing the COUTEN bit (CMCON0<6>). This allows the output of the comparator onto the T0CKI pin, while keeping the T0CKI input active. Therefore, any comparator change on the COUT pin is fed back into the T0CKI input. The T0SE bit (OPTION<4>) determines the source edge. Clearing the T0SE bit selects the rising edge. Restrictions on the external clock input as discussed in **Section 7.1 "Using Timer0 with an External Clock (PIC10F204/206)"**

The prescaler may be used by either the Timer0 module or the Watchdog Timer, but not both. The prescaler assignment is controlled in software by the control bit, PSA (OPTION<3>). Clearing the PSA bit will assign the prescaler to Timer0. The prescaler is not readable or writable. When the prescaler is assigned to the Timer0 module, prescale values of 1:2, 1:4,..., 1:256 are selectable. **Section 7.2 "Prescaler"** details the operation of the prescaler.

A summary of registers associated with the Timer0 module is found in Table 7-1.

FIGURE 7-1: TIMER0 BLOCK DIAGRAM (PIC10F204/206)

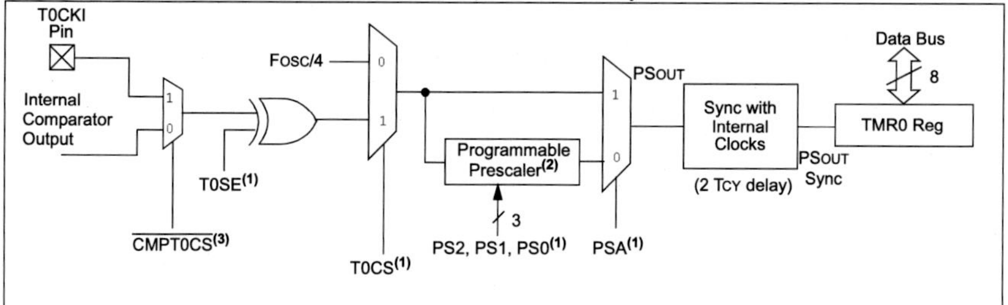

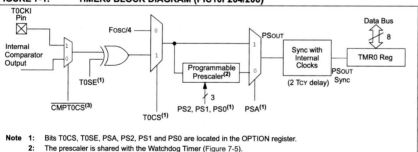

Note 1: Bits T0CS, T0SE, PSA, PS2, PS1 and PS0 are located in the OPTION register.

2: The prescaler is shared with the Watchdog Timer (Figure 7-5).

3: Bit CMPT0CS is located in the CMCON0 register, CMCON0<4>.

FIGURE 7-2: TIMER0 TIMING: INTERNAL CLOCK/NO PRESCALE

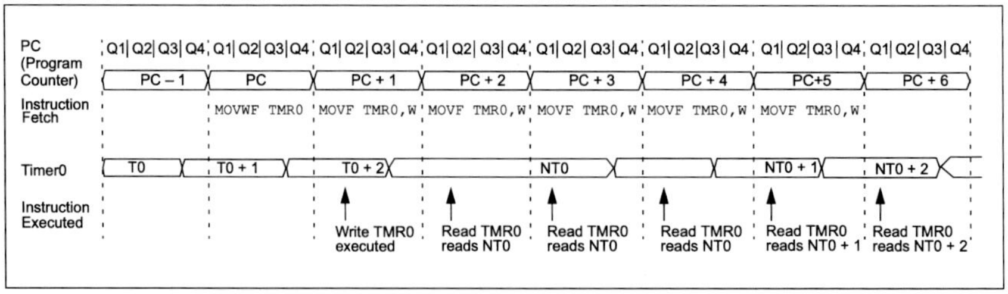

FIGURE 7-3: TIMER0 TIMING: INTERNAL CLOCK/PRESCALE 1:2

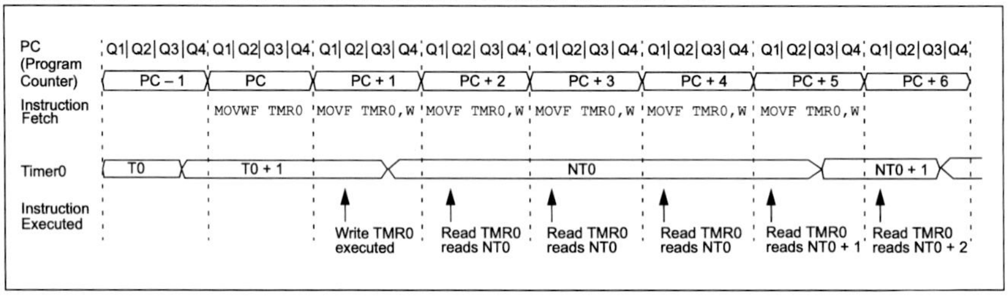

TABLE 7-1: REGISTERS ASSOCIATED WITH TIMER0

Address	Name	Bit 7	Bit 6	Bit 5	Bit 4	Bit 3	Bit 2	Bit 1	Bit 0	Value on Power-On Reset	Value on All Other Resets
01h	TMR0	Timer0 – 8-bit Real-Time Clock/Counter								xxxx xxxx	uuuu uuuu
07h	CMCON0	CMPOUT	COUTEN	POL	CMPT0CS	CMPON	CNREF	CPREF	CWU	1111 1111	uuuu uuuu
N/A	OPTION	GPWU	GPPU	T0CS	T0SE	PSA	PS2	PS1	PS0	1111 1111	1111 1111
N/A	TRISGPIO(1)	—	—	—	—	I/O Control Register				---- 1111	---- 1111

Legend: Shaded cells not used by Timer0. – = unimplemented, x = unknown, u = unchanged.
Note 1: The TRIS of the T0CKI pin is overridden when T0CS = 1.

7.1 Using Timer0 with an External Clock (PIC10F204/206)

When an external clock input is used for Timer0, it must meet certain requirements. The external clock requirement is due to internal phase clock (TOSC) synchronization. Also, there is a delay in the actual incrementing of Timer0 after synchronization.

7.1.1 EXTERNAL CLOCK SYNCHRONIZATION

When no prescaler is used, the external clock input is the same as the prescaler output. The synchronization of an external clock with the internal phase clocks is accomplished by sampling the prescaler output on the Q2 and Q4 cycles of the internal phase clocks (Figure 7-4). Therefore, it is necessary for T0CKI or the comparator output to be high for at least 2 TOSC (and a small RC delay of 2 Tt0H) and low for at least 2 TOSC (and a small RC delay of 2 Tt0H). Refer to the electrical specification of the desired device.

When a prescaler is used, the external clock input is divided by the asynchronous ripple counter type prescaler, so that the prescaler output is symmetrical. For the external clock to meet the sampling requirement, the ripple counter must be taken into account. Therefore, it is necessary for T0CKI or the comparator output to have a period of at least 4 TOSC (and a small RC delay of 4 Tt0H) divided by the prescaler value. The only requirement on T0CKI or the comparator output high and low time is that they do not violate the minimum pulse width requirement of Tt0H. Refer to parameters 40, 41 and 42 in the electrical specification of the desired device.

PIC10F200/202/204/206

7.1.2 TIMER0 INCREMENT DELAY

Since the prescaler output is synchronized with the internal clocks, there is a small delay from the time the external clock edge occurs to the time the Timer0 module is actually incremented. Figure 7-4 shows the delay from the external clock edge to the timer incrementing.

FIGURE 7-4: TIMER0 TIMING WITH EXTERNAL CLOCK

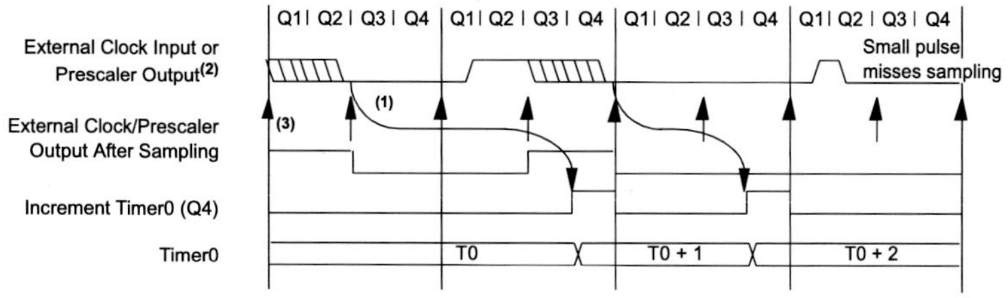

Note 1: Delay from clock input change to Timer0 increment is 3 Tosc to 7 Tosc (Duration of Q = Tosc). Therefore, the error in measuring the interval between two edges on Timer0 input = ±4 Tosc max.

2: External clock if no prescaler selected; prescaler output otherwise.

3: The arrows indicate the points in time where sampling occurs.

7.2 Prescaler

An 8-bit counter is available as a prescaler for the Timer0 module or as a postscaler for the Watchdog Timer (WDT), respectively (see Figure 9-6). For simplicity, this counter is being referred to as "prescaler" throughout this data sheet.

Note: The prescaler may be used by either the Timer0 module or the WDT, but not both. Thus, a prescaler assignment for the Timer0 module means that there is no prescaler for the WDT and vice versa.

The PSA and PS<2:0> bits (OPTION<3:0>) determine prescaler assignment and prescale ratio.

When assigned to the Timer0 module, all instructions writing to the TMR0 register (e.g., `CLRF 1`, `MOVWF 1`, `BSF 1,x`, etc.) will clear the prescaler. When assigned to WDT, a `CLRWDT` instruction will clear the prescaler along with the WDT. The prescaler is neither readable nor writable. On a Reset, the prescaler contains all '0's.

7.2.1 SWITCHING PRESCALER ASSIGNMENT

The prescaler assignment is fully under software control (i.e., it can be changed "on-the-fly" during program execution). To avoid an unintended device Reset, the following instruction sequence (Example 7-1) must be executed when changing the prescaler assignment from Timer0 to the WDT.

EXAMPLE 7-1: CHANGING PRESCALER (TIMER0 → WDT)

```
CLRWDT                 ;Clear WDT
CLRF   TMR0            ;Clear TMR0 & Prescaler
MOVLW  '00xx1111'b     ;These 3 lines (5, 6, 7)
OPTION                 ;are required only if
                       ;desired
CLRWDT                 ;PS<2:0> are 000 or 001
MOVLW  '00xx1xxx'b     ;Set Postscaler to
OPTION                 ;desired WDT rate
```

To change the prescaler from the WDT to the Timer0 module, use the sequence shown in Example 7.2. This sequence must be used even if the WDT is disabled. A `CLRWDT` instruction should be executed before switching the prescaler.

PIC10F200/202/204/206

EXAMPLE 7-2: CHANGING PRESCALER (WDT→TIMER0)

```
CLRWDT                ;Clear WDT and
                      ;prescaler
MOVLW   'xxxx0xxx'    ;Select TMR0, new
                      ;prescale value and
                      ;clock source
OPTION
```

FIGURE 7-5: BLOCK DIAGRAM OF THE TIMER0/WDT PRESCALER

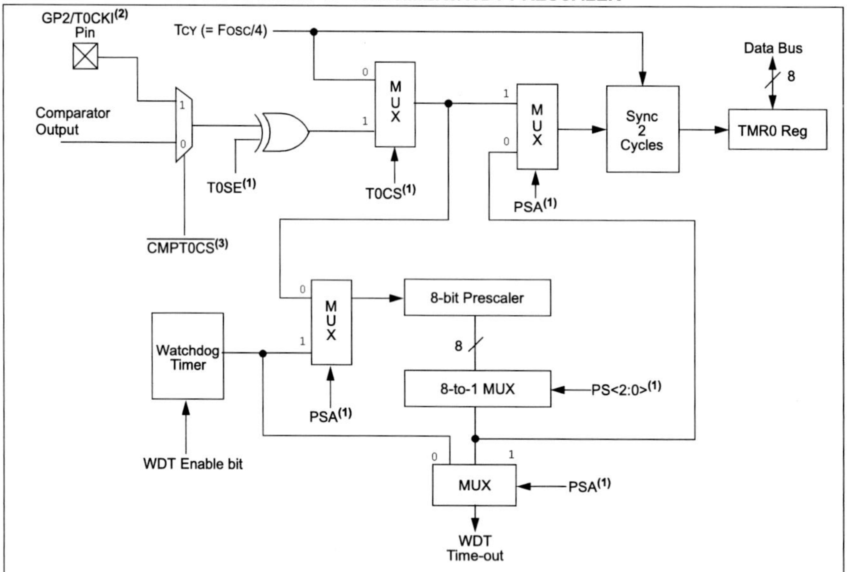

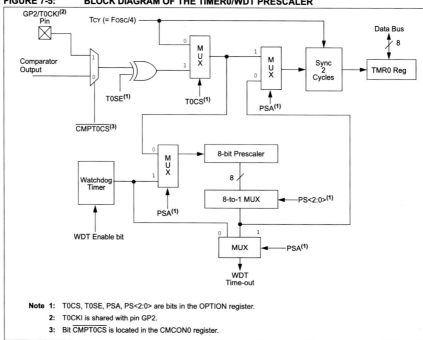

Note 1: T0CS, T0SE, PSA, PS<2:0> are bits in the OPTION register.
2: T0CKI is shared with pin GP2.
3: Bit $\overline{\text{CMPT0CS}}$ is located in the CMCON0 register.

PIC10F200/202/204/206

8.0 COMPARATOR MODULE

The comparator module contains one Analog comparator. The inputs to the comparator are multiplexed with GP0 and GP1 pins. The output of the comparator can be placed on GP2.

The CMCON0 register, shown in Register 8-1, controls the comparator operation. A block diagram of the comparator is shown in Figure 8-1.

REGISTER 8-1: CMCON0 REGISTER

R-1	R/W-1	R/W-1	R/W-1	R/W-1	R/W-1	R/W-1	R/W-1
CMPOUT	$\overline{\text{COUTEN}}$	POL	$\overline{\text{CMPT0CS}}$	CMPON	CNREF	CPREF	$\overline{\text{CWU}}$
bit 7							bit 0

Legend:			
R = Readable bit	W = Writable bit	U = Unimplemented bit, read as ‘0’	
-n = Value at POR	‘1’ = Bit is set	‘0’ = Bit is cleared	x = Bit is unknown

bit 7 **CMPOUT:** Comparator Output bit
1 = $V_{IN+} > V_{IN-}$
0 = $V_{IN+} < V_{IN-}$

bit 6 **$\overline{\text{COUTEN}}$:** Comparator Output Enable bit[1, 2]
1 = Output of comparator is NOT placed on the COUT pin
0 = Output of comparator is placed in the COUT pin

bit 5 **POL:** Comparator Output Polarity bit[2]
1 = Output of comparator not inverted
0 = Output of comparator inverted

bit 4 **$\overline{\text{CMPT0CS}}$:** Comparator TMR0 Clock Source bit[2]
1 = TMR0 clock source selected by T0CS control bit
0 = Comparator output used as TMR0 clock source

bit 3 **CMPON:** Comparator Enable bit
1 = Comparator is on
0 = Comparator is off

bit 2 **CNREF:** Comparator Negative Reference Select bit[2]
1 = CIN- pin[3]
0 = Internal voltage reference

bit 1 **CPREF:** Comparator Positive Reference Select bit[2]
1 = CIN+ pin[3]
0 = CIN- pin[3]

bit 0 **$\overline{\text{CWU}}$:** Comparator Wake-up on Change Enable bit[2]
1 = Wake-up on comparator change is disabled
0 = Wake-up on comparator change is enabled.

Note 1: Overrides T0CS bit for TRIS control of GP2.
2: When the comparator is turned on, these control bits assert themselves. When the comparator is off, these bits have no effect on the device operation and the other control registers have precedence.
3: PIC10F204/206 only.

PIC10F200/202/204/206

8.1 Comparator Configuration

The on-board comparator inputs, (GP0/CIN+, GP1/CIN-), as well as the comparator output (GP2/COUT), are steerable. The CMCON0, OPTION and TRIS registers are used to steer these pins (see Figure 8-1). If the Comparator mode is changed, the comparator output level may not be valid for the specified mode change delay shown in Table 12-1.

Note:	The comparator can have an inverted output (see Figure 8-1).

FIGURE 8-1: BLOCK DIAGRAM OF THE COMPARATOR

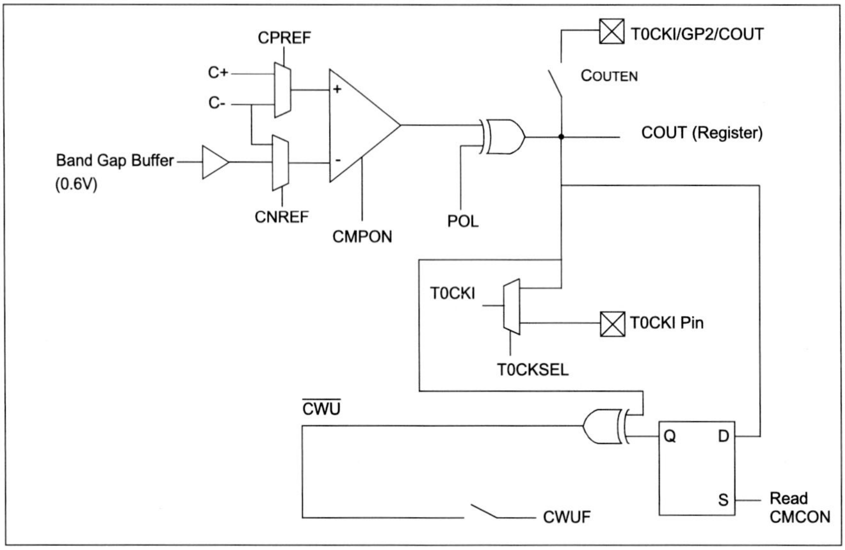

TABLE 8-1: TMR0 CLOCK SOURCE FUNCTION MUXING

T0CS	$\overline{\text{CMPT0CS}}$	$\overline{\text{COUTEN}}$	Source
0	x	x	Internal Instruction Cycle
1	0	0	CMPOUT
1	0	1	CMPOUT
1	1	0	CMPOUT
1	1	1	T0CKI

8.2 Comparator Operation

A single comparator is shown in Figure 8-2 along with the relationship between the analog input levels and the digital output. When the analog input at VIN+ is less than the analog input VIN-, the output of the comparator is a digital low level. When the analog input at VIN+ is greater than the analog input VIN-, the output of the comparator is a digital high level. The shaded areas of the output of the comparator in Figure 8-2 represent the uncertainty due to input offsets and response time. See Table 12-1 for Common Mode Voltage.

FIGURE 8-2: SINGLE COMPARATOR

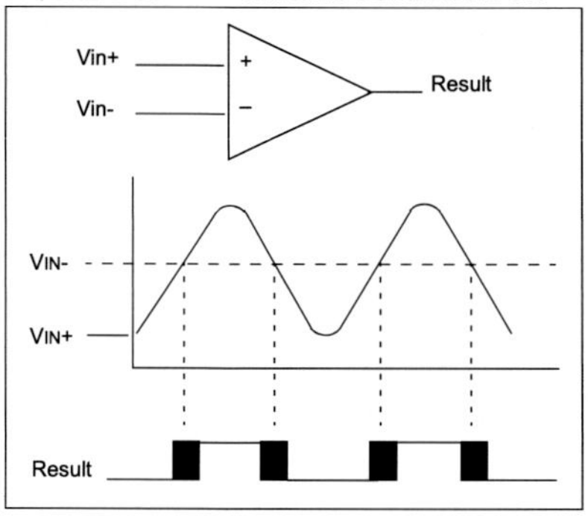

8.3 Comparator Reference

An internal reference signal may be used depending on the Comparator Operating mode. The analog signal that is present at VIN- is compared to the signal at VIN+ and the digital output of the comparator is adjusted accordingly (Figure 8-2). Please see Table 12-1 for internal reference specifications.

8.4 Comparator Response Time

Response time is the minimum time, after selecting a new reference voltage or input source, before the comparator output is to have a valid level. If the comparator inputs are changed, a delay must be used to allow the comparator to settle to its new state. Please see Table 12-1 for comparator response time specifications.

8.5 Comparator Output

The comparator output is read through CMCON0 register. This bit is read-only. The comparator output may also be used internally, see Figure 8-1.

Note:	Analog levels on any pin that is defined as a digital input may cause the input buffer to consume more current than is specified.

8.6 Comparator Wake-up Flag

The comparator wake-up flag is set whenever all of the following conditions are met:

- $\overline{\text{CWU}}$ = `0` (CMCON0<0>)
- CMCON0 has been read to latch the last known state of the CMPOUT bit (`MOVF CMCON0, W`)
- Device is in Sleep
- The output of the comparator has changed state

The wake-up flag may be cleared in software or by another device Reset.

8.7 Comparator Operation During Sleep

When the comparator is active and the device is placed in Sleep mode, the comparator remains active. While the comparator is powered-up, higher Sleep currents than shown in the power-down current specification will occur. To minimize power consumption while in Sleep mode, turn off the comparator before entering Sleep.

8.8 Effects of a Reset

A Power-on Reset (POR) forces the CMCON0 register to its Reset state. This forces the comparator module to be in the comparator Reset mode. This ensures that all potential inputs are analog inputs. Device current is minimized when analog inputs are present at Reset time. The comparator will be powered-down during the Reset interval.

8.9 Analog Input Connection Considerations

A simplified circuit for an analog input is shown in Figure 8-3. Since the analog pins are connected to a digital output, they have reverse biased diodes to VDD and VSS. The analog input therefore, must be between VSS and VDD. If the input voltage deviates from this range by more than 0.6V in either direction, one of the diodes is forward biased and a latch-up may occur. A maximum source impedance of 10 kΩ is recommended for the analog sources. Any external component connected to an analog input pin, such as a capacitor or a Zener diode, should have very little leakage current.

PIC10F200/202/204/206

FIGURE 8-3: ANALOG INPUT MODE

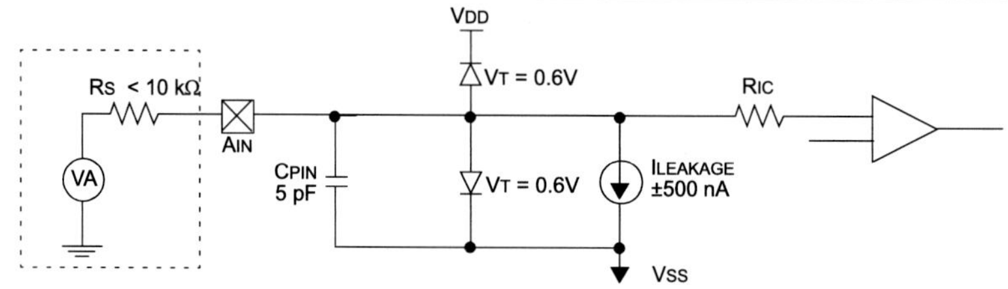

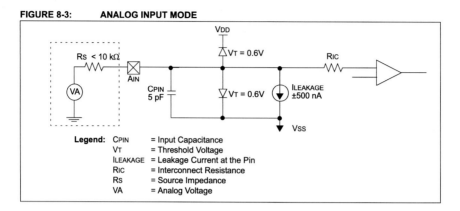

Legend:
CPIN = Input Capacitance
VT = Threshold Voltage
ILEAKAGE = Leakage Current at the Pin
RIC = Interconnect Resistance
RS = Source Impedance
VA = Analog Voltage

TABLE 8-2: REGISTERS ASSOCIATED WITH COMPARATOR MODULE

Address	Name	Bit 7	Bit 6	Bit 5	Bit 4	Bit 3	Bit 2	Bit 1	Bit 0	Value on POR	Value on All Other Resets
03h	STATUS	GPWUF	CWUF	—	$\overline{TO}$	$\overline{PD}$	Z	DC	C	00-1 1xxx	qq0q quuu
07h	CMCON0	CMPOUT	$\overline{COUTEN}$	POL	$\overline{CMPT0CS}$	CMPON	CNREF	CPREF	$\overline{CWU}$	1111 1111	uuuu uuuu
N/A	TRISGPIO	—	—	—	—	I/O Control Register				---- 1111	---- 1111

Legend: x = Unknown, u = Unchanged, – = Unimplemented, read as '0', q = Depends on condition.

PIC10F200/202/204/206

9.0 SPECIAL FEATURES OF THE CPU

What sets a microcontroller apart from other processors are special circuits that deal with the needs of real-time applications. The PIC10F200/202/204/206 microcontrollers have a host of such features intended to maximize system reliability, minimize cost through elimination of external components, provide power-saving operating modes and offer code protection. These features are:

- Reset:
  - Power-on Reset (POR)
  - Device Reset Timer (DRT)
  - Watchdog Timer (WDT)
  - Wake-up from Sleep on pin change
  - Wake-up from Sleep on comparator change
- Sleep
- Code Protection
- ID Locations
- In-Circuit Serial Programming™
- Clock Out

The PIC10F200/202/204/206 devices have a Watchdog Timer, which can be shut off only through Configuration bit WDTE. It runs off of its own RC oscillator for added reliability. When using INTRC, there is an 18 ms delay only on VDD power-up. With this timer on-chip, most applications need no external Reset circuitry.

The Sleep mode is designed to offer a very low-current Power-Down mode. The user can wake-up from Sleep through a change on input pins, wake-up from comparator change, or through a Watchdog Timer time-out.

9.1 Configuration Bits

The PIC10F200/202/204/206 Configuration Words consist of 12 bits. Configuration bits can be programmed to select various device configurations. One bit is the Watchdog Timer enable bit, one bit is the $\overline{MCLR}$ enable bit and one bit is for code protection (see Register 9-1).

REGISTER 9-1: CONFIGURATION WORD FOR PIC10F200/202/204/206(1,2)

R/P-1	R/P-1	R/P-1	R/P-1	R/P-1	R/P-1	R/P-1	R/P-1	R/P-1	R/P-1	R/P-1	R/P-1
—	—	—	—	—	—	—	MCLRE	$\overline{CP}$	WDTE	—	—
bit 11											bit 0

Legend:			
R = Readable bit	W = Writable bit	U = Unimplemented bit, read as '0'	
-n = Value at POR	'1' = Bit is set	'0' = Bit is cleared	x = Bit is unknown

bit 11-5 **Unimplemented:** Read as '0'

bit 4 **MCLRE:** GP3/$\overline{MCLR}$ Pin Function Select bit
1 = GP3/$\overline{MCLR}$ pin function is $\overline{MCLR}$
0 = GP3/$\overline{MCLR}$ pin function is digital I/O, $\overline{MCLR}$ internally tied to VDD

bit 3 **$\overline{CP}$:** Code Protection bit
1 = Code protection off
0 = Code protection on

bit 2 **WDTE:** Watchdog Timer Enable bit
1 = WDT enabled
0 = WDT disabled

bit 1-0 **Reserved:** Read as '0'

Note 1: Refer to the "*PIC10F200/202/204/206 Memory Programming Specifications*" (DS41228) to determine how to access the Configuration Word. The Configuration Word is not user addressable during device operation.

2: INTRC is the only oscillator mode offered on the PIC10F200/202/204/206.

9.2 Oscillator Configurations

9.2.1 OSCILLATOR TYPES

The PIC10F200/202/204/206 devices are offered with Internal Oscillator mode only.

- INTOSC: Internal 4 MHz Oscillator

9.2.2 INTERNAL 4 MHz OSCILLATOR

The internal oscillator provides a 4 MHz (nominal) system clock (see **Section 12.0 "Electrical Characteristics"** for information on variation over voltage and temperature).

In addition, a calibration instruction is programmed into the last address of memory, which contains the calibration value for the internal oscillator. This location is always uncode protected, regardless of the code-protect settings. This value is programmed as a `MOVLW xx` instruction where `xx` is the calibration value and is placed at the Reset vector. This will load the W register with the calibration value upon Reset and the PC will then roll over to the users program at address 0x000. The user then has the option of writing the value to the OSCCAL Register (05h) or ignoring it.

OSCCAL, when written to with the calibration value, will "trim" the internal oscillator to remove process variation from the oscillator frequency.

Note:	Erasing the device will also erase the pre-programmed internal calibration value for the internal oscillator. The calibration value must be read prior to erasing the part so it can be reprogrammed correctly later.

9.3 Reset

The device differentiates between various kinds of Reset:

- Power-on Reset (POR)
- $\overline{\text{MCLR}}$ Reset during normal operation
- $\overline{\text{MCLR}}$ Reset during Sleep
- WDT time-out Reset during normal operation
- WDT time-out Reset during Sleep
- Wake-up from Sleep on pin change
- Wake-up from Sleep on comparator change

Some registers are not reset in any way, they are unknown on POR and unchanged in any other Reset. Most other registers are reset to "Reset state" on Power-on Reset (POR), $\overline{\text{MCLR}}$, WDT or Wake-up on pin change Reset during normal operation. They are not affected by a WDT Reset during Sleep or $\overline{\text{MCLR}}$ Reset during Sleep, since these Resets are viewed as resumption of normal operation. The exceptions to this are $\overline{\text{TO}}$, $\overline{\text{PD}}$, GPWUF and CWUF bits. They are set or cleared differently in different Reset situations. These bits are used in software to determine the nature of Reset. See Table 9-1 for a full description of Reset states of all registers.

TABLE 9-1: RESET CONDITIONS FOR REGISTERS – PIC10F200/202/204/206

Register	Address	Power-on Reset	$\overline{\text{MCLR}}$ Reset, WDT Time-out, Wake-up On Pin Change, Wake on Comparator Change
W	—	qqqq qqqu(1)	qqqq qqqu(1)
INDF	00h	xxxx xxxx	uuuu uuuu
TMR0	01h	xxxx xxxx	uuuu uuuu
PCL	02h	1111 1111	1111 1111
STATUS	03h	00-1 1xxx	q00q quuu(2)
STATUS(3)	03h	00-1 1xxx	qq0q quuu(2)
FSR	04h	111x xxxx	111u uuuu
OSCCAL	05h	1111 1110	uuuu uuuu
GPIO	06h	---- xxxx	---- uuuu
CMCON(3)	07h	1111 1111	uuuu uuuu
OPTION	—	1111 1111	1111 1111
TRISGPIO	—	---- 1111	---- 1111

Legend: u = unchanged, x = unknown, – = unimplemented bit, read as '0', q = value depends on condition.
Note 1: Bits <7:2> of W register contain oscillator calibration values due to `MOVLW XX` instruction at top of memory.
2: See Table 9-2 for Reset value for specific conditions.
3: PIC10F204/206 only.

TABLE 9-2: RESET CONDITION FOR SPECIAL REGISTERS

—	STATUS Address: 03h	PCL Address: 02h
Power-on Reset	00-1 1xxx	1111 1111
MCLR Reset during normal operation	000u uuuu	1111 1111
MCLR Reset during Sleep	0001 0uuu	1111 1111
WDT Reset during Sleep	0000 0uuu	1111 1111
WDT Reset normal operation	0000 uuuu	1111 1111
Wake-up from Sleep on pin change	1001 0uuu	1111 1111
Wake-up from Sleep on comparator change	0101 0uuu	1111 1111

Legend: u = unchanged, x = unknown, – = unimplemented bit, read as '0'.

9.3.1 MCLR ENABLE

This Configuration bit, when unprogrammed (left in the '1' state), enables the external MCLR function. When programmed, the MCLR function is tied to the internal VDD and the pin is assigned to be a I/O. See Figure 9-1.

FIGURE 9-1: MCLR SELECT

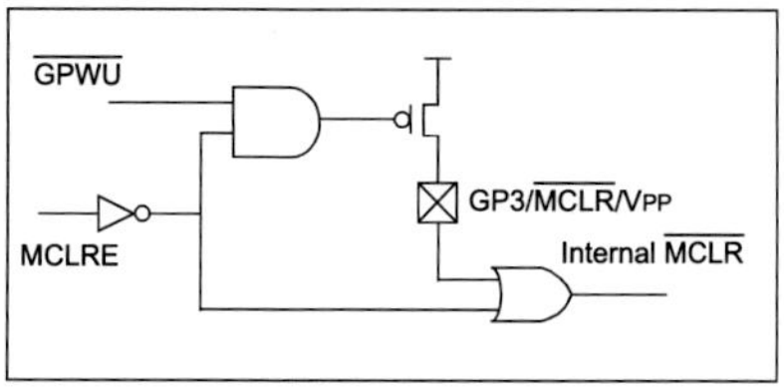

9.4 Power-on Reset (POR)

The PIC10F200/202/204/206 devices incorporate an on-chip Power-on Reset (POR) circuitry, which provides an internal chip Reset for most power-up situations.

The on-chip POR circuit holds the chip in Reset until VDD has reached a high enough level for proper operation. To take advantage of the internal POR, program the GP3/MCLR/VPP pin as MCLR and tie through a resistor to VDD, or program the pin as GP3. An internal weak pull-up resistor is implemented using a transistor (refer to Table 12-2 for the pull-up resistor ranges). This will eliminate external RC components usually needed to create a Power-on Reset. A maximum rise time for VDD is specified. See **Section 12.0 "Electrical Characteristics"** for details.

When the devices start normal operation (exit the Reset condition), device operating parameters (voltage, frequency, temperature,...) must be met to ensure operation. If these conditions are not met, the devices must be held in Reset until the operating parameters are met.

A simplified block diagram of the on-chip Power-on Reset circuit is shown in Figure 9-2.

The Power-on Reset circuit and the Device Reset Timer (see **Section 9.5 "Device Reset Timer (DRT)"**) circuit are closely related. On power-up, the Reset latch is set and the DRT is reset. The DRT timer begins counting once it detects MCLR to be high. After the time-out period, which is typically 18 ms, it will reset the Reset latch and thus end the on-chip Reset signal.

A power-up example where MCLR is held low is shown in Figure 9-3. VDD is allowed to rise and stabilize before bringing MCLR high. The chip will actually come out of Reset TDRT msec after MCLR goes high.

In Figure 9-4, the on-chip Power-on Reset feature is being used (MCLR and VDD are tied together or the pin is programmed to be GP3). The VDD is stable before the Start-up Timer times out and there is no problem in getting a proper Reset. However, Figure 9-5 depicts a problem situation where VDD rises too slowly. The time between when the DRT senses that MCLR is high and when MCLR and VDD actually reach their full value, is too long. In this situation, when the Start-up Timer times out, VDD has not reached the VDD (min) value and the chip may not function correctly. For such situations, we recommend that external RC circuits be used to achieve longer POR delay times (Figure 9-4).

> **Note:** When the devices start normal operation (exit the Reset condition), device operating parameters (voltage, frequency, temperature, etc.) must be met to ensure operation. If these conditions are not met, the device must be held in Reset until the operating conditions are met.

For additional information, refer to Application Notes AN522 *"Power-up Considerations"*, (DS00522) and AN607 *"Power-up Trouble Shooting"*, (DS00000607).

PIC10F200/202/204/206

FIGURE 9-2: SIMPLIFIED BLOCK DIAGRAM OF ON-CHIP RESET CIRCUIT

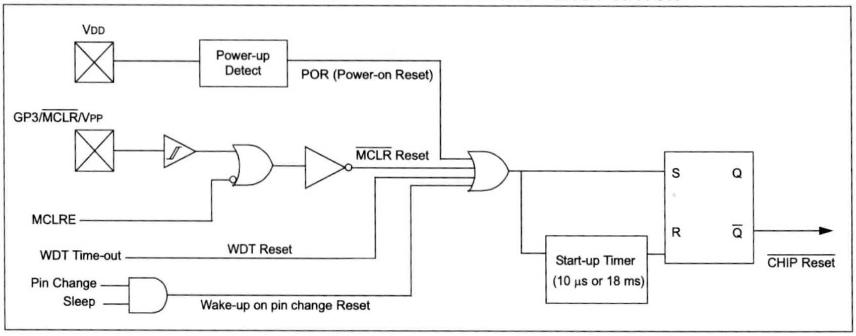

FIGURE 9-3: TIME-OUT SEQUENCE ON POWER-UP (MCLR PULLED LOW)

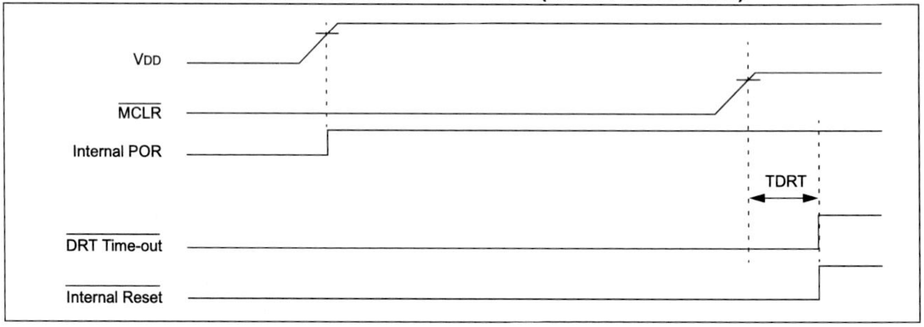

FIGURE 9-4: TIME-OUT SEQUENCE ON POWER-UP (MCLR TIED TO VDD): FAST VDD RISE TIME

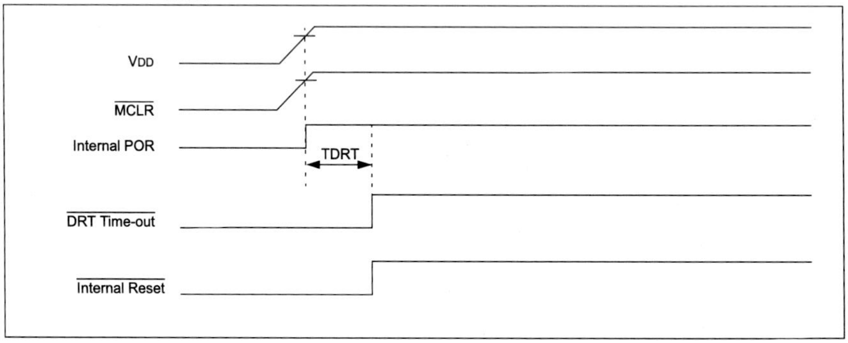

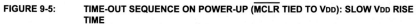

PIC10F200/202/204/206

FIGURE 9-5: TIME-OUT SEQUENCE ON POWER-UP ($\overline{MCLR}$ TIED TO VDD): SLOW VDD RISE TIME

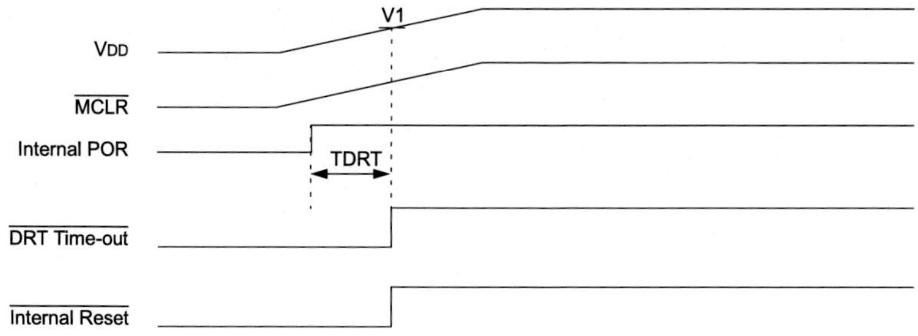

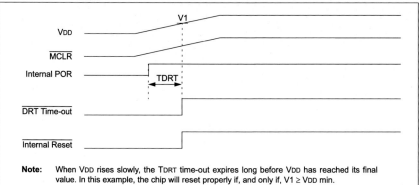

Note: When VDD rises slowly, the TDRT time-out expires long before VDD has reached its final value. In this example, the chip will reset properly if, and only if, V1 ≥ VDD min.

PIC10F200/202/204/206

9.5 Device Reset Timer (DRT)

On the PIC10F200/202/204/206 devices, the DRT runs any time the device is powered-up.

The DRT operates on an internal oscillator. The processor is kept in Reset as long as the DRT is active. The DRT delay allows V_{DD} to rise above V_{DD} min. and for the oscillator to stabilize.

The on-chip DRT keeps the devices in a Reset condition for approximately 18 ms after $\overline{MCLR}$ has reached a logic high (V_{IH} $\overline{MCLR}$) level. Programming GP3/$\overline{MCLR}$/V_{PP} as $\overline{MCLR}$ and using an external RC network connected to the $\overline{MCLR}$ input is not required in most cases. This allows savings in cost-sensitive and/ or space restricted applications, as well as allowing the use of the GP3/$\overline{MCLR}$/V_{PP} pin as a general purpose input.

The Device Reset Time delays will vary from chip-to-chip due to V_{DD}, temperature and process variation. See AC parameters for details.

Reset sources are POR, $\overline{MCLR}$, WDT time-out and wake-up on pin change. See **Section 9.9.2 "Wake-up from Sleep"**, Notes 1, 2 and 3.

TABLE 9-3: DRT PERIOD

Oscillator	POR Reset	Subsequent Resets
INTOSC	18 ms (typical)	10 μs (typical)

9.6 Watchdog Timer (WDT)

The Watchdog Timer (WDT) is a free running on-chip RC oscillator, which does not require any external components. This RC oscillator is separate from the internal 4 MHz oscillator. This means that the WDT will run even if the main processor clock has been stopped, for example, by execution of a `SLEEP` instruction. During normal operation or Sleep, a WDT Reset or wake-up Reset, generates a device Reset.

The $\overline{TO}$ bit (STATUS<4>) will be cleared upon a Watchdog Timer Reset.

The WDT can be permanently disabled by programming the configuration WDTE as a '0' (see **Section 9.1 "Configuration Bits"**). Refer to the PIC10F200/202/204/206 Programming Specifications to determine how to access the Configuration Word.

9.6.1 WDT PERIOD

The WDT has a nominal time-out period of 18 ms, (with no prescaler). If a longer time-out period is desired, a prescaler with a division ratio of up to 1:128 can be assigned to the WDT (under software control) by writing to the OPTION register. Thus, a time-out period of a nominal 2.3 seconds can be realized. These periods vary with temperature, V_{DD} and part-to-part process variations (see DC specs).

Under worst-case conditions (V_{DD} = Min., Temperature = Max., max. WDT prescaler), it may take several seconds before a WDT time-out occurs.

9.6.2 WDT PROGRAMMING CONSIDERATIONS

The `CLRWDT` instruction clears the WDT and the postscaler, if assigned to the WDT, and prevents it from timing out and generating a device Reset.

The `SLEEP` instruction resets the WDT and the postscaler, if assigned to the WDT. This gives the maximum Sleep time before a WDT wake-up Reset.

PIC10F200/202/204/206

FIGURE 9-6: WATCHDOG TIMER BLOCK DIAGRAM

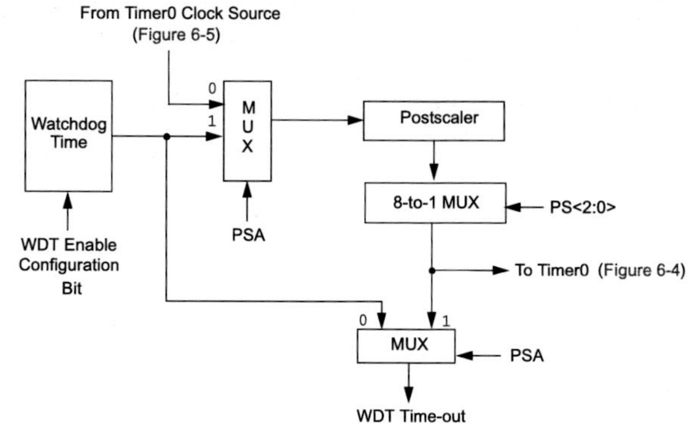

TABLE 9-4: SUMMARY OF REGISTERS ASSOCIATED WITH THE WATCHDOG TIMER

Address	Name	Bit 7	Bit 6	Bit 5	Bit 4	Bit 3	Bit 2	Bit 1	Bit 0	Value on Power-On Reset	Value on All Other Resets
N/A	OPTION	$\overline{GPWU}$	$\overline{GPPU}$	T0CS	T0SE	PSA	PS2	PS1	PS0	1111 1111	1111 1111

Legend: Shaded boxes = Not used by Watchdog Timer, – = unimplemented, read as '0', u = unchanged.

9.7 Time-out Sequence, Power-down and Wake-up from Sleep Status Bits ($\overline{\text{TO}}$, $\overline{\text{PD}}$, GPWUF, CWUF)

The $\overline{\text{TO}}$, $\overline{\text{PD}}$, GPWUF and CWUF bits in the STATUS register can be tested to determine if a Reset condition has been caused by a power-up condition, a $\overline{\text{MCLR}}$, Watchdog Timer (WDT) Reset, wake-up on comparator change or wake-up on pin change.

TABLE 9-5: $\overline{\text{TO}}$, $\overline{\text{PD}}$, GPWUF, CWUF STATUS AFTER RESET

CWUF	GPWUF	$\overline{\text{TO}}$	$\overline{\text{PD}}$	Reset Caused By
0	0	0	0	WDT wake-up from Sleep
0	0	0	u	WDT time-out (not from Sleep)
0	0	1	0	$\overline{\text{MCLR}}$ wake-up from Sleep
0	0	1	1	Power-up
0	0	u	u	$\overline{\text{MCLR}}$ not during Sleep
0	1	1	0	Wake-up from Sleep on pin change
1	0	1	0	Wake-up from Sleep on comparator change

Legend: u = unchanged, x = unknown, – = unimplemented bit, read as '0', q = value depends on condition.

Note 1: The $\overline{\text{TO}}$, $\overline{\text{PD}}$, GPWUF and CWUF bits maintain their status (u) until a Reset occurs. A low-pulse on the $\overline{\text{MCLR}}$ input does not change the $\overline{\text{TO}}$, $\overline{\text{PD}}$, GPWUF or CWUF Status bits.

9.8 Reset on Brown-out

A Brown-out Reset is a condition where device power (VDD) dips below its minimum value, but not to zero, and then recovers. The device should be reset in the event of a brown-out.

To reset PIC10F200/202/204/206 devices when a Brown-out Reset occurs, external brown-out protection circuits may be built, as shown in Figure 9-7 and Figure 9-8.

FIGURE 9-7: BROWN-OUT PROTECTION CIRCUIT 1

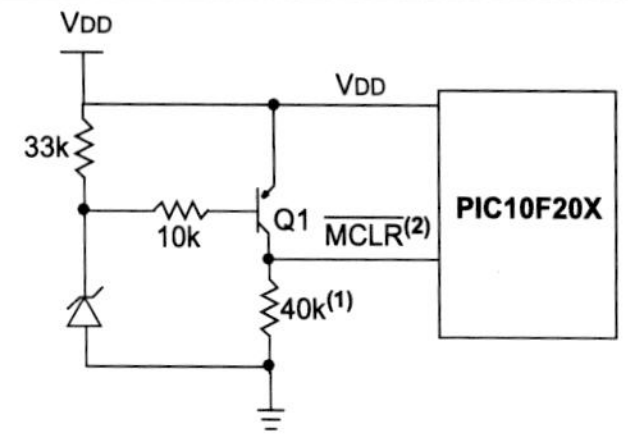

Note 1: This circuit will activate Reset when VDD goes below Vz + 0.7V (where Vz = Zener voltage).

2: Pin must be confirmed as $\overline{\text{MCLR}}$.

FIGURE 9-8: BROWN-OUT PROTECTION CIRCUIT 2

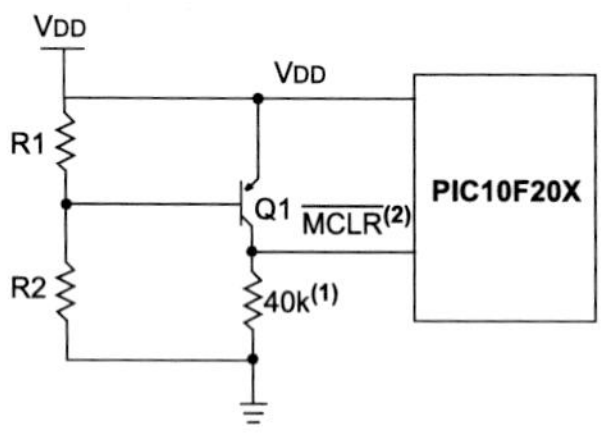

Note 1: This brown-out circuit is less expensive, although less accurate. Transistor Q1 turns off when VDD is below a certain level such that:

$$V_{DD} \cdot \frac{R1}{R1 + R2} = 0.7V$$

2: Pin must be confirmed as $\overline{\text{MCLR}}$.

FIGURE 9-9: BROWN-OUT PROTECTION CIRCUIT 3

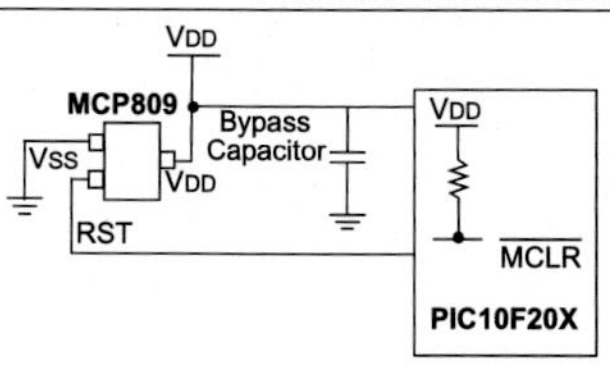

Note: This brown-out protection circuit employs Microchip Technology's MCP809 microcontroller supervisor. There are seven different trip point selections to accommodate 5V to 3V systems.

9.9 Power-down Mode (Sleep)

A device may be powered-down (Sleep) and later powered-up (wake-up from Sleep).

9.9.1 SLEEP

The Power-down mode is entered by executing a `SLEEP` instruction.

If enabled, the Watchdog Timer will be cleared but keeps running, the $\overline{\text{TO}}$ bit (STATUS<4>) is set, the $\overline{\text{PD}}$ bit (STATUS<3>) is cleared and the oscillator driver is turned off. The I/O ports maintain the status they had before the `SLEEP` instruction was executed (driving high, driving low or high-impedance).

> **Note:** A Reset generated by a WDT time-out does not drive the $\overline{\text{MCLR}}$ pin low.

For lowest current consumption while powered-down, the T0CKI input should be at VDD or VSS and the GP3/$\overline{\text{MCLR}}$/VPP pin must be at a logic high level if $\overline{\text{MCLR}}$ is enabled.

9.9.2 WAKE-UP FROM SLEEP

The device can wake-up from Sleep through one of the following events:

1. An external Reset input on GP3/$\overline{\text{MCLR}}$/VPP pin, when configured as $\overline{\text{MCLR}}$.
2. A Watchdog Timer time-out Reset (if WDT was enabled).
3. A change on input pin GP0, GP1 or GP3 when wake-up on change is enabled.
4. A comparator output change has occurred when wake-up on comparator change is enabled.

These events cause a device Reset. The $\overline{\text{TO}}$, $\overline{\text{PD}}$ GPWUF and CWUF bits can be used to determine the cause of device Reset. The $\overline{\text{TO}}$ bit is cleared if a WDT time-out occurred (and caused wake-up). The $\overline{\text{PD}}$ bit, which is set on power-up, is cleared when `SLEEP` is invoked. The GPWUF bit indicates a change in state while in Sleep at pins GP0, GP1 or GP3 (since the last file or bit operation on GP port). The CWUF bit indicates a change in the state while in Sleep of the comparator output.

> **Caution:** Right before entering Sleep, read the input pins. When in Sleep, wake-up occurs when the values at the pins change from the state they were in at the last reading. If a wake-up on change occurs and the pins are not read before re-entering Sleep, a wake-up will occur immediately even if no pins change while in Sleep mode.

> **Note:** The WDT is cleared when the device wakes from Sleep, regardless of the wake-up source.

PIC10F200/202/204/206

9.10 Program Verification/Code Protection

If the code protection bit has not been programmed, the on-chip program memory can be read out for verification purposes.

The first 64 locations and the last location (Reset vector) can be read, regardless of the code protection bit setting.

9.11 ID Locations

Four memory locations are designated as ID locations where the user can store checksum or other code identification numbers. These locations are not accessible during normal execution, but are readable and writable during Program/Verify.

Use only the lower four bits of the ID locations and always program the upper eight bits as '0's.

9.12 In-Circuit Serial Programming™

The PIC10F200/202/204/206 microcontrollers can be serially programmed while in the end application circuit. This is simply done with two lines for clock and data, and three other lines for power, ground and the programming voltage. This allows customers to manufacture boards with unprogrammed devices and then program the microcontroller just before shipping the product. This also allows the most recent firmware or a custom firmware, to be programmed.

The devices are placed into a Program/Verify mode by holding the GP1 and GP0 pins low while raising the $\overline{MCLR}$ (V_{PP}) pin from V_{IL} to V_{IHH} (see programming specification). GP1 becomes the programming clock and GP0 becomes the programming data. Both GP1 and GP0 are Schmitt Trigger inputs in this mode.

After Reset, a 6-bit command is then supplied to the device. Depending on the command, 16 bits of program data are then supplied to or from the device, depending if the command was a Load or a Read. For complete details of serial programming, please refer to the PIC10F200/202/204/206 Programming Specifications.

A typical In-Circuit Serial Programming connection is shown in Figure 9-10.

FIGURE 9-10: TYPICAL IN-CIRCUIT SERIAL PROGRAMMING™ CONNECTION

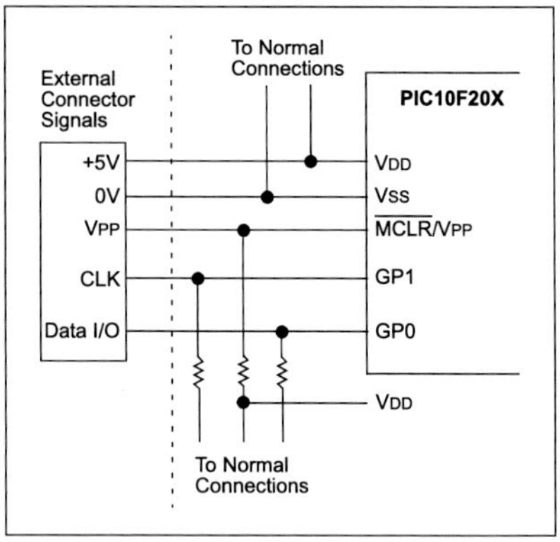

PIC10F200/202/204/206

10.0 INSTRUCTION SET SUMMARY

The PIC16 instruction set is highly orthogonal and is comprised of three basic categories.

- **Byte-oriented** operations
- **Bit-oriented** operations
- **Literal and control** operations

Each PIC16 instruction is a 12-bit word divided into an **opcode**, which specifies the instruction type and one or more **operands** which further specify the operation of the instruction. The formats for each of the categories is presented in Figure 10-1, while the various opcode fields are summarized in Table 10-1.

For **byte-oriented** instructions, 'f' represents a file register designator and 'd' represents a destination designator. The file register designator specifies which file register is to be used by the instruction.

The destination designator specifies where the result of the operation is to be placed. If 'd' is '0', the result is placed in the W register. If 'd' is '1', the result is placed in the file register specified in the instruction.

For **bit-oriented** instructions, 'b' represents a bit field designator which selects the number of the bit affected by the operation, while 'f' represents the number of the file in which the bit is located.

For **literal and control** operations, 'k' represents an 8 or 9-bit constant or literal value.

TABLE 10-1: OPCODE FIELD DESCRIPTIONS

Field	Description
f	Register file address (0x00 to 0x7F)
W	Working register (accumulator)
b	Bit address within an 8-bit file register
k	Literal field, constant data or label
x	Don't care location (= 0 or 1) The assembler will generate code with x = 0. It is the recommended form of use for compatibility with all Microchip software tools.
d	Destination select; d = 0 (store result in W) d = 1 (store result in file register 'f') Default is d = 1
label	Label name
TOS	Top-of-Stack
PC	Program Counter
WDT	Watchdog Timer counter
$\overline{TO}$	Time-out bit
$\overline{PD}$	Power-down bit
dest	Destination, either the W register or the specified register file location
[ ]	Options
( )	Contents
→	Assigned to
< >	Register bit field
∈	In the set of
italics	User defined term (font is courier)

All instructions are executed within a single instruction cycle, unless a conditional test is true or the program counter is changed as a result of an instruction. In this case, the execution takes two instruction cycles. One instruction cycle consists of four oscillator periods. Thus, for an oscillator frequency of 4 MHz, the normal instruction execution time is 1 μs. If a conditional test is true or the program counter is changed as a result of an instruction, the instruction execution time is 2 μs.

Figure 10-1 shows the three general formats that the instructions can have. All examples in the figure use the following format to represent a hexadecimal number:

```
0xhhh
```

where 'h' signifies a hexadecimal digit.

FIGURE 10-1: GENERAL FORMAT FOR INSTRUCTIONS

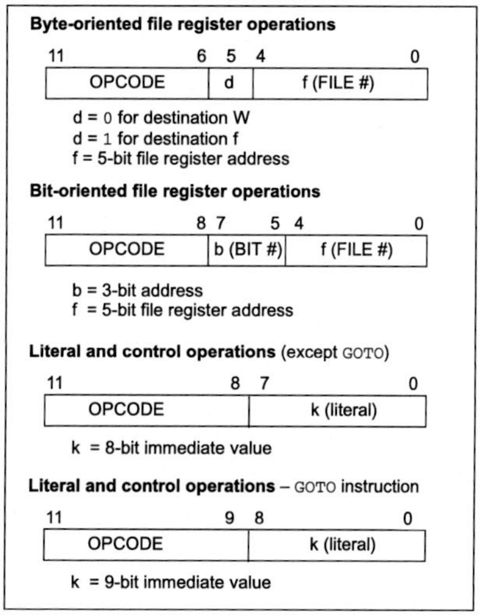

PIC10F200/202/204/206

TABLE 10-2: INSTRUCTION SET SUMMARY

Mnemonic, Operands		Description	Cycles	12-Bit Opcode MSb		LSb	Status Affected	Notes
ADDWF	f, d	Add W and f	1	0001	11df	ffff	C, DC, Z	1, 2, 4
ANDWF	f, d	AND W with f	1	0001	01df	ffff	Z	2, 4
CLRF	f	Clear f	1	0000	011f	ffff	Z	4
CLRW	—	Clear W	1	0000	0100	0000	Z	
COMF	f, d	Complement f	1	0010	01df	ffff	Z	
DECF	f, d	Decrement f	1	0000	11df	ffff	Z	2, 4
DECFSZ	f, d	Decrement f, Skip if 0	1(2)	0010	11df	ffff	None	2, 4
INCF	f, d	Increment f	1	0010	10df	ffff	Z	2, 4
INCFSZ	f, d	Increment f, Skip if 0	1(2)	0011	11df	ffff	None	2, 4
IORWF	f, d	Inclusive OR W with f	1	0001	00df	ffff	Z	2, 4
MOVF	f, d	Move f	1	0010	00df	ffff	Z	2, 4
MOVWF	f	Move W to f	1	0000	001f	ffff	None	1, 4
NOP	—	No Operation	1	0000	0000	0000	None	
RLF	f, d	Rotate left f through Carry	1	0011	01df	ffff	C	2, 4
RRF	f, d	Rotate right f through Carry	1	0011	00df	ffff	C	2, 4
SUBWF	f, d	Subtract W from f	1	0000	10df	ffff	C, DC, Z	1, 2, 4
SWAPF	f, d	Swap f	1	0011	10df	ffff	None	2, 4
XORWF	f, d	Exclusive OR W with f	1	0001	10df	ffff	Z	2, 4
BIT-ORIENTED FILE REGISTER OPERATIONS								
BCF	f, b	Bit Clear f	1	0100	bbbf	ffff	None	2, 4
BSF	f, b	Bit Set f	1	0101	bbbf	ffff	None	2, 4
BTFSC	f, b	Bit Test f, Skip if Clear	1(2)	0110	bbbf	ffff	None	
BTFSS	f, b	Bit Test f, Skip if Set	1(2)	0111	bbbf	ffff	None	
LITERAL AND CONTROL OPERATIONS								
ANDLW	k	AND literal with W	1	1110	kkkk	kkkk	Z	
CALL	k	Call Subroutine	2	1001	kkkk	kkkk	None	1
CLRWDT		Clear Watchdog Timer	1	0000	0000	0100	$\overline{TO}$, $\overline{PD}$	
GOTO	k	Unconditional branch	2	101k	kkkk	kkkk	None	
IORLW	k	Inclusive OR literal with W	1	1101	kkkk	kkkk	Z	
MOVLW	k	Move literal to W	1	1100	kkkk	kkkk	None	
OPTION	—	Load OPTION register	1	0000	0000	0010	None	
RETLW	k	Return, place Literal in W	2	1000	kkkk	kkkk	None	
SLEEP	—	Go into Standby mode	1	0000	0000	0011	$\overline{TO}$, $\overline{PD}$	
TRIS	f	Load TRIS register	1	0000	0000	0fff	None	3
XORLW	k	Exclusive OR literal to W	1	1111	kkkk	kkkk	Z	

Note 1: The 9th bit of the program counter will be forced to a '0' by any instruction that writes to the PC except for `GOTO`. See **Section 4.7 "Program Counter"**.

2: When an I/O register is modified as a function of itself (e.g. `MOVF PORTB, 1`), the value used will be that value present on the pins themselves. For example, if the data latch is '1' for a pin configured as input and is driven low by an external device, the data will be written back with a '0'.

3: The instruction `TRIS f`, where f = 6, causes the contents of the W register to be written to the tri-state latches of PORTB. A '1' forces the pin to a high-impedance state and disables the output buffers.

4: If this instruction is executed on the TMR0 register (and where applicable, d = 1), the prescaler will be cleared (if assigned to TMR0).

PIC10F200/202/204/206

ADDWF	Add W and f
Syntax:	[*label*] ADDWF f,d
Operands:	$0 \le f \le 31$ $d \in [0,1]$
Operation:	(W) + (f) → (dest)
Status Affected:	C, DC, Z
Description:	Add the contents of the W register and register 'f'. If 'd' is '0', the result is stored in the W register. If 'd' is '1', the result is stored back in register 'f'.

ANDLW	AND literal with W
Syntax:	[*label*] ANDLW k
Operands:	$0 \le k \le 255$
Operation:	(W).AND. (k) → (W)
Status Affected:	Z
Description:	The contents of the W register are AND'ed with the 8-bit literal 'k'. The result is placed in the W register.

ANDWF	AND W with f
Syntax:	[*label*] ANDWF f,d
Operands:	$0 \le f \le 31$ $d \in [0,1]$
Operation:	(W) .AND. (f) → (dest)
Status Affected:	Z
Description:	The contents of the W register are AND'ed with register 'f'. If 'd' is '0', the result is stored in the W register. If 'd' is '1', the result is stored back in register 'f'.

BCF	Bit Clear f
Syntax:	[*label*] BCF f,b
Operands:	$0 \le f \le 31$ $0 \le b \le 7$
Operation:	0 → (f<b>)
Status Affected:	None
Description:	Bit 'b' in register 'f' is cleared.

BSF	Bit Set f
Syntax:	[*label*] BSF f,b
Operands:	$0 \le f \le 31$ $0 \le b \le 7$
Operation:	1 → (f<b>)
Status Affected:	None
Description:	Bit 'b' in register 'f' is set.

BTFSC	Bit Test f, Skip if Clear
Syntax:	[*label*] BTFSC f,b
Operands:	$0 \le f \le 31$ $0 \le b \le 7$
Operation:	skip if (f<b>) = 0
Status Affected:	None
Description:	If bit 'b' in register 'f' is '0', then the next instruction is skipped. If bit 'b' is '0', then the next instruction fetched during the current instruction execution is discarded, and a NOP is executed instead, making this a 2-cycle instruction.

PIC10F200/202/204/206

BTFSS	**Bit Test f, Skip if Set**
Syntax:	[*label*] BTFSS f,b
Operands:	$0 \le f \le 31$ $0 \le b < 7$
Operation:	skip if (f<b>) = `1`
Status Affected:	None
Description:	If bit ‘b’ in register ‘f’ is ‘`1`’, then the next instruction is skipped. If bit ‘b’ is ‘`1`’, then the next instruction fetched during the current instruction execution, is discarded and a `NOP` is executed instead, making this a 2-cycle instruction.

CLRW	**Clear W**
Syntax:	[*label*] CLRW
Operands:	None
Operation:	00h → (W); 1 → Z
Status Affected:	Z
Description:	The W register is cleared. Zero bit (Z) is set.

CALL	**Subroutine Call**
Syntax:	[*label*] CALL k
Operands:	$0 \le k \le 255$
Operation:	(PC) + 1→ Top-of-Stack; k → PC<7:0>; (STATUS<6:5>) → PC<10:9>; 0 → PC<8>
Status Affected:	None
Description:	Subroutine call. First, return address (PC + 1) is PUSHed onto the stack. The 8-bit immediate address is loaded into PC bits <7:0>. The upper bits PC<10:9> are loaded from STATUS<6:5>, PC<8> is cleared. `CALL` is a 2-cycle instruction.

CLRWDT	**Clear Watchdog Timer**
Syntax:	[*label*] CLRWDT
Operands:	None
Operation:	00h → WDT; 0 → WDT prescaler (if assigned); 1 → $\overline{TO}$; 1 → $\overline{PD}$
Status Affected:	$\overline{TO}$, $\overline{PD}$
Description:	The `CLRWDT` instruction resets the WDT. It also resets the prescaler, if the prescaler is assigned to the WDT and not Timer0. Status bits $\overline{TO}$ and $\overline{PD}$ are set.

CLRF	**Clear f**
Syntax:	[*label*] CLRF f
Operands:	$0 \le f \le 31$
Operation:	00h → (f); 1 → Z
Status Affected:	Z
Description:	The contents of register ‘f’ are cleared and the Z bit is set.

COMF	**Complement f**
Syntax:	[*label*] COMF f,d
Operands:	$0 \le f \le 31$ $d \in$ [`0`,`1`]
Operation:	$(\bar{f})$ → (dest)
Status Affected:	Z
Description:	The contents of register ‘f’ are complemented. If ‘d’ is ‘`0`’, the result is stored in the W register. If ‘d’ is ‘`1`’, the result is stored back in register ‘f’.

PIC10F200/202/204/206

DECF	Decrement f
Syntax:	[*label*] DECF f,d
Operands:	$0 \leq f \leq 31$ $d \in [0,1]$
Operation:	(f) – 1 → (dest)
Status Affected:	Z
Description:	Decrement register 'f'. If 'd' is '0', the result is stored in the W register. If 'd' is '1', the result is stored back in register 'f'.

DECFSZ	Decrement f, Skip if 0
Syntax:	[*label*] DECFSZ f,d
Operands:	$0 \leq f \leq 31$ $d \in [0,1]$
Operation:	(f) – 1 → d; skip if result = 0
Status Affected:	None
Description:	The contents of register 'f' are decremented. If 'd' is '0', the result is placed in the W register. If 'd' is '1', the result is placed back in register 'f'. If the result is '0', the next instruction, which is already fetched, is discarded and a NOP is executed instead making it a 2-cycle instruction.

GOTO	Unconditional Branch
Syntax:	[*label*] GOTO k
Operands:	$0 \leq k \leq 511$
Operation:	k → PC<8:0>; STATUS<6:5> → PC<10:9>
Status Affected:	None
Description:	GOTO is an unconditional branch. The 9-bit immediate value is loaded into PC bits <8:0>. The upper bits of PC are loaded from STATUS<6:5>. GOTO is a 2-cycle instruction.

INCF	Increment f
Syntax:	[*label*] INCF f,d
Operands:	$0 \leq f \leq 31$ $d \in [0,1]$
Operation:	(f) + 1 → (dest)
Status Affected:	Z
Description:	The contents of register 'f' are incremented. If 'd' is '0', the result is placed in the W register. If 'd' is '1', the result is placed back in register 'f'.

INCFSZ	Increment f, Skip if 0
Syntax:	[*label*] INCFSZ f,d
Operands:	$0 \leq f \leq 31$ $d \in [0,1]$
Operation:	(f) + 1 → (dest), skip if result = 0
Status Affected:	None
Description:	The contents of register 'f' are incremented. If 'd' is '0', the result is placed in the W register. If 'd' is '1', the result is placed back in register 'f'. If the result is '0', then the next instruction, which is already fetched, is discarded and a NOP is executed instead making it a 2-cycle instruction.

IORLW	Inclusive OR literal with W
Syntax:	[*label*] IORLW k
Operands:	$0 \leq k \leq 255$
Operation:	(W) .OR. (k) → (W)
Status Affected:	Z
Description:	The contents of the W register are OR'ed with the 8-bit literal 'k'. The result is placed in the W register.

PIC10F200/202/204/206

IORWF	Inclusive OR W with f
Syntax:	[*label*] IORWF f,d
Operands:	0 ≤ f ≤ 31 d ∈ [0,1]
Operation:	(W).OR. (f) → (dest)
Status Affected:	Z
Description:	Inclusive OR the W register with register 'f'. If 'd' is '0', the result is placed in the W register. If 'd' is '1', the result is placed back in register 'f'.

MOVWF	Move W to f
Syntax:	[*label*] MOVWF f
Operands:	0 ≤ f ≤ 31
Operation:	(W) → (f)
Status Affected:	None
Description:	Move data from the W register to register 'f'.

MOVF	Move f
Syntax:	[*label*] MOVF f,d
Operands:	0 ≤ f ≤ 31 d ∈ [0,1]
Operation:	(f) → (dest)
Status Affected:	Z
Description:	The contents of register 'f' are moved to destination 'd'. If 'd' is '0', destination is the W register. If 'd' is '1', the destination is file register 'f'. 'd' = 1 is useful as a test of a file register, since status flag Z is affected.

NOP	No Operation
Syntax:	[*label*] NOP
Operands:	None
Operation:	No operation
Status Affected:	None
Description:	No operation.

MOVLW	Move literal to W
Syntax:	[*label*] MOVLW k
Operands:	0 ≤ k ≤ 255
Operation:	k → (W)
Status Affected:	None
Description:	The 8-bit literal 'k' is loaded into the W register. The "don't cares" will assembled as '0's.

OPTION	Load OPTION Register
Syntax:	[*label*] OPTION
Operands:	None
Operation:	(W) → Option
Status Affected:	None
Description:	The content of the W register is loaded into the OPTION register.

RETLW	Return with literal in W
Syntax:	[*label*] RETLW k
Operands:	$0 \le k \le 255$
Operation:	k → (W); TOS → PC
Status Affected:	None
Description:	The W register is loaded with the 8-bit literal 'k'. The program counter is loaded from the top of the stack (the return address). This is a 2-cycle instruction.

RLF	Rotate Left f through Carry
Syntax:	[*label*] RLF f,d
Operands:	$0 \le f \le 31$ d ∈ [0,1]
Operation:	See description below
Status Affected:	C
Description:	The contents of register 'f' are rotated one bit to the left through the Carry flag. If 'd' is '0', the result is placed in the W register. If 'd' is '1', the result is stored back in register 'f'.

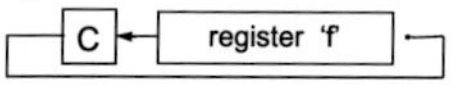

RRF	Rotate Right f through Carry
Syntax:	[*label*] RRF f,d
Operands:	$0 \le f \le 31$ d ∈ [0,1]
Operation:	See description below
Status Affected:	C
Description:	The contents of register 'f' are rotated one bit to the right through the Carry flag. If 'd' is '0', the result is placed in the W register. If 'd' is '1', the result is placed back in register 'f'.

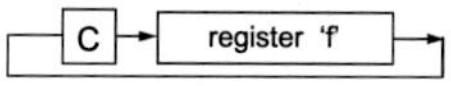

SLEEP	Enter SLEEP Mode
Syntax:	[*label*] SLEEP
Operands:	None
Operation:	00h → WDT; 0 → WDT prescaler; 1 → $\overline{TO}$; 0 → $\overline{PD}$
Status Affected:	$\overline{TO}$, $\overline{PD}$, RBWUF
Description:	Time-out Status bit ($\overline{TO}$) is set. The Power-down Status bit ($\overline{PD}$) is cleared. RBWUF is unaffected. The WDT and its prescaler are cleared. The processor is put into Sleep mode with the oscillator stopped. See **Section 9.9 "Power-down Mode (Sleep)"** for more details.

SUBWF	Subtract W from f
Syntax:	[*label*] SUBWF f,d
Operands:	$0 \le f \le 31$ d ∈ [0,1]
Operation:	(f) – (W) → (dest)
Status Affected:	C, DC, Z
Description:	Subtract (2's complement method) the W register from register 'f'. If 'd' is '0', the result is stored in the W register. If 'd' is '1', the result is stored back in register 'f'.

SWAPF	Swap Nibbles in f
Syntax:	[*label*] SWAPF f,d
Operands:	$0 \le f \le 31$ d ∈ [0,1]
Operation:	(f<3:0>) → (dest<7:4>); (f<7:4>) → (dest<3:0>)
Status Affected:	None
Description:	The upper and lower nibbles of register 'f' are exchanged. If 'd' is '0', the result is placed in W register. If 'd' is '1', the result is placed in register 'f'.

PIC10F200/202/204/206

TRIS	**Load TRIS Register**
Syntax:	[*label*] TRIS f
Operands:	f = 6
Operation:	(W) → TRIS register f
Status Affected:	None
Description:	TRIS register 'f' (f = 6 or 7) is loaded with the contents of the W register

XORLW	**Exclusive OR literal with W**
Syntax:	[*label*] XORLW k
Operands:	$0 \leq k \leq 255$
Operation:	(W) .XOR. k → (W)
Status Affected:	Z
Description:	The contents of the W register are XOR'ed with the 8-bit literal 'k'. The result is placed in the W register.

XORWF	**Exclusive OR W with f**
Syntax:	[*label*] XORWF f,d
Operands:	$0 \leq f \leq 31$ $d \in [0,1]$
Operation:	(W) .XOR. (f) → (dest)
Status Affected:	Z
Description:	Exclusive OR the contents of the W register with register 'f'. If 'd' is '0', the result is stored in the W register. If 'd' is '1', the result is stored back in register 'f'.

11.0 DEVELOPMENT SUPPORT

The PIC® microcontrollers (MCU) and dsPIC® digital signal controllers (DSC) are supported with a full range of software and hardware development tools:

- Integrated Development Environment
  - MPLAB® X IDE Software
- Compilers/Assemblers/Linkers
  - MPLAB XC Compiler
  - MPASM™ Assembler
  - MPLINK™ Object Linker/ MPLIB™ Object Librarian
  - MPLAB Assembler/Linker/Librarian for Various Device Families
- Simulators
  - MPLAB X SIM Software Simulator
- Emulators
  - MPLAB REAL ICE™ In-Circuit Emulator
- In-Circuit Debuggers/Programmers
  - MPLAB ICD 3
  - PICkit™ 3
- Device Programmers
  - MPLAB PM3 Device Programmer
- Low-Cost Demonstration/Development Boards, Evaluation Kits and Starter Kits
- Third-party development tools

11.1 MPLAB X Integrated Development Environment Software

The MPLAB X IDE is a single, unified graphical user interface for Microchip and third-party software, and hardware development tool that runs on Windows®, Linux and Mac OS® X. Based on the NetBeans IDE, MPLAB X IDE is an entirely new IDE with a host of free software components and plug-ins for high-performance application development and debugging. Moving between tools and upgrading from software simulators to hardware debugging and programming tools is simple with the seamless user interface.

With complete project management, visual call graphs, a configurable watch window and a feature-rich editor that includes code completion and context menus, MPLAB X IDE is flexible and friendly enough for new users. With the ability to support multiple tools on multiple projects with simultaneous debugging, MPLAB X IDE is also suitable for the needs of experienced users.

Feature-Rich Editor:

- Color syntax highlighting
- Smart code completion makes suggestions and provides hints as you type
- Automatic code formatting based on user-defined rules
- Live parsing

User-Friendly, Customizable Interface:

- Fully customizable interface: toolbars, toolbar buttons, windows, window placement, etc.
- Call graph window

Project-Based Workspaces:

- Multiple projects
- Multiple tools
- Multiple configurations
- Simultaneous debugging sessions

File History and Bug Tracking:

- Local file history feature
- Built-in support for Bugzilla issue tracker

PIC10F200/202/204/206

11.2 MPLAB XC Compilers

The MPLAB XC Compilers are complete ANSI C compilers for all of Microchip's 8, 16, and 32-bit MCU and DSC devices. These compilers provide powerful integration capabilities, superior code optimization and ease of use. MPLAB XC Compilers run on Windows, Linux or MAC OS X.

For easy source level debugging, the compilers provide debug information that is optimized to the MPLAB X IDE.

The free MPLAB XC Compiler editions support all devices and commands, with no time or memory restrictions, and offer sufficient code optimization for most applications.

MPLAB XC Compilers include an assembler, linker and utilities. The assembler generates relocatable object files that can then be archived or linked with other relocatable object files and archives to create an executable file. MPLAB XC Compiler uses the assembler to produce its object file. Notable features of the assembler include:

- Support for the entire device instruction set
- Support for fixed-point and floating-point data
- Command-line interface
- Rich directive set
- Flexible macro language
- MPLAB X IDE compatibility

11.3 MPASM Assembler

The MPASM Assembler is a full-featured, universal macro assembler for PIC10/12/16/18 MCUs.

The MPASM Assembler generates relocatable object files for the MPLINK Object Linker, Intel® standard HEX files, MAP files to detail memory usage and symbol reference, absolute LST files that contain source lines and generated machine code, and COFF files for debugging.

The MPASM Assembler features include:

- Integration into MPLAB X IDE projects
- User-defined macros to streamline assembly code
- Conditional assembly for multipurpose source files
- Directives that allow complete control over the assembly process

11.4 MPLINK Object Linker/ MPLIB Object Librarian

The MPLINK Object Linker combines relocatable objects created by the MPASM Assembler. It can link relocatable objects from precompiled libraries, using directives from a linker script.

The MPLIB Object Librarian manages the creation and modification of library files of precompiled code. When a routine from a library is called from a source file, only the modules that contain that routine will be linked in with the application. This allows large libraries to be used efficiently in many different applications.

The object linker/library features include:

- Efficient linking of single libraries instead of many smaller files
- Enhanced code maintainability by grouping related modules together
- Flexible creation of libraries with easy module listing, replacement, deletion and extraction

11.5 MPLAB Assembler, Linker and Librarian for Various Device Families

MPLAB Assembler produces relocatable machine code from symbolic assembly language for PIC24, PIC32 and dsPIC DSC devices. MPLAB XC Compiler uses the assembler to produce its object file. The assembler generates relocatable object files that can then be archived or linked with other relocatable object files and archives to create an executable file. Notable features of the assembler include:

- Support for the entire device instruction set
- Support for fixed-point and floating-point data
- Command-line interface
- Rich directive set
- Flexible macro language
- MPLAB X IDE compatibility

11.6 MPLAB X SIM Software Simulator

The MPLAB X SIM Software Simulator allows code development in a PC-hosted environment by simulating the PIC MCUs and dsPIC DSCs on an instruction level. On any given instruction, the data areas can be examined or modified and stimuli can be applied from a comprehensive stimulus controller. Registers can be logged to files for further run-time analysis. The trace buffer and logic analyzer display extend the power of the simulator to record and track program execution, actions on I/O, most peripherals and internal registers.

The MPLAB X SIM Software Simulator fully supports symbolic debugging using the MPLAB XC Compilers, and the MPASM and MPLAB Assemblers. The software simulator offers the flexibility to develop and debug code outside of the hardware laboratory environment, making it an excellent, economical software development tool.

11.7 MPLAB REAL ICE In-Circuit Emulator System

The MPLAB REAL ICE In-Circuit Emulator System is Microchip's next generation high-speed emulator for Microchip Flash DSC and MCU devices. It debugs and programs all 8, 16 and 32-bit MCU, and DSC devices with the easy-to-use, powerful graphical user interface of the MPLAB X IDE.

The emulator is connected to the design engineer's PC using a high-speed USB 2.0 interface and is connected to the target with either a connector compatible with in-circuit debugger systems (RJ-11) or with the new high-speed, noise tolerant, Low-Voltage Differential Signal (LVDS) interconnection (CAT5).

The emulator is field upgradable through future firmware downloads in MPLAB X IDE. MPLAB REAL ICE offers significant advantages over competitive emulators including full-speed emulation, run-time variable watches, trace analysis, complex breakpoints, logic probes, a ruggedized probe interface and long (up to three meters) interconnection cables.

11.8 MPLAB ICD 3 In-Circuit Debugger System

The MPLAB ICD 3 In-Circuit Debugger System is Microchip's most cost-effective, high-speed hardware debugger/programmer for Microchip Flash DSC and MCU devices. It debugs and programs PIC Flash microcontrollers and dsPIC DSCs with the powerful, yet easy-to-use graphical user interface of the MPLAB IDE.

The MPLAB ICD 3 In-Circuit Debugger probe is connected to the design engineer's PC using a high-speed USB 2.0 interface and is connected to the target with a connector compatible with the MPLAB ICD 2 or MPLAB REAL ICE systems (RJ-11). MPLAB ICD 3 supports all MPLAB ICD 2 headers.

11.9 PICkit 3 In-Circuit Debugger/ Programmer

The MPLAB PICkit 3 allows debugging and programming of PIC and dsPIC Flash microcontrollers at a most affordable price point using the powerful graphical user interface of the MPLAB IDE. The MPLAB PICkit 3 is connected to the design engineer's PC using a full-speed USB interface and can be connected to the target via a Microchip debug (RJ-11) connector (compatible with MPLAB ICD 3 and MPLAB REAL ICE). The connector uses two device I/O pins and the Reset line to implement in-circuit debugging and In-Circuit Serial Programming™ (ICSP™).

11.10 MPLAB PM3 Device Programmer

The MPLAB PM3 Device Programmer is a universal, CE compliant device programmer with programmable voltage verification at VDDMIN and VDDMAX for maximum reliability. It features a large LCD display (128 x 64) for menus and error messages, and a modular, detachable socket assembly to support various package types. The ICSP cable assembly is included as a standard item. In Stand-Alone mode, the MPLAB PM3 Device Programmer can read, verify and program PIC devices without a PC connection. It can also set code protection in this mode. The MPLAB PM3 connects to the host PC via an RS-232 or USB cable. The MPLAB PM3 has high-speed communications and optimized algorithms for quick programming of large memory devices, and incorporates an MMC card for file storage and data applications.

PIC10F200/202/204/206

11.11 Demonstration/Development Boards, Evaluation Kits, and Starter Kits

A wide variety of demonstration, development and evaluation boards for various PIC MCUs and dsPIC DSCs allows quick application development on fully functional systems. Most boards include prototyping areas for adding custom circuitry and provide application firmware and source code for examination and modification.

The boards support a variety of features, including LEDs, temperature sensors, switches, speakers, RS-232 interfaces, LCD displays, potentiometers and additional EEPROM memory.

The demonstration and development boards can be used in teaching environments, for prototyping custom circuits and for learning about various microcontroller applications.

In addition to the PICDEM™ and dsPICDEM™ demonstration/development board series of circuits, Microchip has a line of evaluation kits and demonstration software for analog filter design, KEELOQ® security ICs, CAN, IrDA®, PowerSmart battery management, SEEVAL® evaluation system, Sigma-Delta ADC, flow rate sensing, plus many more.

Also available are starter kits that contain everything needed to experience the specified device. This usually includes a single application and debug capability, all on one board.

Check the Microchip web page (www.microchip.com) for the complete list of demonstration, development and evaluation kits.

11.12 Third-Party Development Tools

Microchip also offers a great collection of tools from third-party vendors. These tools are carefully selected to offer good value and unique functionality.

- Device Programmers and Gang Programmers from companies, such as SoftLog and CCS
- Software Tools from companies, such as Gimpel and Trace Systems
- Protocol Analyzers from companies, such as Saleae and Total Phase
- Demonstration Boards from companies, such as MikroElektronika, Digilent® and Olimex
- Embedded Ethernet Solutions from companies, such as EZ Web Lynx, WIZnet and IPLogika®

PIC10F200/202/204/206

12.0 ELECTRICAL CHARACTERISTICS

Absolute Maximum Ratings(†)

Ambient temperature under bias	-40°C to +125°C
Storage temperature	-65°C to +150°C
Voltage on V_{DD} with respect to V_{SS}	0 to +6.5V
Voltage on $\overline{MCLR}$ with respect to V_{SS}	0 to +13.5V
Voltage on all other pins with respect to V_{SS}	-0.3V to (V_{DD} + 0.3V)
Total power dissipation[1]	800 mW
Max. current out of V_{SS} pin	80 mA
Max. current into V_{DD} pin	80 mA
Input clamp current, I_{IK} ($V_I < 0$ or $V_I > V_{DD}$)	±20 mA
Output clamp current, I_{OK} ($V_O < 0$ or $V_O > V_{DD}$)	±20 mA
Max. output current sunk by any I/O pin	25 mA
Max. output current sourced by any I/O pin	25 mA
Max. output current sourced by I/O port	75 mA
Max. output current sunk by I/O port	75 mA

Note 1: Power dissipation is calculated as follows: $P_{DIS} = V_{DD} \times \{I_{DD} - \sum I_{OH}\} + \sum \{(V_{DD} - V_{OH}) \times I_{OH}\} + \sum(V_{OL} \times I_{OL})$

†NOTICE: Stresses above those listed under "Absolute Maximum Ratings" may cause permanent damage to the device. This is a stress rating only and functional operation of the device at those or any other conditions above those indicated in the operation listings of this specification is not implied. Exposure to maximum rating conditions for extended periods may affect device reliability.

PIC10F200/202/204/206

FIGURE 12-1: PIC10F200/202/204/206 VOLTAGE-FREQUENCY GRAPH, -40°C ≤ T_A ≤ +125°C

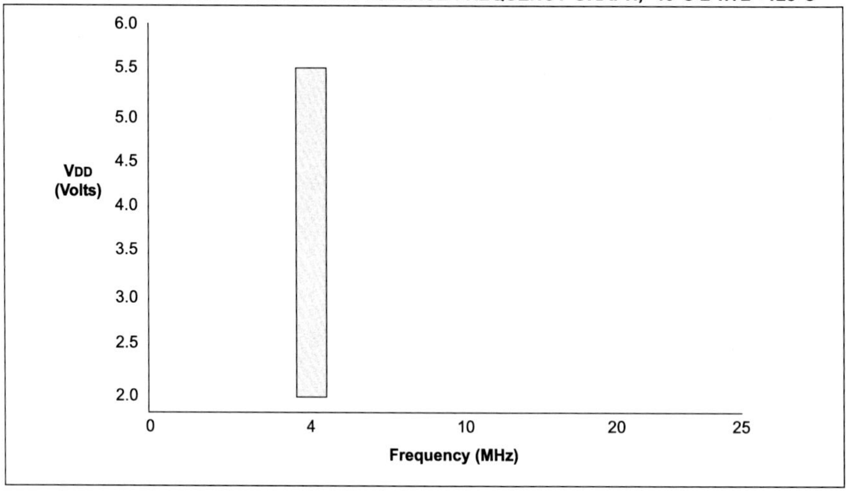

PIC10F200/202/204/206

12.1 DC Characteristics: PIC10F200/202/204/206 (Industrial)

DC CHARACTERISTICS			Standard Operating Conditions (unless otherwise specified) Operating Temperature -40°C ≤ TA ≤ +85°C (industrial)				
Param. No.	Sym.	Characteristic	Min.	Typ.(1)	Max.	Units	Conditions
D001	VDD	**Supply Voltage**	2.0		5.5	V	See Figure 12-1
D002	VDR	**RAM Data Retention Voltage(2)**	1.5*	—	—	V	Device in Sleep mode
D003	VPOR	**VDD Start Voltage** to ensure Power-on Reset	—	Vss	—	V	
D004	SVDD	**VDD Rise Rate** to ensure Power-on Reset	0.05*	—	—	V/ms	
D010	IDD	**Supply Current(3)**					
			—	175	275	μA	VDD = 2.0V
			—	0.63	1.1	mA	VDD = 5.0V
D020	IPD	**Power-down Current(4)**					
			—	0.1	1.2	μA	VDD = 2.0V
			—	0.35	2.4	μA	VDD = 5.0V
D022	IWDT	**WDT Current(5)**					
			—	1.0	3	μA	VDD = 2.0V
			—	7	16	μA	VDD = 5.0V
D023	ICMP	**Comparator Current(5)**					
			—	12	23	μA	VDD = 2.0V
			—	44	80	μA	VDD = 5.0V
D024	IVREF	**Internal Reference Current(5,6)**					
			—	85	115	μA	VDD = 2.0V
				175	195	μA	VDD = 5.0V

* These parameters are characterized but not tested.

Note 1: Data in the Typical ("Typ.") column is based on characterization results at 25°C. This data is for design guidance only and is not tested.

2: This is the limit to which VDD can be lowered in Sleep mode without losing RAM data.

3: The supply current is mainly a function of the operating voltage and frequency. Other factors such as bus loading, bus rate, internal code execution pattern and temperature also have an impact on the current consumption.

a) The test conditions for all IDD measurements in active operation mode are:
All I/O pins tri-stated, pulled to Vss, T0CKI = VDD, $\overline{MCLR}$ = VDD; WDT enabled/disabled as specified.

b) For standby current measurements, the conditions are the same, except that the device is in Sleep mode.

4: Power-down current is measured with the part in Sleep mode, with all I/O pins in high-impedance state and tied to VDD or VSS.

5: The peripheral current is the sum of the base IDD or IPD and the additional current consumed when this peripheral is enabled.

6: Measured with the comparator enabled.

PIC10F200/202/204/206

12.2 DC Characteristics: PIC10F200/202/204/206 (Extended)

DC CHARACTERISTICS			Standard Operating Conditions (unless otherwise specified) Operating Temperature -40°C ≤ TA ≤ +125°C (extended)				
Param. No.	**Sym.**	**Characteristic**	**Min.**	**Typ.(1)**	**Max.**	**Units**	**Conditions**
D001	VDD	**Supply Voltage**	2.0		5.5	V	See Figure 12-1
D002	VDR	**RAM Data Retention Voltage(2)**	1.5*		—	V	Device in Sleep mode
D003	VPOR	**VDD Start Voltage** to ensure Power-on Reset	—	Vss	—	V	
D004	SVDD	**VDD Rise Rate** to ensure Power-on Reset	0.05*	—	—	V/ms	
D010	IDD	**Supply Current(3)**					
			—	175	275	μA	VDD = 2.0V
			—	0.63	1.1	mA	VDD = 5.0V
D020	IPD	**Power-down Current(4)**					
			—	0.1	9	μA	VDD = 2.0V
			—	0.35	15	μA	VDD = 5.0V
D022	IWDT	**WDT Current(5)**					
			—	1.0	18	μA	VDD = 2.0V
			—	7	22	μA	VDD = 5.0V
D023	ICMP	**Comparator Current(5)**					
			—	12	27	μA	VDD = 2.0V
			—	42	85	μA	VDD = 5.0V
D024	VREF	**Internal Reference Current(5,6)**					
			—	85	120	μA	VDD = 2.0V
				175	200	μA	VDD = 5.0V

* These parameters are characterized but not tested.

Note 1: Data in the Typical ("Typ.") column is based on characterization results at 25°C. This data is for design guidance only and is not tested.

2: This is the limit to which VDD can be lowered in Sleep mode without losing RAM data.

3: The supply current is mainly a function of the operating voltage and frequency. Other factors such as bus loading, bus rate, internal code execution pattern and temperature also have an impact on the current consumption.

a) The test conditions for all IDD measurements in active operation mode are: All I/O pins tri-stated, pulled to VSS, T0CKI = VDD, $\overline{MCLR}$ = VDD; WDT enabled/disabled as specified.

b) For standby current measurements, the conditions are the same, except that the device is in Sleep mode.

4: Power-down current is measured with the part in Sleep mode, with all I/O pins in high-impedance state and tied to VDD or VSS.

5: The peripheral current is the sum of the base IDD or IPD and the additional current consumed when this peripheral is enabled.

6: Measured with the Comparator enabled.

PIC10F200/202/204/206

12.3 DC Characteristics: PIC10F200/202/204/206 (Industrial, Extended)

DC CHARACTERISTICS			Standard Operating Conditions (unless otherwise specified) Operating temperature -40°C ≤ TA ≤ +85°C (industrial) -40°C ≤ TA ≤ +125°C (extended) Operating voltage VDD range as described in DC specification				
Param. No.	**Sym.**	**Characteristic**	**Min.**	**Typ.†**	**Max.**	**Units**	**Conditions**
	VIL	**Input Low Voltage**					
		I/O ports:					
D030		with TTL buffer	Vss	—	0.8	V	For all 4.5V ≤ VDD ≤ 5.5V
D030A			Vss	—	0.15 VDD	V	
D031		with Schmitt Trigger buffer	Vss	—	0.2 VDD	V	
D032		$\overline{MCLR}$, T0CKI	Vss	—	0.2 VDD	V	
	VIH	**Input High Voltage**					
		I/O ports:		—			
D040		with TTL buffer	2.0	—	VDD	V	4.5V ≤ VDD ≤ 5.5V
D040A			0.25 VDD + 0.8	—	VDD	V	Otherwise
D041		with Schmitt Trigger buffer	0.8VDD	—	VDD	V	For entire VDD range
D042		$\overline{MCLR}$, T0CKI	0.8VDD	—	VDD	V	
D070	IPUR	**GPIO weak pull-up current**(3)	50	250	400	μA	VDD = 5V, VPIN = VSS
	IIL	**Input Leakage Current**(1, 2)					
D060		I/O ports	—	±0.1	± 1	μA	Vss ≤ VPIN ≤ VDD, Pin at high-impedance
D061		GP3/$\overline{MCLR}$(3)	—	±0.7	± 5	μA	Vss ≤ VPIN ≤ VDD
		Output Low Voltage					
D080		I/O ports	—	—	0.6	V	IOL = 8.5 mA, VDD = 4.5V, -40°C to +85°C
D080A			—	—	0.6	V	IOL = 7.0 mA, VDD = 4.5V, -40°C to +125°C
		Output High Voltage					
D090		I/O ports(2)	VDD – 0.7	—	—	V	IOH = -3.0 mA, VDD = 4.5V, -40°C to +85°C
D090A			VDD – 0.7	—	—	V	IOH = -2.5 mA, VDD = 4.5V, -40°C to +125°C
		Capacitive Loading Specs on Output Pins					
D101		All I/O pins	—	—	50*	pF	

† Data in "Typ." column is at 5V, 25°C unless otherwise stated. These parameters are for design guidance only and are not tested.

* These parameters are for design guidance only and are not tested.

Note 1: The leakage current on the $\overline{MCLR}$ pin is strongly dependent on the applied voltage level. The specified levels represent normal operating conditions. Higher leakage current may be measured at different input voltages.

2: Negative current is defined as coming out of the pin.

3: This specification applies when GP3/$\overline{MCLR}$ is configured as an input with pull-up disabled. The leakage current of the $\overline{MCLR}$ circuit is higher than the standard I/O logic.

PIC10F200/202/204/206

TABLE 12-1: COMPARATOR SPECIFICATIONS

Standard Operating Conditions (unless otherwise stated)
Operating Temperature -40°C ≤ TA ≤ +125°C

Param. No.	Sym.	Characteristics		Min.	Typ.†	Max.	Units	Comments
D300	VOS	Input Offset Voltage		—	± 5.0	± 10	mV	(VDD - 1.5)/2
D301	VCM	Input Common Mode Voltage		0	—	VDD–1.5*	V	
D302	CMRR	Common Mode Rejection Ratio		55*	—	—	dB	
D303*	TRT	Response Time	Falling	—	150	600	ns	**(Note 1)**
			Rising	—	200	1000	ns	
D304*	TMC2COV	Comparator Mode Change to Output Valid		—	—	10*	μs	
D305	VIVRF	Internal Reference Voltage		0.55	0.6	0.65	V	2.0V ≤ VDD ≤ 5.5V -40°C ≤ TA ≤ ±125°C (extended)

* These parameters are characterized but not tested.

† Data in 'Typ.' column is at 5V, 25°C unless otherwise stated. These parameters are for design guidance only and are not tested.

Note 1: Response time is measured with one comparator input at (VDD - 1.5)/2 - 100 mV to (VDD - 1.5)/2 + 20 mV.

TABLE 12-2: PULL-UP RESISTOR RANGES

VDD (Volts)	Temperature (°C)	Min.	Typ.	Max.	Units
GP0/GP1					
2.0	-40	73K	105K	186K	Ω
	25	73K	113K	187K	Ω
	85	82K	123K	190K	Ω
	125	86K	132k	190K	Ω
5.5	-40	15K	21K	33K	Ω
	25	15K	22K	34K	Ω
	85	19K	26k	35K	Ω
	125	23K	29K	35K	Ω
GP3					
2.0	-40	63K	81K	96K	Ω
	25	77K	93K	116K	Ω
	85	82K	96k	116K	Ω
	125	86K	100K	119K	Ω
5.5	-40	16K	20k	22K	Ω
	25	16K	21K	23K	Ω
	85	24K	25k	28K	Ω
	125	26K	27K	29K	Ω

PIC10F200/202/204/206

12.4 Timing Parameter Symbology and Load Conditions – PIC10F200/202/204/206

The timing parameter symbols have been created following one of the following formats:

1. TppS2ppS
2. TppS

T			
F	Frequency	T	Time

Lowercase subscripts (pp) and their meanings:

pp			
2	to	mc	$\overline{\text{MCLR}}$
ck	CLKOUT	osc	Oscillator
cy	Cycle time	t0	T0CKI
drt	Device Reset Timer	wdt	Watchdog Timer
io	I/O port	wdt	Watchdog Timer

Uppercase letters and their meanings:

S			
F	Fall	P	Period
H	High	R	Rise
I	Invalid (high-impedance)	V	Valid
L	Low	Z	High-impedance

FIGURE 12-2: LOAD CONDITIONS – PIC10F200/202/204/206

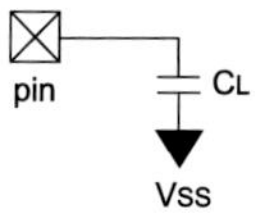

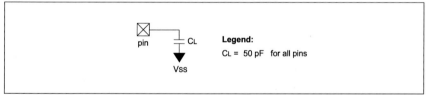

Legend:
C_L = 50 pF for all pins

PIC10F200/202/204/206

TABLE 12-3: CALIBRATED INTERNAL RC FREQUENCIES – PIC10F200/202/204/206

AC CHARACTERISTICS			Standard Operating Conditions (unless otherwise specified) Operating Temperature -40°C ≤ TA ≤ +85°C (industrial), -40°C ≤ TA ≤ +125°C (extended) Operating Voltage VDD range is described in **Section 12.1 "DC Characteristics: PIC10F200/202/204/206 (Industrial)"**					
Param. No.	**Sym.**	**Characteristic**	**Freq. Tolerance**	**Min.**	**Typ.†**	**Max.**	**Units**	**Conditions**
F10	FOSC	Internal Calibrated INTOSC Frequency(1,2)	± 1%	3.96	4.00	4.04	MHz	VDD=3.5V @ 25°C
			± 2%	3.92	4.00	4.08	MHz	2.5V ≤ VDD ≤ 5.5V 0°C ≤ TA ≤ +85°C (industrial)
			± 5%	3.80	4.00	4.20	MHz	2.0V ≤ VDD ≤ 5.5V -40°C ≤ TA ≤ +85°C (industrial) -40°C ≤ TA ≤ +125°C (extended)

* These parameters are characterized but not tested.

† Data in the Typical ("Typ.") column is at 5V, 25°C unless otherwise stated. These parameters are for design guidance only and are not tested.

Note 1: To ensure these oscillator frequency tolerances, VDD and VSS must be capacitively decoupled as close to the device as possible. 0.1 μF and 0.01 μF values in parallel are recommended.

2: Under stable VDD conditions.

FIGURE 12-3: RESET, WATCHDOG TIMER AND DEVICE RESET TIMER TIMING – PIC10F200/202/204/206

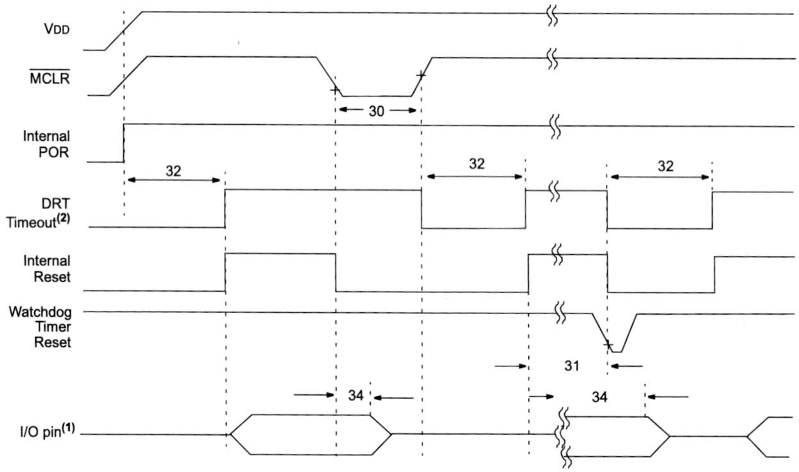

Note 1: I/O pins must be taken out of High-Impedance mode by enabling the output drivers in software.

2: Runs on POR only.

PIC10F200/202/204/206

TABLE 12-4: RESET, WATCHDOG TIMER AND DEVICE RESET TIMER – PIC10F200/202/204/206

AC CHARACTERISTICS			Standard Operating Conditions (unless otherwise specified) Operating Temperature -40°C ≤ TA ≤ +85°C (industrial) -40°C ≤ TA ≤ +125°C (extended) Operating Voltage VDD range is described in Section 12.1 "DC Characteristics: PIC10F200/202/204/206 (Industrial)"				
Param. No.	**Sym.**	**Characteristic**	**Min.**	**Typ.(1)**	**Max.**	**Units**	**Conditions**
30	T_{MCL}	MCLR Pulse Width (low)	2* 5*	— —	— —	μs μs	VDD = 5V, -40°C to +85°C VDD = 5.0V
31	TWDT	Watchdog Timer Time-out Period (no prescaler)	10 10	16 16	29 31	ms ms	VDD = 5.0V (industrial) VDD = 5.0V (extended)
32	TDRT	Device Reset Timer Period (standard)	10 10	16 16	29 31	ms ms	VDD = 5.0V (industrial) VDD = 5.0V (extended)
34	TIOZ	I/O High-impedance from MCLR low	—	—	2*	μs	

* These parameters are characterized but not tested.

Note 1: Data in the Typical ("Typ.") column is at 5V, 25°C unless otherwise stated. These parameters are for design guidance only and are not tested.

FIGURE 12-4: TIMER0 CLOCK TIMINGS – PIC10F200/202/204/206

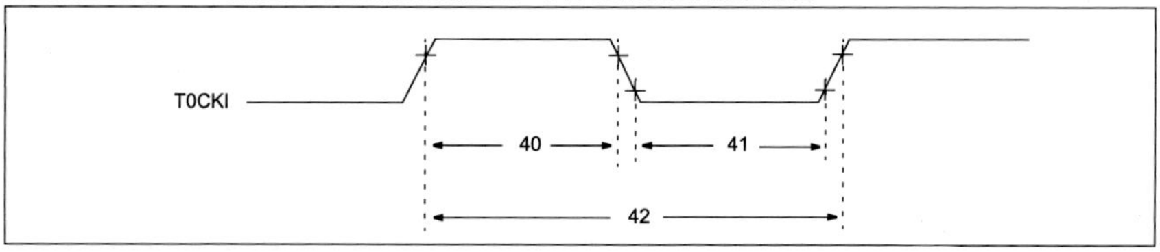

TABLE 12-5: TIMER0 CLOCK REQUIREMENTS – PIC10F200/202/204/206

AC CHARACTERISTICS				Standard Operating Conditions (unless otherwise specified) Operating Temperature -40°C ≤ TA ≤ +85°C (industrial) -40°C ≤ TA ≤ +125°C (extended) Operating Voltage VDD range is described in Section 12.1 "DC Characteristics: PIC10F200/202/204/206 (Industrial)".				
Param. No.	**Sym.**	**Characteristic**		**Min.**	**Typ.(1)**	**Max.**	**Units**	**Conditions**
40	Tt0H	T0CKI High Pulse Width	No Prescaler	0.5 TCY + 20*	—	—	ns	
			With Prescaler	10*	—	—	ns	
41	Tt0L	T0CKI Low Pulse Width	No Prescaler	0.5 TCY + 20*	—	—	ns	
			With Prescaler	10*	—	—	ns	
42	Tt0P	T0CKI Period		20 or $\frac{T_{CY} + 40*}{N}$	—	—	ns	Whichever is greater. N = Prescale Value (1, 2, 4,..., 256)

* These parameters are characterized but not tested.

Note 1: Data in the Typical ("Typ.") column is at 5V, 25°C unless otherwise stated. These parameters are for design guidance only and are not tested.

PIC10F200/202/204/206

TABLE 12-6: THERMAL CONSIDERATIONS

Standard Operating Conditions (unless otherwise specified)					
Param. No.	**Sym.**	**Characteristic**	**Typ.**	**Units**	**Conditions**
TH01	θ_{JA}	Thermal Resistance Junction to Ambient	60	°C/W	6-pin SOT-23 package
			80	°C/W	8-pin PDIP package
			90	°C/W	8-pin DFN package
TH02	θ_{JC}	Thermal Resistance Junction to Case	31.4	°C/W	6-pin SOT-23 package
			24	°C/W	8-pin PDIP package
			24	°C/W	8-pin DFN package
TH03	T_{JMAX}	Maximum Junction Temperature	150	°C	
TH04	PD	Power Dissipation	—	W	$PD = P_{INTERNAL} + P_{I/O}$
TH05	$P_{INTERNAL}$	Internal Power Dissipation	—	W	$P_{INTERNAL} = I_{DD} \times V_{DD}$(1)
TH06	$P_{I/O}$	I/O Power Dissipation	—	W	$P_{I/O} = \Sigma (I_{OL} * V_{OL}) + \Sigma (I_{OH} * (V_{DD} - V_{OH}))$
TH07	P_{DER}	Derated Power	—	W	$P_{DER} = PD_{MAX} (T_J - T_A)/\theta_{JA}$(2)

Note 1: I_{DD} is current to run the chip alone without driving any load on the output pins.
2: T_A = Ambient Temperature; T_J = Junction Temperature.

PIC10F200/202/204/206

13.0 DC AND AC CHARACTERISTICS GRAPHS AND TABLES

The graphs and tables provided in this section are for **design guidance** and are **not tested**.

In some graphs or tables, the data presented are **outside specified operating range** (i.e., outside specified VDD range). This is for **information only** and devices are ensured to operate properly only within the specified range.

Note:	The graphs and tables provided following this note are a statistical summary based on a limited number of samples and are provided for informational purposes only. The performance characteristics listed herein are not tested or guaranteed. In some graphs or tables, the data presented may be outside the specified operating range (e.g., outside specified power supply range) and therefore, outside the warranted range.

"Typical" represents the mean of the distribution at 25°C. "MAXIMUM", "Max.", "MINIMUM" or "Min." represents (mean + 3σ) or (mean - 3σ) respectively, where σ is a standard deviation, over each temperature range.

FIGURE 13-1: IDD vs. VDD OVER FOSC

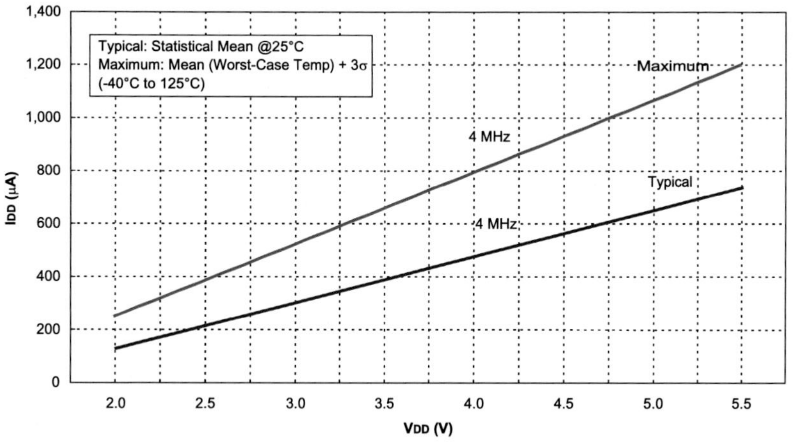

PIC10F200/202/204/206

FIGURE 13-2: TYPICAL IPD vs. VDD (SLEEP MODE, ALL PERIPHERALS DISABLED)

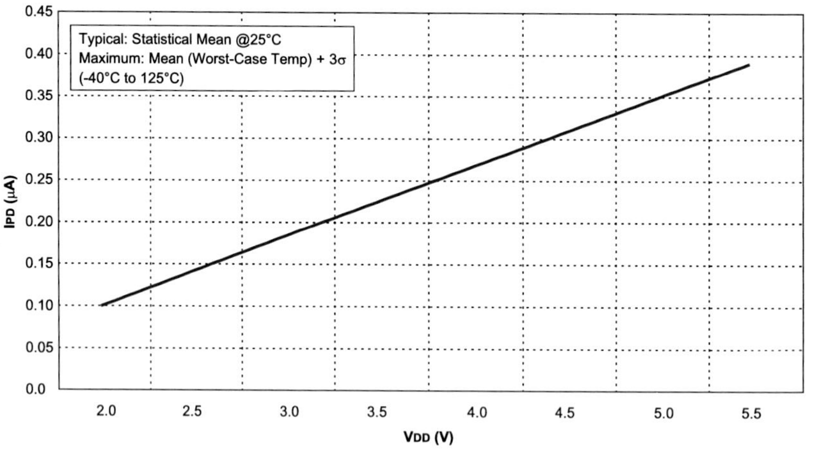

FIGURE 13-3: MAXIMUM IPD vs. VDD (SLEEP MODE, ALL PERIPHERALS DISABLED)

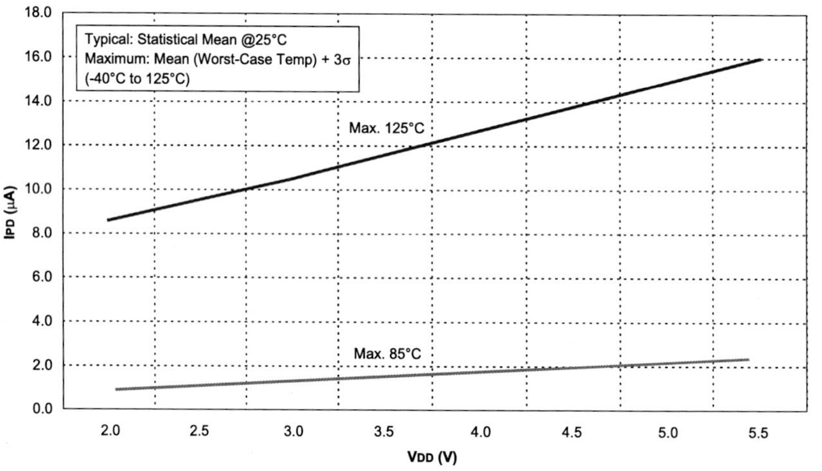

PIC10F200/202/204/206

FIGURE 13-4: COMPARATOR IPD vs. VDD (COMPARATOR ENABLED)

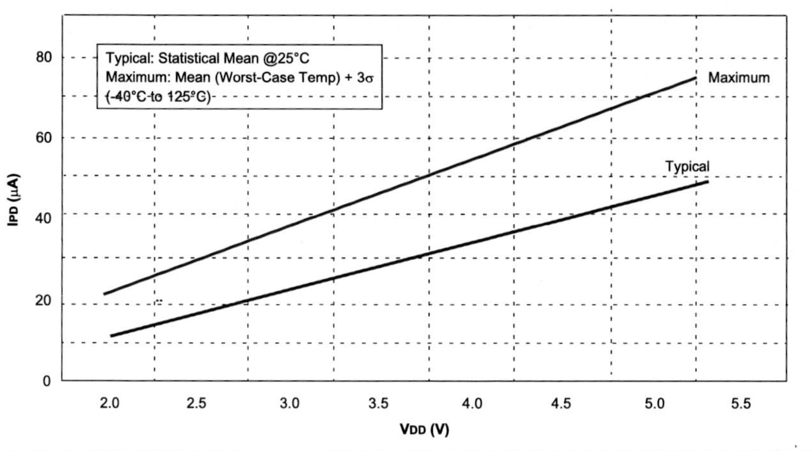

FIGURE 13-5: TYPICAL WDT IPD vs. VDD

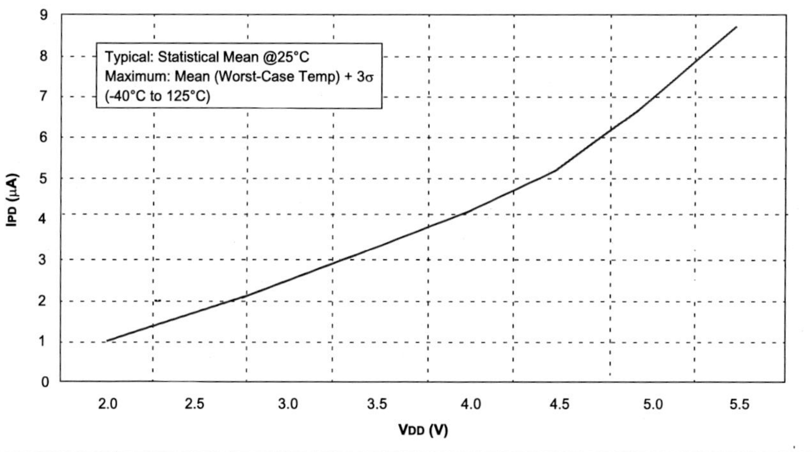

PIC10F200/202/204/206

FIGURE 13-6: MAXIMUM WDT IPD vs. VDD OVER TEMPERATURE

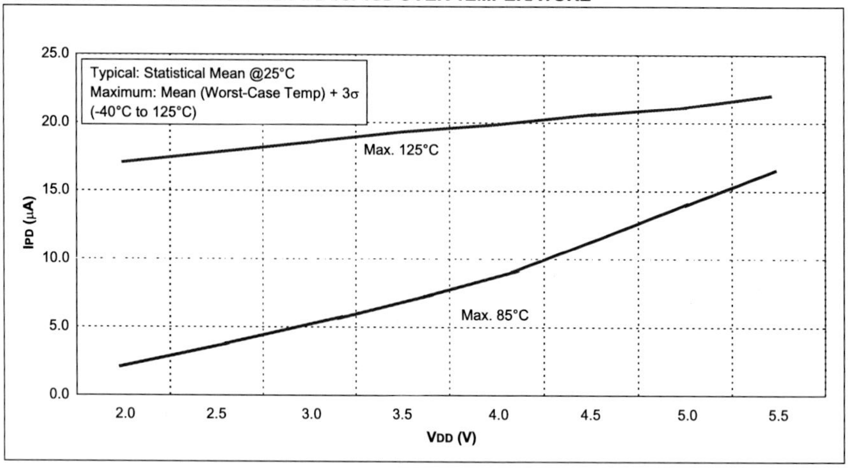

FIGURE 13-7: WDT TIME-OUT vs. VDD OVER TEMPERATURE (NO PRESCALER)

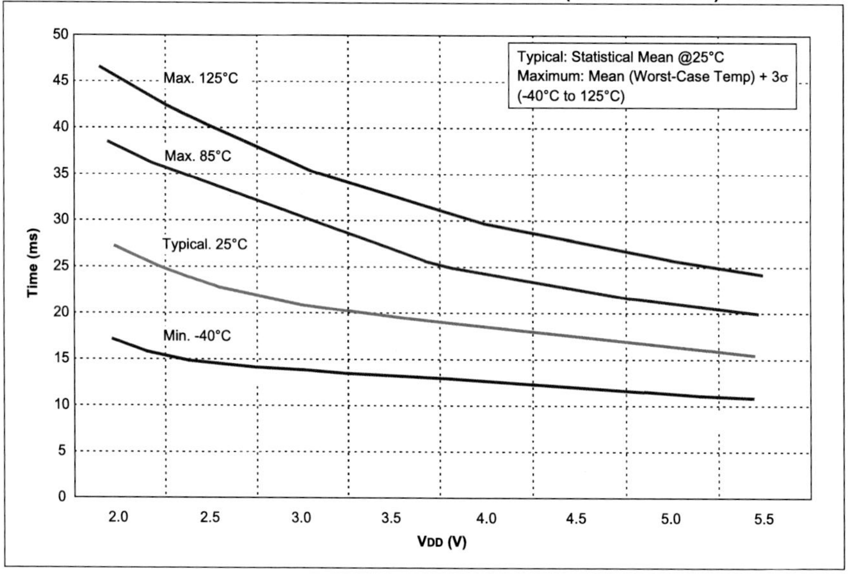

FIGURE 13-8: V_{OL} vs. I_{OL} OVER TEMPERATURE (V_{DD} = 3.0V)

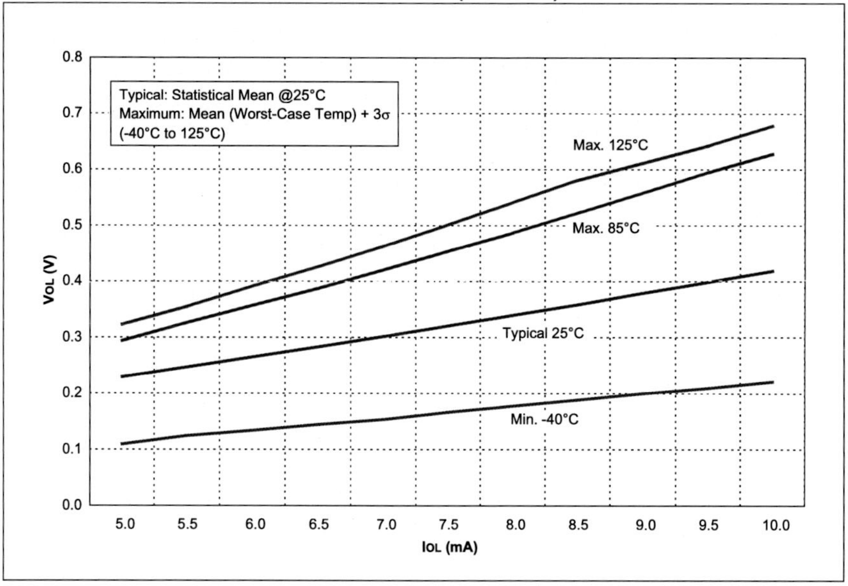

FIGURE 13-9: V_{OL} vs. I_{OL} OVER TEMPERATURE (V_{DD} = 5.0V)

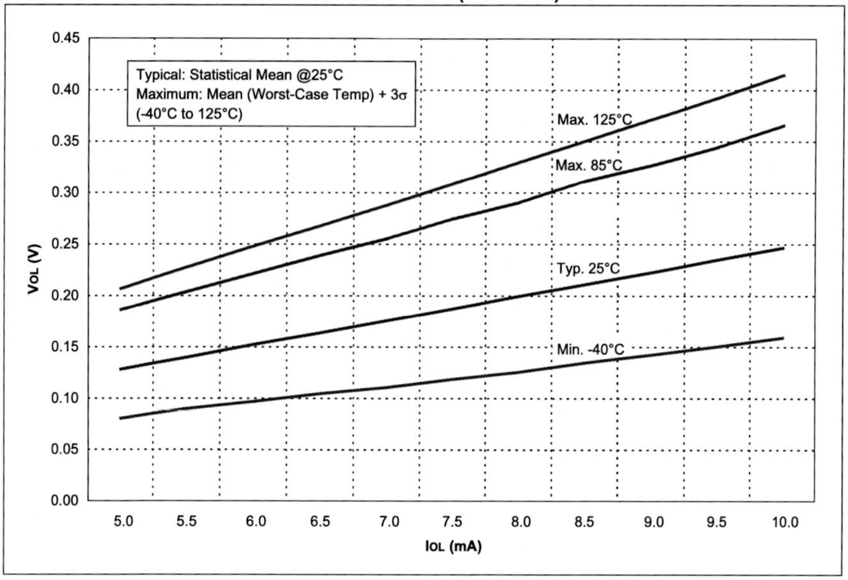

PIC10F200/202/204/206

FIGURE 13-10: VOH vs. IOH OVER TEMPERATURE (VDD = 3.0V)

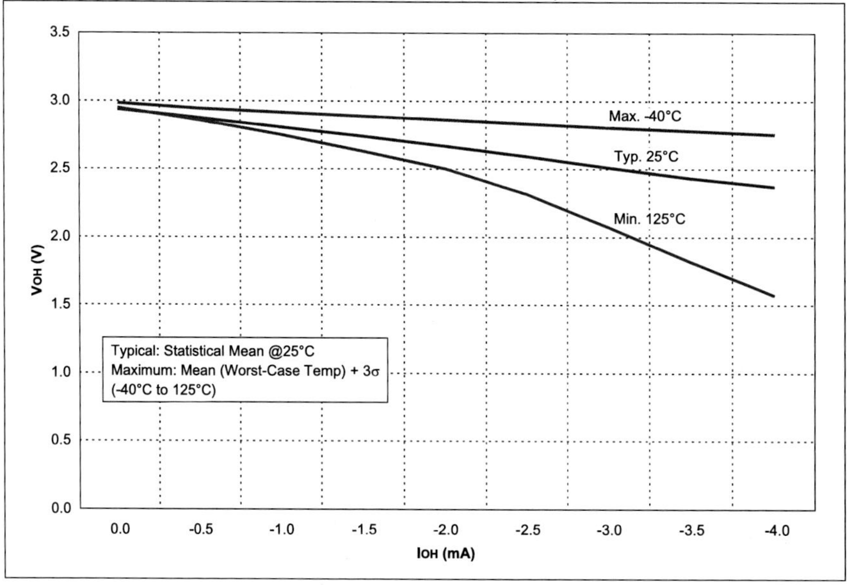

FIGURE 13-11: VOH vs. IOH OVER TEMPERATURE (VDD = 5.0V)

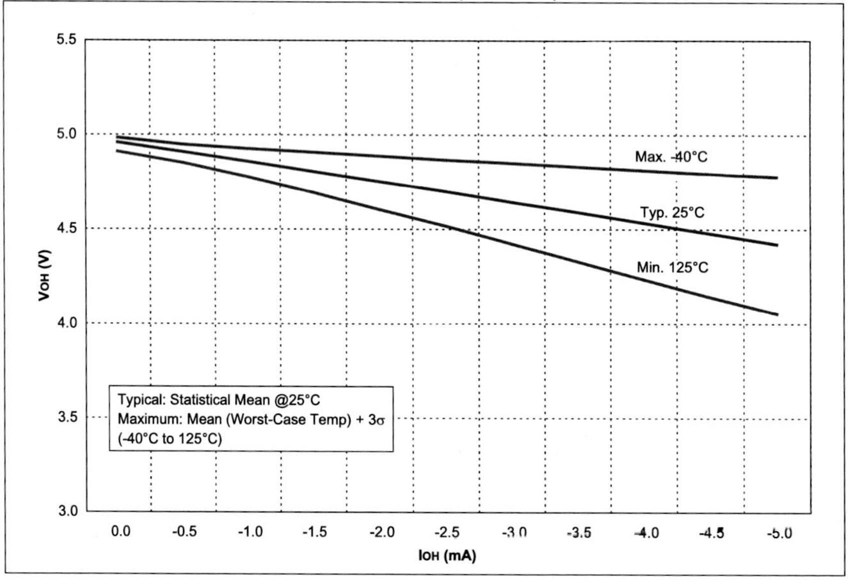

PIC10F200/202/204/206

FIGURE 13-12: TTL INPUT THRESHOLD VIN vs. VDD

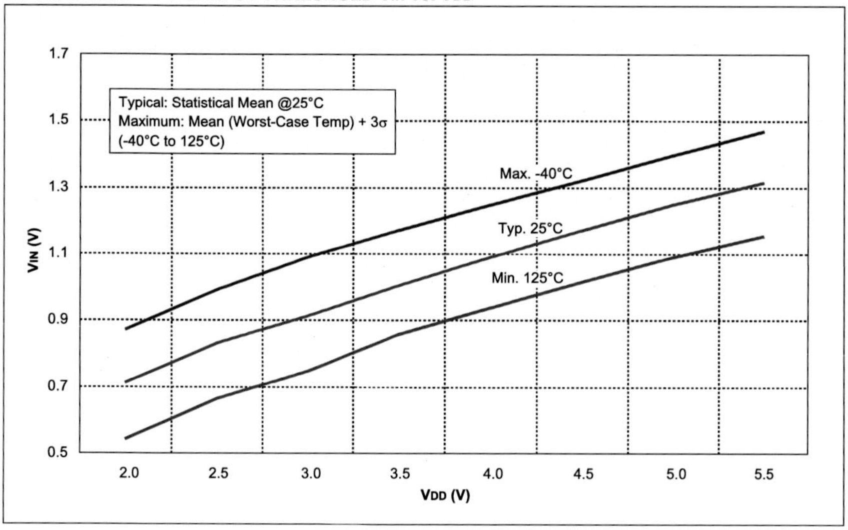

FIGURE 13-13: SCHMITT TRIGGER INPUT THRESHOLD VIN vs. VDD

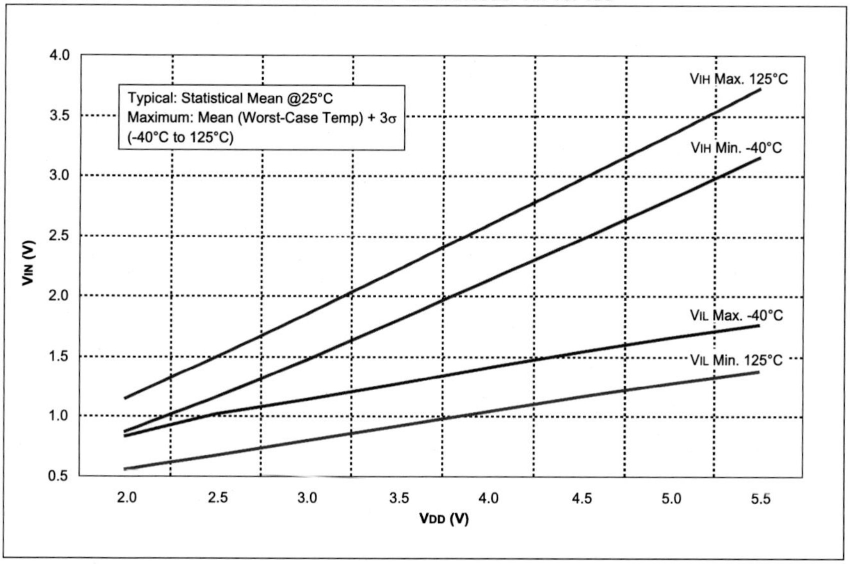

PIC10F200/202/204/206

FIGURE 13-14: INTOSC (INTERNAL OSCILLATOR) POWER-UP TIMES vs. VDD

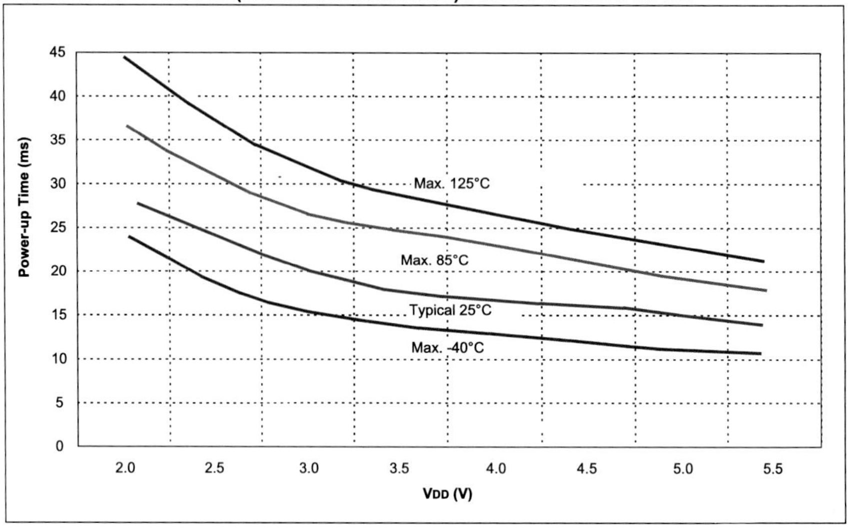

14.0 PACKAGING INFORMATION

14.1 Package Marking Information

6-Lead SOT-23

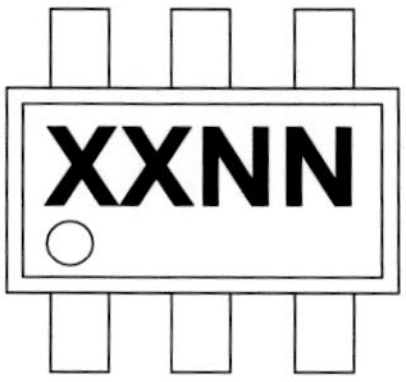

Example

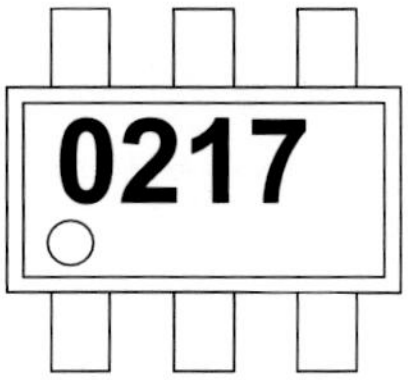

8-Lead PDIP (300 mil)

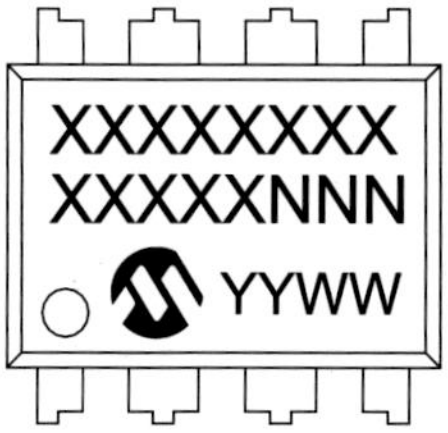

Example

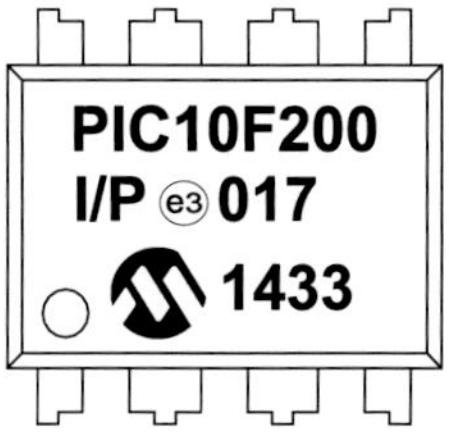

Legend:	XX...X	Customer-specific information
	Y	Year code (last digit of calendar year)
	YY	Year code (last 2 digits of calendar year)
	WW	Week code (week of January 1 is week '01')
	NNN	Alphanumeric traceability code
	(e3)	Pb-free JEDEC® designator for Matte Tin (Sn)
	*	This package is Pb-free. The Pb-free JEDEC designator ((e3)) can be found on the outer packaging for this package.
Note:		In the event the full Microchip part number cannot be marked on one line, it will be carried over to the next line, thus limiting the number of available characters for customer-specific information.

* Standard PIC® device marking consists of Microchip part number, year code, week code, and traceability code. For PIC device marking beyond this, certain price adders apply. Please check with your Microchip Sales Office. For QTP devices, any special marking adders are included in QTP price.

PIC10F200/202/204/206

Package Marking Information (Continued)

8-Lead DFN (2x3x0.9 mm)

Example

Legend:	XX...X	Customer-specific information
	Y	Year code (last digit of calendar year)
	YY	Year code (last 2 digits of calendar year)
	WW	Week code (week of January 1 is week '01')
	NNN	Alphanumeric traceability code
	(e3)	Pb-free JEDEC® designator for Matte Tin (Sn)
	*	This package is Pb-free. The Pb-free JEDEC designator ((e3)) can be found on the outer packaging for this package.

Note: In the event the full Microchip part number cannot be marked on one line, it will be carried over to the next line, thus limiting the number of available characters for customer-specific information.

* Standard PIC® device marking consists of Microchip part number, year code, week code, and traceability code. For PIC device marking beyond this, certain price adders apply. Please check with your Microchip Sales Office. For QTP devices, any special marking adders are included in QTP price.

PIC10F200/202/204/206

TABLE 14-1: 8-LEAD 2x3 DFN (MC) PACKAGE TOP MARKING

Part Number	Marking
PIC10F200-I/MC	BA0
PIC10F200-E/MC	BB0
PIC10F202-I/MC	BC0
PIC10F202-E/MC	BD0
PIC10F204-I/MC	BE0
PIC10F204-E/MC	BF0
PIC10F206-I/MC	BG0
PIC10F206-E/MC	BH0

TABLE 14-2: 6-LEAD SOT-23 (OT) PACKAGE TOP MARKING

Part Number	Marking
PIC10F200-I/OT	00NN
PIC10F200-E/OT	00NN
PIC10F202-I/OT	02NN
PIC10F202-E/OT	02NN
PIC10F204-I/OT	04NN
PIC10F204-E/OT	04NN
PIC10F206-I/OT	06NN
PIC10F206-E/OT	06NN

Note: NN represents the alphanumeric traceability code.

PIC10F200/202/204/206

14.2 Package Details

The following sections give the technical details of the packages.

6-Lead Plastic Small Outline Transistor (OT) [SOT-23]

Note: For the most current package drawings, please see the Microchip Packaging Specification located at http://www.microchip.com/packaging

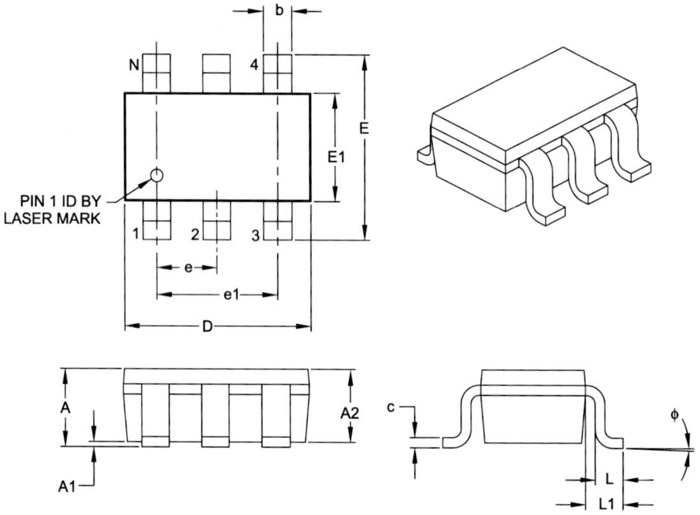

		Units	MILLIMETERS		
		Dimension Limits	MIN	NOM	MAX
Number of Pins	N			6	
Pitch	e			0.95 BSC	
Outside Lead Pitch	e1			1.90 BSC	
Overall Height	A		0.90	–	1.45
Molded Package Thickness	A2		0.89	–	1.30
Standoff	A1		0.00	–	0.15
Overall Width	E		2.20	–	3.20
Molded Package Width	E1		1.30	–	1.80
Overall Length	D		2.70	–	3.10
Foot Length	L		0.10	–	0.60
Footprint	L1		0.35	–	0.80
Foot Angle	ϕ		0°	–	30°
Lead Thickness	c		0.08	–	0.26
Lead Width	b		0.20	–	0.51

Notes:

1. Dimensions D and E1 do not include mold flash or protrusions. Mold flash or protrusions shall not exceed 0.127 mm per side.
2. Dimensioning and tolerancing per ASME Y14.5M.

   BSC: Basic Dimension. Theoretically exact value shown without tolerances.

Microchip Technology Drawing C04-028B

6-Lead Plastic Small Outline Transistor (OT) [SOT-23]

Note: For the most current package drawings, please see the Microchip Packaging Specification located at http://www.microchip.com/packaging

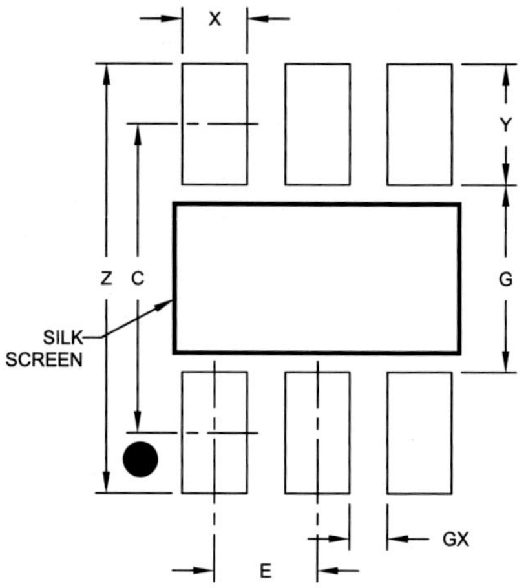

RECOMMENDED LAND PATTERN

	Units	MILLIMETERS		
	Dimension Limits	MIN	NOM	MAX
Contact Pitch	E	0.95 BSC		
Contact Pad Spacing	C		2.80	
Contact Pad Width (X6)	X			0.60
Contact Pad Length (X6)	Y			1.10
Distance Between Pads	G	1.70		
Distance Between Pads	GX	0.35		
Overall Width	Z			3.90

Notes:

1. Dimensioning and tolerancing per ASME Y14.5M

BSC: Basic Dimension. Theoretically exact value shown without tolerances.

Microchip Technology Drawing No. C04-2028A

PIC10F200/202/204/206

8-Lead Plastic Dual In-Line (P) - 300 mil Body [PDIP]

Note:	For the most current package drawings, please see the Microchip Packaging Specification located at http://www.microchip.com/packaging

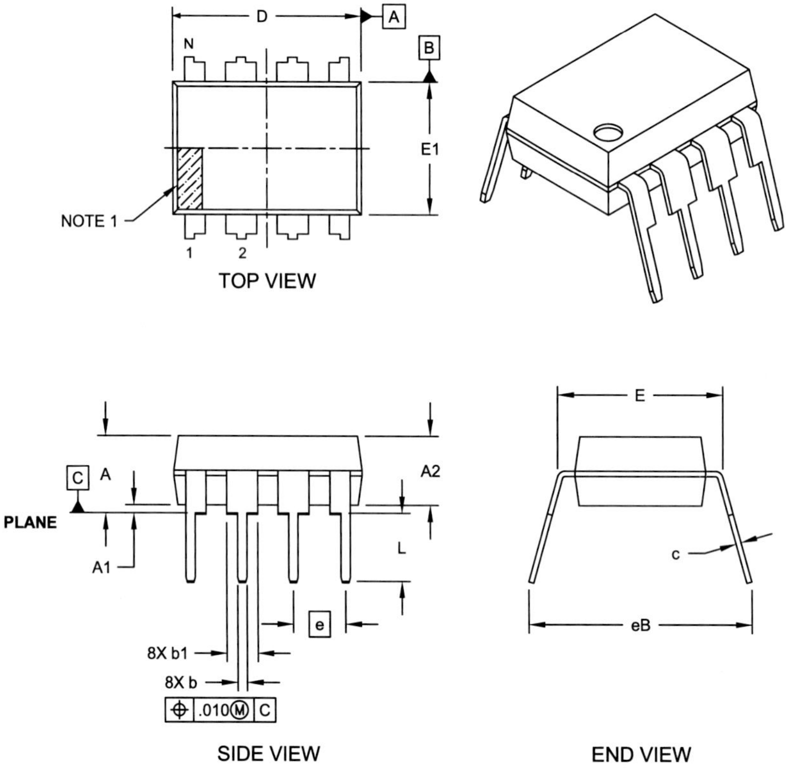

Microchip Technology Drawing No. C04-018D Sheet 1 of 2

PIC10F200/202/204/206

8-Lead Plastic Dual In-Line (P) - 300 mil Body [PDIP]

Note: For the most current package drawings, please see the Microchip Packaging Specification located at http://www.microchip.com/packaging

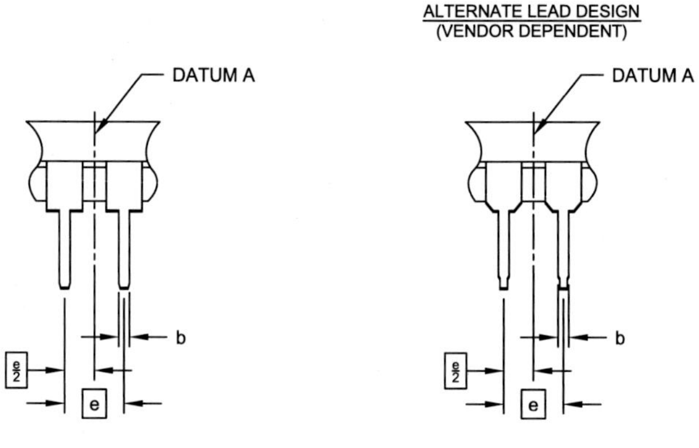

Units		INCHES		
Dimension Limits		MIN	NOM	MAX
Number of Pins	N		8	
Pitch	e		.100 BSC	
Top to Seating Plane	A	-	-	.210
Molded Package Thickness	A2	.115	.130	.195
Base to Seating Plane	A1	.015	-	-
Shoulder to Shoulder Width	E	.290	.310	.325
Molded Package Width	E1	.240	.250	.280
Overall Length	D	.348	.365	.400
Tip to Seating Plane	L	.115	.130	.150
Lead Thickness	c	.008	.010	.015
Upper Lead Width	b1	.040	.060	.070
Lower Lead Width	b	.014	.018	.022
Overall Row Spacing §	eB	-	-	.430

Notes:

1. Pin 1 visual index feature may vary, but must be located within the hatched area.
2. § Significant Characteristic
3. Dimensions D and E1 do not include mold flash or protrusions. Mold flash or protrusions shall not exceed .010" per side.
4. Dimensioning and tolerancing per ASME Y14.5M

   BSC: Basic Dimension. Theoretically exact value shown without tolerances.

Microchip Technology Drawing No. C04-018D Sheet 2 of 2

PIC10F200/202/204/206

8-Lead Plastic Dual Flat, No Lead Package (MC) – 2x3x0.9 mm Body [DFN]

Note: For the most current package drawings, please see the Microchip Packaging Specification located at http://www.microchip.com/packaging

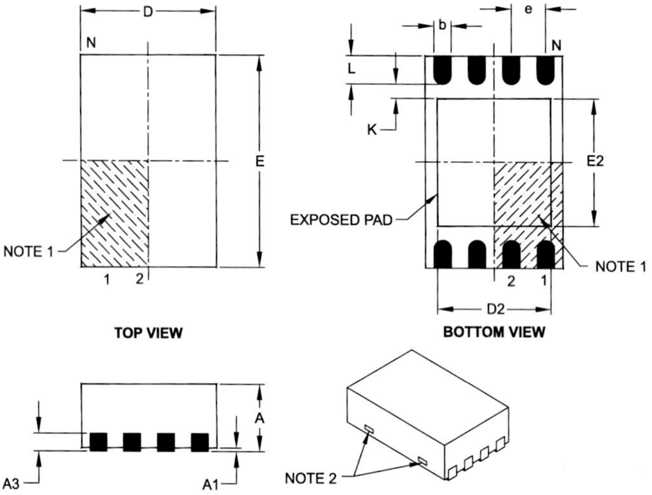

Units		MILLIMETERS		
Dimension Limits		MIN	NOM	MAX
Number of Pins	N		8	
Pitch	e		0.50 BSC	
Overall Height	A	0.80	0.90	1.00
Standoff	A1	0.00	0.02	0.05
Contact Thickness	A3		0.20 REF	
Overall Length	D		2.00 BSC	
Overall Width	E		3.00 BSC	
Exposed Pad Length	D2	1.30	–	1.55
Exposed Pad Width	E2	1.50	–	1.75
Contact Width	b	0.20	0.25	0.30
Contact Length	L	0.30	0.40	0.50
Contact-to-Exposed Pad	K	0.20	–	–

Notes:

1. Pin 1 visual index feature may vary, but must be located within the hatched area.
2. Package may have one or more exposed tie bars at ends.
3. Package is saw singulated.
4. Dimensioning and tolerancing per ASME Y14.5M.
   BSC: Basic Dimension. Theoretically exact value shown without tolerances.
   REF: Reference Dimension, usually without tolerance, for information purposes only.

Microchip Technology Drawing C04-123C

8-Lead Plastic Dual Flat, No Lead Package (MC) - 2x3x0.9mm Body [DFN]

Note: For the most current package drawings, please see the Microchip Packaging Specification located at http://www.microchip.com/packaging

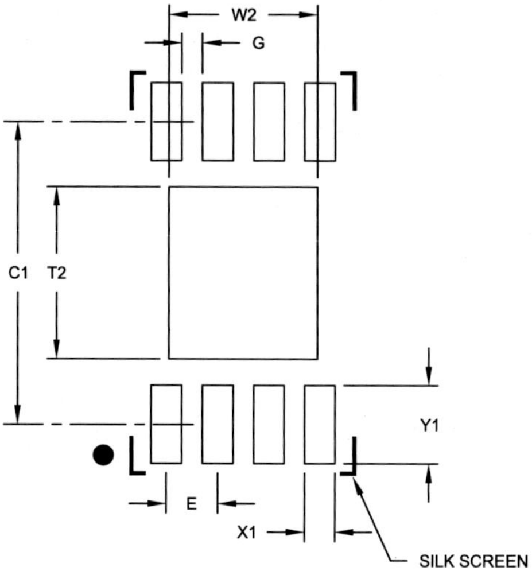

RECOMMENDED LAND PATTERN

Units		MILLIMETERS		
Dimension Limits		MIN	NOM	MAX
Contact Pitch	E		0.50 BSC	
Optional Center Pad Width	W2			1.45
Optional Center Pad Length	T2			1.75
Contact Pad Spacing	C1		2.90	
Contact Pad Width (X8)	X1			0.30
Contact Pad Length (X8)	Y1			0.75
Distance Between Pads	G	0.20		

Notes:

1. Dimensioning and tolerancing per ASME Y14.5M

BSC: Basic Dimension. Theoretically exact value shown without tolerances.

Microchip Technology Drawing No. C04-2123B

PIC10F200/202/204/206

APPENDIX A: REVISION HISTORY

Revision C (August 2006)

Added 8-Pin DFN Pin Diagram; Revised Table 1-1; Reformatted all Registers; Revised Section 4.8 and added note; Section 5.3 (changed Figure reference to Figure 5-1); Tables 6-1 and 7-1 (removed shading from TRISGPIO (I/O Control Register); Sections 8.1-8.4 (changed Table reference to Table 12-2); Section 14.1 Revised and replaced Package Marking Information and drawings, Added Tables 14-1 & 14-2, Added DFN Package drawing.

Revision D (April 2007)

Revised section 12.1, 12.2, 12.3, Table 1-1, 12-1, 12-3, 12-4. Added Section 13.0. Replaced Package Drawings (Rev. AP); Removed instances of PICmicro® and replaced it with PIC®.

Revision E (October 2013)

Revised Figure 8-1 (deleted OSCCAL); Revised Packaging Legend.

Revision F (September 2014)

Added Table 12-6 (Thermal Considerations); Updated Register 4-1, Register 9-1 and Chapter 14 (Packaging Information); Other minor corrections.

THE MICROCHIP WEB SITE

Microchip provides online support via our WWW site at www.microchip.com. This web site is used as a means to make files and information easily available to customers. Accessible by using your favorite Internet browser, the web site contains the following information:

- **Product Support** – Data sheets and errata, application notes and sample programs, design resources, user's guides and hardware support documents, latest software releases and archived software
- **General Technical Support** – Frequently Asked Questions (FAQ), technical support requests, online discussion groups, Microchip consultant program member listing
- **Business of Microchip** – Product selector and ordering guides, latest Microchip press releases, listing of seminars and events, listings of Microchip sales offices, distributors and factory representatives

CUSTOMER CHANGE NOTIFICATION SERVICE

Microchip's customer notification service helps keep customers current on Microchip products. Subscribers will receive e-mail notification whenever there are changes, updates, revisions or errata related to a specified product family or development tool of interest.

To register, access the Microchip web site at www.microchip.com. Under "Support", click on "Customer Change Notification" and follow the registration instructions.

CUSTOMER SUPPORT

Users of Microchip products can receive assistance through several channels:

- Distributor or Representative
- Local Sales Office
- Field Application Engineer (FAE)
- Technical Support

Customers should contact their distributor, representative or Field Application Engineer (FAE) for support. Local sales offices are also available to help customers. A listing of sales offices and locations is included in the back of this document.

Technical support is available through the web site at: http://microchip.com/support

PIC10F200/202/204/206

PRODUCT IDENTIFICATION SYSTEM

To order or obtain information, e.g., on pricing or delivery, refer to the factory or the listed sales office.

PART NO.	[X][(1)]	-	X	/XX	XXX
Device	Tape and Reel Option		Temperature Range	Package	Pattern

Device:	PIC10F200 PIC10F202 PIC10F204 PIC10F206 PIC10F200T (Tape & Reel) PIC10F202T (Tape & Reel) PIC10F204T (Tape & Reel) PIC10F206T (Tape & Reel)
Tape and Reel Option:	Blank = Standard packaging (tube or tray) T = Tape and Reel[(1)]
Temperature Range:	I = -40°C to +85°C (Industrial) E = -40°C to +125°C (Extended)
Package:	P = 300 mil PDIP (Pb-free) OT = SOT-23, 6-LD (Pb-free) MC = DFN, 8-LD 2x3 (Pb-free)
Pattern:	QTP, SQTP, Code or Special Requirements (blank otherwise)

Examples:

a) PIC10F202T - E/OT
Tape and Reel
Extended temperature
SOT-23 package (Pb-free)

b) PIC10F200 - I/P
Industrial temperature,
PDIP package (Pb-free)

c) PIC10F204 - I/MC
Industrial temperature
DFN package (Pb-free)

Note 1: Tape and Reel identifier only appears in the catalog part number description. This identifier is used for ordering purposes and is not printed on the device package. Check with your Microchip Sales Office for package availability with the Tape and Reel option.

Note the following details of the code protection feature on Microchip devices:

- Microchip products meet the specification contained in their particular Microchip Data Sheet.
- Microchip believes that its family of products is one of the most secure families of its kind on the market today, when used in the intended manner and under normal conditions.
- There are dishonest and possibly illegal methods used to breach the code protection feature. All of these methods, to our knowledge, require using the Microchip products in a manner outside the operating specifications contained in Microchip's Data Sheets. Most likely, the person doing so is engaged in theft of intellectual property.
- Microchip is willing to work with the customer who is concerned about the integrity of their code.
- Neither Microchip nor any other semiconductor manufacturer can guarantee the security of their code. Code protection does not mean that we are guaranteeing the product as "unbreakable."

Code protection is constantly evolving. We at Microchip are committed to continuously improving the code protection features of our products. Attempts to break Microchip's code protection feature may be a violation of the Digital Millennium Copyright Act. If such acts allow unauthorized access to your software or other copyrighted work, you may have a right to sue for relief under that Act.

Trademarks

The Microchip name and logo, the Microchip logo, dsPIC, FlashFlex, flexPWR, JukeBlox, KEELOQ, KEELOQ logo, Kleer, LANCheck, MediaLB, MOST, MOST logo, MPLAB, OptoLyzer, PIC, PICSTART, PIC[32] logo, RightTouch, SpyNIC, SST, SST Logo, SuperFlash and UNI/O are registered trademarks of Microchip Technology Incorporated in the U.S.A. and other countries.

The Embedded Control Solutions Company and mTouch are registered trademarks of Microchip Technology Incorporated in the U.S.A.

Analog-for-the-Digital Age, BodyCom, chipKIT, chipKIT logo, CodeGuard, dsPICDEM, dsPICDEM.net, ECAN, In-Circuit Serial Programming, ICSP, Inter-Chip Connectivity, KleerNet, KleerNet logo, MiWi, MPASM, MPF, MPLAB Certified logo, MPLIB, MPLINK, MultiTRAK, NetDetach, Omniscient Code Generation, PICDEM, PICDEM.net, PICkit, PICtail, RightTouch logo, REAL ICE, SQI, Serial Quad I/O, Total Endurance, TSHARC, USBCheck, VariSense, ViewSpan, WiperLock, Wireless DNA, and ZENA are trademarks of Microchip Technology Incorporated in the U.S.A. and other countries.

SQTP is a service mark of Microchip Technology Incorporated in the U.S.A.

Silicon Storage Technology is a registered trademark of Microchip Technology Inc. in other countries.

GestIC is a registered trademarks of Microchip Technology Germany II GmbH & Co. KG, a subsidiary of Microchip Technology Inc., in other countries.

All other trademarks mentioned herein are property of their respective companies.

ISBN: 978-1-63276-597-0

QUALITY MANAGEMENT SYSTEM
CERTIFIED BY DNV
= ISO/TS 16949 =

Microchip received ISO/TS-16949:2009 certification for its worldwide headquarters, design and wafer fabrication facilities in Chandler and Tempe, Arizona; Gresham, Oregon and design centers in California and India. The Company's quality system processes and procedures are for its PIC® MCUs and dsPIC® DSCs, KEELOQ® code hopping devices, Serial EEPROMs, microperipherals, nonvolatile memory and analog products. In addition, Microchip's quality system for the design and manufacture of development systems is ISO 9001:2000 certified.

MICROCHIP

Worldwide Sales and Service

AMERICAS

Corporate Office
2355 West Chandler Blvd.
Chandler, AZ 85224-6199
Tel: 480-792-7200
Fax: 480-792-7277
Technical Support:
http://www.microchip.com/support
Web Address:
www.microchip.com

Atlanta
Duluth, GA
Tel: 678-957-9614
Fax: 678-957-1455

Austin, TX
Tel: 512-257-3370

Boston
Westborough, MA
Tel: 774-760-0087
Fax: 774-760-0088

Chicago
Itasca, IL
Tel: 630-285-0071
Fax: 630-285-0075

Cleveland
Independence, OH
Tel: 216-447-0464
Fax: 216-447-0643

Dallas
Addison, TX
Tel: 972-818-7423
Fax: 972-818-2924

Detroit
Novi, MI
Tel: 248-848-4000

Houston, TX
Tel: 281-894-5983

Indianapolis
Noblesville, IN
Tel: 317-773-8323
Fax: 317-773-5453

Los Angeles
Mission Viejo, CA
Tel: 949-462-9523
Fax: 949-462-9608

New York, NY
Tel: 631-435-6000

San Jose, CA
Tel: 408-735-9110

Canada - Toronto
Tel: 905-673-0699
Fax: 905-673-6509

ASIA/PACIFIC

Asia Pacific Office
Suites 3707-14, 37th Floor
Tower 6, The Gateway
Harbour City, Kowloon
Hong Kong
Tel: 852-2943-5100
Fax: 852-2401-3431

Australia - Sydney
Tel: 61-2-9868-6733
Fax: 61-2-9868-6755

China - Beijing
Tel: 86-10-8569-7000
Fax: 86-10-8528-2104

China - Chengdu
Tel: 86-28-8665-5511
Fax: 86-28-8665-7889

China - Chongqing
Tel: 86-23-8980-9588
Fax: 86-23-8980-9500

China - Hangzhou
Tel: 86-571-8792-8115
Fax: 86-571-8792-8116

China - Hong Kong SAR
Tel: 852-2943-5100
Fax: 852-2401-3431

China - Nanjing
Tel: 86-25-8473-2460
Fax: 86-25-8473-2470

China - Qingdao
Tel: 86-532-8502-7355
Fax: 86-532-8502-7205

China - Shanghai
Tel: 86-21-5407-5533
Fax: 86-21-5407-5066

China - Shenyang
Tel: 86-24-2334-2829
Fax: 86-24-2334-2393

China - Shenzhen
Tel: 86-755-8864-2200
Fax: 86-755-8203-1760

China - Wuhan
Tel: 86-27-5980-5300
Fax: 86-27-5980-5118

China - Xian
Tel: 86-29-8833-7252
Fax: 86-29-8833-7256

China - Xiamen
Tel: 86-592-2388138
Fax: 86-592-2388130

China - Zhuhai
Tel: 86-756-3210040
Fax: 86-756-3210049

ASIA/PACIFIC

India - Bangalore
Tel: 91-80-3090-4444
Fax: 91-80-3090-4123

India - New Delhi
Tel: 91-11-4160-8631
Fax: 91-11-4160-8632

India - Pune
Tel: 91-20-3019-1500

Japan - Osaka
Tel: 81-6-6152-7160
Fax: 81-6-6152-9310

Japan - Tokyo
Tel: 81-3-6880- 3770
Fax: 81-3-6880-3771

Korea - Daegu
Tel: 82-53-744-4301
Fax: 82-53-744-4302

Korea - Seoul
Tel: 82-2-554-7200
Fax: 82-2-558-5932 or
82-2-558-5934

Malaysia - Kuala Lumpur
Tel: 60-3-6201-9857
Fax: 60-3-6201-9859

Malaysia - Penang
Tel: 60-4-227-8870
Fax: 60-4-227-4068

Philippines - Manila
Tel: 63-2-634-9065
Fax: 63-2-634-9069

Singapore
Tel: 65-6334-8870
Fax: 65-6334-8850

Taiwan - Hsin Chu
Tel: 886-3-5778-366
Fax: 886-3-5770-955

Taiwan - Kaohsiung
Tel: 886-7-213-7830

Taiwan - Taipei
Tel: 886-2-2508-8600
Fax: 886-2-2508-0102

Thailand - Bangkok
Tel: 66-2-694-1351
Fax: 66-2-694-1350

EUROPE

Austria - Wels
Tel: 43-7242-2244-39
Fax: 43-7242-2244-393

Denmark - Copenhagen
Tel: 45-4450-2828
Fax: 45-4485-2829

France - Paris
Tel: 33-1-69-53-63-20
Fax: 33-1-69-30-90-79

Germany - Dusseldorf
Tel: 49-2129-3766400

Germany - Munich
Tel: 49-89-627-144-0
Fax: 49-89-627-144-44

Germany - Pforzheim
Tel: 49-7231-424750

Italy - Milan
Tel: 39-0331-742611
Fax: 39-0331-466781

Italy - Venice
Tel: 39-049-7625286

Netherlands - Drunen
Tel: 31-416-690399
Fax: 31-416-690340

Poland - Warsaw
Tel: 48-22-3325737

Spain - Madrid
Tel: 34-91-708-08-90
Fax: 34-91-708-08-91

Sweden - Stockholm
Tel: 46-8-5090-4654

UK - Wokingham
Tel: 44-118-921-5800
Fax: 44-118-921-5820

03/25/14

MICROCHIP **PIC16F882/883/884/886/887**

28/40/44-Pin Flash-Based, 8-Bit CMOS Microcontrollers with nanoWatt Technology

High-Performance RISC CPU:

- Only 35 Instructions to Learn:
  - All single-cycle instructions except branches
- Operating Speed:
  - DC – 20 MHz oscillator/clock input
  - DC – 200 ns instruction cycle
- Interrupt Capability
- 8-Level Deep Hardware Stack
- Direct, Indirect and Relative Addressing modes

Special Microcontroller Features:

- Precision Internal Oscillator:
  - Factory calibrated to ±1%
  - Software selectable frequency range of 8 MHz to 31 kHz
  - Software tunable
  - Two-Speed Start-up mode
  - Crystal fail detect for critical applications
  - Clock mode switching during operation for power savings
- Power-Saving Sleep mode
- Wide Operating Voltage Range (2.0V-5.5V)
- Industrial and Extended Temperature Range
- Power-on Reset (POR)
- Power-up Timer (PWRT) and Oscillator Start-up Timer (OST)
- Brown-out Reset (BOR) with Software Control Option
- Enhanced Low-Current Watchdog Timer (WDT) with On-Chip Oscillator (software selectable nominal 268 seconds with full prescaler) with software enable
- Multiplexed Master Clear with Pull-up/Input Pin
- Programmable Code Protection
- High Endurance Flash/EEPROM Cell:
  - 100,000 write Flash endurance
  - 1,000,000 write EEPROM endurance
  - Flash/Data EEPROM retention: > 40 years
- Program Memory Read/Write during run time
- In-Circuit Debugger (on board)

Low-Power Features:

- Standby Current:
  - 50 nA @ 2.0V, typical
- Operating Current:
  - 11 μA @ 32 kHz, 2.0V, typical
  - 220 μA @ 4 MHz, 2.0V, typical
- Watchdog Timer Current:
  - 1 μA @ 2.0V, typical

Peripheral Features:

- 24/35 I/O Pins with Individual Direction Control:
  - High current source/sink for direct LED drive
  - Interrupt-on-Change pin
  - Individually programmable weak pull-ups
  - Ultra Low-Power Wake-up (ULPWU)
- Analog Comparator Module with:
  - Two analog comparators
  - Programmable on-chip voltage reference (CVREF) module (% of VDD)
  - Fixed voltage reference (0.6V)
  - Comparator inputs and outputs externally accessible
  - SR Latch mode
  - External Timer1 Gate (count enable)
- A/D Converter:
  - 10-bit resolution and 11/14 channels
- Timer0: 8-bit Timer/Counter with 8-bit Programmable Prescaler
- Enhanced Timer1:
  - 16-bit timer/counter with prescaler
  - External Gate Input mode
  - Dedicated low-power 32 kHz oscillator
- Timer2: 8-bit Timer/Counter with 8-bit Period Register, Prescaler and Postscaler
- Enhanced Capture, Compare, PWM+ Module:
  - 16-bit Capture, max. resolution 12.5 ns
  - Compare, max. resolution 200 ns
  - 10-bit PWM with 1, 2 or 4 output channels, programmable "dead time", max. frequency 20 kHz
  - PWM output steering control
- Capture, Compare, PWM Module:
  - 16-bit Capture, max. resolution 12.5 ns
  - 16-bit Compare, max. resolution 200 ns
  - 10-bit PWM, max. frequency 20 kHz
- Enhanced USART Module:
  - Supports RS-485, RS-232, and LIN 2.0
  - Auto-Baud Detect
  - Auto-Wake-Up on Start bit
- In-Circuit Serial Programming™ (ICSP™) via Two Pins
- Master Synchronous Serial Port (MSSP) Module supporting 3-wire SPI (all 4 modes) and I²C™ Master and Slave Modes with I²C Address Mask

PIC16F882/883/884/886/887

Device	Program Memory	Data Memory		I/O	10-bit A/D (ch)	ECCP/ CCP	EUSART	MSSP	Comparators	Timers 8/16-bit
	Flash (words)	SRAM (bytes)	EEPROM (bytes)							
PIC16F882	2048	128	128	24	11	1/1	1	1	2	2/1
PIC16F883	4096	256	256	24	11	1/1	1	1	2	2/1
PIC16F884	4096	256	256	35	14	1/1	1	1	2	2/1
PIC16F886	8192	368	256	24	11	1/1	1	1	2	2/1
PIC16F887	8192	368	256	35	14	1/1	1	1	2	2/1

PIC16F882/883/884/886/887

Pin Diagrams – PIC16F882/883/886, 28-Pin PDIP, SOIC, SSOP

28-pin PDIP, SOIC, SSOP

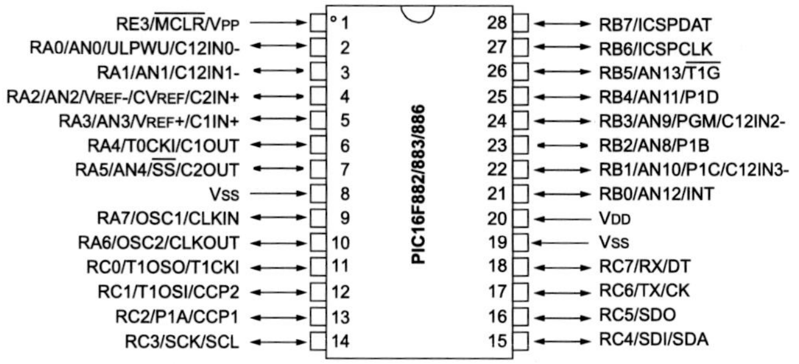

TABLE 1: PIC16F882/883/886 28-PIN SUMMARY (PDIP, SOIC, SSOP)

I/O	Pin	Analog	Comparators	Timers	ECCP	EUSART	MSSP	Interrupt	Pull-up	Basic
RA0	2	AN0/ULPWU	C12IN0-	—	—	—	—	—	—	—
RA1	3	AN1	C12IN1-	—	—	—	—	—	—	—
RA2	4	AN2	C2IN+	—	—	—	—	—	—	VREF-/CVREF
RA3	5	AN3	C1IN+	—	—	—	—	—	—	VREF+
RA4	6	—	C1OUT	T0CKI	—	—	—	—	—	—
RA5	7	AN4	C2OUT	—	—	—	SS	—	—	—
RA6	10	—	—	—	—	—	—	—	—	OSC2/CLKOUT
RA7	9	—	—	—	—	—	—	—	—	OSC1/CLKIN
RB0	21	AN12	—	—	—	—	—	IOC/INT	Y	—
RB1	22	AN10	C12IN3-	—	P1C	—	—	IOC	Y	—
RB2	23	AN8	—	—	P1B	—	—	IOC	Y	—
RB3	24	AN9	C12IN2-	—	—	—	—	IOC	Y	PGM
RB4	25	AN11	—	—	P1D	—	—	IOC	Y	—
RB5	26	AN13	—	T1G	—	—	—	IOC	Y	—
RB6	27	—	—	—	—	—	—	IOC	Y	ICSPCLK
RB7	28	—	—	—	—	—	—	IOC	Y	ICSPDAT
RC0	11	—	—	T1OSO/T1CKI	—	—	—	—	—	—
RC1	12	—	—	T1OSI	CCP2	—	—	—	—	—
RC2	13	—	—	—	CCP1/P1A	—	—	—	—	—
RC3	14	—	—	—	—	—	SCK/SCL	—	—	—
RC4	15	—	—	—	—	—	SDI/SDA	—	—	—
RC5	16	—	—	—	—	—	SDO	—	—	—
RC6	17	—	—	—	—	TX/CK	—	—	—	—
RC7	18	—	—	—	—	RX/DT	—	—	—	—
RE3	1	—	—	—	—	—	—	—	Y(1)	MCLR/VPP
—	20	—	—	—	—	—	—	—	—	VDD
—	8	—	—	—	—	—	—	—	—	VSS
—	19	—	—	—	—	—	—	—	—	VSS

Note 1: Pull-up activated only with external MCLR configuration.

PIC16F882/883/884/886/887

Table of Contents

TO OUR VALUED CUSTOMERS

It is our intention to provide our valued customers with the best documentation possible to ensure successful use of your Microchip products. To this end, we will continue to improve our publications to better suit your needs. Our publications will be refined and enhanced as new volumes and updates are introduced.

If you have any questions or comments regarding this publication, please contact the Marketing Communications Department via E-mail at **docerrors@microchip.com** or fax the **Reader Response Form** in the back of this data sheet to (480) 792-4150. We welcome your feedback.

Most Current Data Sheet

To obtain the most up-to-date version of this data sheet, please register at our Worldwide Web site at:

http://www.microchip.com

You can determine the version of a data sheet by examining its literature number found on the bottom outside corner of any page. The last character of the literature number is the version number, (e.g., DS30000A is version A of document DS30000).

Errata

An errata sheet, describing minor operational differences from the data sheet and recommended workarounds, may exist for current devices. As device/documentation issues become known to us, we will publish an errata sheet. The errata will specify the revision of silicon and revision of document to which it applies.

To determine if an errata sheet exists for a particular device, please check with one of the following:

- Microchip's Worldwide Web site; http://www.microchip.com
- Your local Microchip sales office (see last page)

When contacting a sales office, please specify which device, revision of silicon and data sheet (include literature number) you are using.

Customer Notification System

Register on our web site at **www.microchip.com** to receive the most current information on all of our products.

PIC16F882/883/884/886/887

FIGURE 1-1: PIC16F882/883/886 BLOCK DIAGRAM

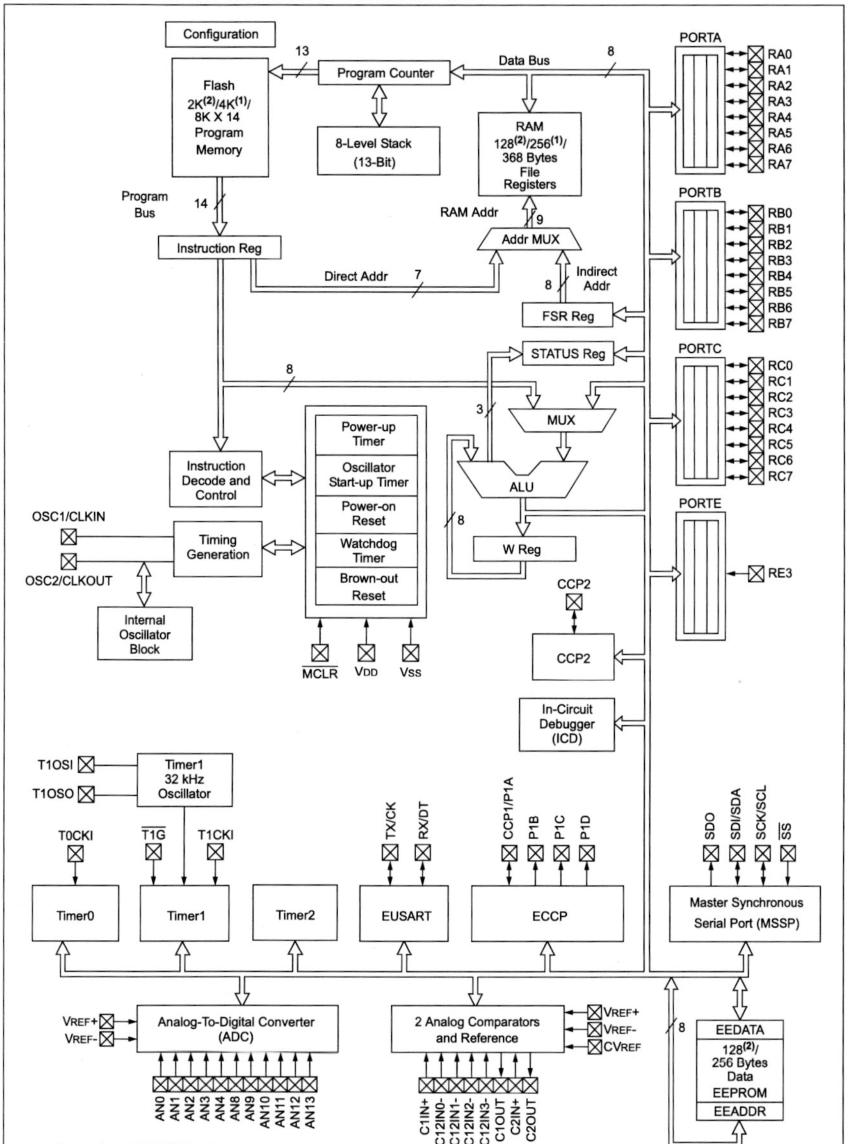

Note 1: PIC16F883 only.
2: PIC16F882 only.

FIGURE 1-2: PIC16F884/PIC16F887 BLOCK DIAGRAM

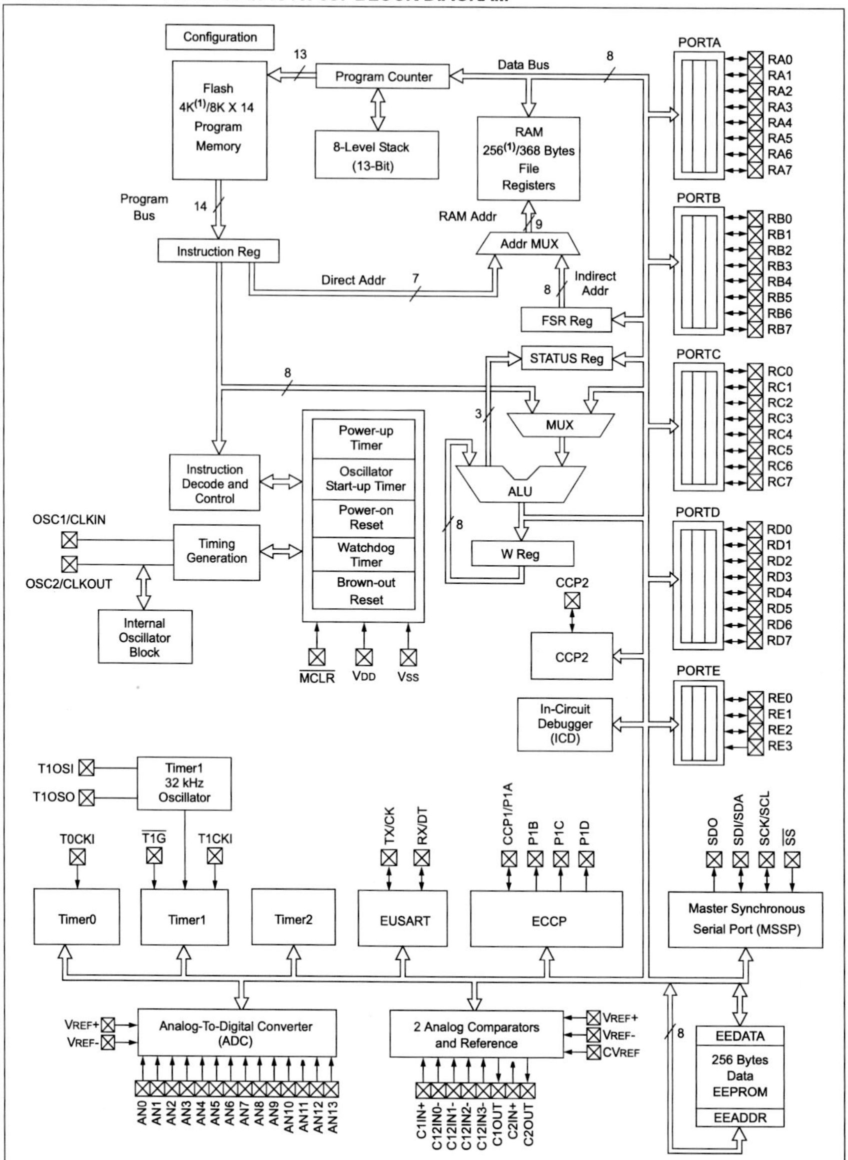

Note 1: PIC16F884 only.

PIC16F882/883/884/886/887

TABLE 1-1: PIC16F882/883/886 PINOUT DESCRIPTION

Name	Function	Input Type	Output Type	Description
RA0/AN0/ULPWU/C12IN0-	RA0	TTL	CMOS	General purpose I/O.
	AN0	AN	—	A/D Channel 0 input.
	ULPWU	AN	—	Ultra Low-Power Wake-up input.
	C12IN0-	AN	—	Comparator C1 or C2 negative input.
RA1/AN1/C12IN1-	RA1	TTL	CMOS	General purpose I/O.
	AN1	AN	—	A/D Channel 1 input.
	C12IN1-	AN	—	Comparator C1 or C2 negative input.
RA2/AN2/VREF-/CVREF/C2IN+	RA2	TTL	CMOS	General purpose I/O.
	AN2	AN	—	A/D Channel 2.
	VREF-	AN	—	A/D Negative Voltage Reference input.
	CVREF	—	AN	Comparator Voltage Reference output.
	C2IN+	AN	—	Comparator C2 positive input.
RA3/AN3/VREF+/C1IN+	RA3	TTL	—	General purpose I/O.
	AN3	AN	—	A/D Channel 3.
	VREF+	AN	—	Programming voltage.
	C1IN+	AN	—	Comparator C1 positive input.
RA4/T0CKI/C1OUT	RA4	TTL	CMOS	General purpose I/O.
	T0CKI	ST	—	Timer0 clock input.
	C1OUT	—	CMOS	Comparator C1 output.
RA5/AN4/$\overline{SS}$/C2OUT	RA5	TTL	CMOS	General purpose I/O.
	AN4	AN	—	A/D Channel 4.
	$\overline{SS}$	ST	—	Slave Select input.
	C2OUT	—	CMOS	Comparator C2 output.
RA6/OSC2/CLKOUT	RA6	TTL	CMOS	General purpose I/O.
	OSC2	—	XTAL	Master Clear with internal pull-up.
	CLKOUT	—	CMOS	FOSC/4 output.
RA7/OSC1/CLKIN	RA7	TTL	CMOS	General purpose I/O.
	OSC1	XTAL	—	Crystal/Resonator.
	CLKIN	ST	—	External clock input/RC oscillator connection.
RB0/AN12/INT	RB0	TTL	CMOS	General purpose I/O. Individually controlled interrupt-on-change. Individually enabled pull-up.
	AN12	AN	—	A/D Channel 12.
	INT	ST	—	External interrupt.
RB1/AN10/P1C/C12IN3-	RB1	TTL	CMOS	General purpose I/O. Individually controlled interrupt-on-change. Individually enabled pull-up.
	AN10	AN	—	A/D Channel 10.
	P1C	—	CMOS	PWM output.
	C12IN3-	AN	—	Comparator C1 or C2 negative input.
RB2/AN8/P1B	RB2	TTL	CMOS	General purpose I/O. Individually controlled interrupt-on-change. Individually enabled pull-up.
	AN8	AN	—	A/D Channel 8.
	P1B	—	CMOS	PWM output.

Legend: AN = Analog input or output; CMOS = CMOS compatible input or output; OD = Open Drain
TTL = TTL compatible input; ST = Schmitt Trigger input with CMOS levels
HV = High Voltage; XTAL = Crystal

TABLE 1-1: PIC16F882/883/886 PINOUT DESCRIPTION (CONTINUED)

Name	Function	Input Type	Output Type	Description
RB3/AN9/PGM/C12IN2-	RB3	TTL	CMOS	General purpose I/O. Individually controlled interrupt-on-change. Individually enabled pull-up.
	AN9	AN	—	A/D Channel 9.
	PGM	ST	—	Low-voltage ICSP™ Programming enable pin.
	C12IN2-	AN	—	Comparator C1 or C2 negative input.
RB4/AN11/P1D	RB4	TTL	CMOS	General purpose I/O. Individually controlled interrupt-on-change. Individually enabled pull-up.
	AN11	AN	—	A/D Channel 11.
	P1D	—	CMOS	PWM output.
RB5/AN13/$\overline{\text{T1G}}$	RB5	TTL	CMOS	General purpose I/O. Individually controlled interrupt-on-change. Individually enabled pull-up.
	AN13	AN	—	A/D Channel 13.
	$\overline{\text{T1G}}$	ST	—	Timer1 Gate input.
RB6/ICSPCLK	RB6	TTL	CMOS	General purpose I/O. Individually controlled interrupt-on-change. Individually enabled pull-up.
	ICSPCLK	ST	—	Serial Programming Clock.
RB7/ICSPDAT	RB7	TTL	CMOS	General purpose I/O. Individually controlled interrupt-on-change. Individually enabled pull-up.
	ICSPDAT	ST	CMOS	ICSP™ Data I/O.
RC0/T1OSO/T1CKI	RC0	ST	CMOS	General purpose I/O.
	T1OSO	—	CMOS	Timer1 oscillator output.
	T1CKI	ST	—	Timer1 clock input.
RC1/T1OSI/CCP2	RC1	ST	CMOS	General purpose I/O.
	T1OSI	ST	—	Timer1 oscillator input.
	CCP2	ST	CMOS	Capture/Compare/PWM2.
RC2/P1A/CCP1	RC2	ST	CMOS	General purpose I/O.
	P1A	—	CMOS	PWM output.
	CCP1	ST	CMOS	Capture/Compare/PWM1.
RC3/SCK/SCL	RC3	ST	CMOS	General purpose I/O.
	SCK	ST	CMOS	SPI clock.
	SCL	ST	OD	I^{2}C™ clock.
RC4/SDI/SDA	RC4	ST	CMOS	General purpose I/O.
	SDI	ST	—	SPI data input.
	SDA	ST	OD	I^{2}C data input/output.
RC5/SDO	RC5	ST	CMOS	General purpose I/O.
	SDO	—	CMOS	SPI data output.
RC6/TX/CK	RC6	ST	CMOS	General purpose I/O.
	TX	—	CMOS	EUSART asynchronous transmit.
	CK	ST	CMOS	EUSART synchronous clock.
RC7/RX/DT	RC7	ST	CMOS	General purpose I/O.
	RX	ST	—	EUSART asynchronous input.
	DT	ST	CMOS	EUSART synchronous data.
RE3/$\overline{\text{MCLR}}$/VPP	RE3	TTL	—	General purpose input.
	$\overline{\text{MCLR}}$	ST	—	Master Clear with internal pull-up.
	VPP	HV	—	Programming voltage.
VSS	VSS	Power	—	Ground reference.
VDD	VDD	Power	—	Positive supply.

Legend: AN = Analog input or output; CMOS = CMOS compatible input or output; OD = Open Drain
TTL = TTL compatible input; ST = Schmitt Trigger input with CMOS levels
HV = High Voltage; XTAL = Crystal

PIC16F882/883/884/886/887

TABLE 1-2: PIC16F884/887 PINOUT DESCRIPTION

Name	Function	Input Type	Output Type	Description
RA0/AN0/ULPWU/C12IN0-	RA0	TTL	CMOS	General purpose I/O.
	AN0	AN	—	A/D Channel 0 input.
	ULPWU	AN	—	Ultra Low-Power Wake-up input.
	C12IN0-	AN	—	Comparator C1 or C2 negative input.
RA1/AN1/C12IN1-	RA1	TTL	CMOS	General purpose I/O.
	AN1	AN	—	A/D Channel 1 input.
	C12IN1-	AN	—	Comparator C1 or C2 negative input.
RA2/AN2/VREF-/CVREF/C2IN+	RA2	TTL	CMOS	General purpose I/O.
	AN2	AN	—	A/D Channel 2.
	VREF-	AN	—	A/D Negative Voltage Reference input.
	CVREF	—	AN	Comparator Voltage Reference output.
	C2IN+	AN	—	Comparator C2 positive input.
RA3/AN3/VREF+/C1IN+	RA3	TTL	CMOS	General purpose I/O.
	AN3	AN	—	A/D Channel 3.
	VREF+	AN	—	A/D Positive Voltage Reference input.
	C1IN+	AN	—	Comparator C1 positive input.
RA4/T0CKI/C1OUT	RA4	TTL	CMOS	General purpose I/O.
	T0CKI	ST	—	Timer0 clock input.
	C1OUT	—	CMOS	Comparator C1 output.
RA5/AN4/$\overline{SS}$/C2OUT	RA5	TTL	CMOS	General purpose I/O.
	AN4	AN	—	A/D Channel 4.
	$\overline{SS}$	ST	—	Slave Select input.
	C2OUT	—	CMOS	Comparator C2 output.
RA6/OSC2/CLKOUT	RA6	TTL	CMOS	General purpose I/O.
	OSC2	—	XTAL	Crystal/Resonator.
	CLKOUT	—	CMOS	FOSC/4 output.
RA7/OSC1/CLKIN	RA7	TTL	CMOS	General purpose I/O.
	OSC1	XTAL	—	Crystal/Resonator.
	CLKIN	ST	—	External clock input/RC oscillator connection.
RB0/AN12/INT	RB0	TTL	CMOS	General purpose I/O. Individually controlled interrupt-on-change. Individually enabled pull-up.
	AN12	AN	—	A/D Channel 12.
	INT	ST	—	External interrupt.
RB1/AN10/C12IN3-	RB1	TTL	CMOS	General purpose I/O. Individually controlled interrupt-on-change. Individually enabled pull-up.
	AN10	AN	—	A/D Channel 10.
	C12IN3-	AN	—	Comparator C1 or C2 negative input.
RB2/AN8	RB2	TTL	CMOS	General purpose I/O. Individually controlled interrupt-on-change. Individually enabled pull-up.
	AN8	AN	—	A/D Channel 8.
RB3/AN9/PGM/C12IN2-	RB3	TTL	CMOS	General purpose I/O. Individually controlled interrupt-on-change. Individually enabled pull-up.
	AN9	AN	—	A/D Channel 9.
	PGM	ST	—	Low-voltage ICSP™ Programming enable pin.
	C12IN2-	AN	—	Comparator C1 or C2 negative input.

Legend: AN = Analog input or output; CMOS = CMOS compatible input or output; OD = Open Drain
TTL = TTL compatible input; ST = Schmitt Trigger input with CMOS levels
HV = High Voltage; XTAL = Crystal

TABLE 1-2: PIC16F884/887 PINOUT DESCRIPTION (CONTINUED)

Name	Function	Input Type	Output Type	Description
RB4/AN11	RB4	TTL	CMOS	General purpose I/O. Individually controlled interrupt-on-change. Individually enabled pull-up.
	AN11	AN	—	A/D Channel 11.
RB5/AN13/$\overline{T1G}$	RB5	TTL	CMOS	General purpose I/O. Individually controlled interrupt-on-change. Individually enabled pull-up.
	AN13	AN	—	A/D Channel 13.
	$\overline{T1G}$	ST	—	Timer1 Gate input.
RB6/ICSPCLK	RB6	TTL	CMOS	General purpose I/O. Individually controlled interrupt-on-change. Individually enabled pull-up.
	ICSPCLK	ST	—	Serial Programming Clock.
RB7/ICSPDAT	RB7	TTL	CMOS	General purpose I/O. Individually controlled interrupt-on-change. Individually enabled pull-up.
	ICSPDAT	ST	TTL	ICSP™ Data I/O.
RC0/T1OSO/T1CKI	RC0	ST	CMOS	General purpose I/O.
	T1OSO	—	XTAL	Timer1 oscillator output.
	T1CKI	ST	—	Timer1 clock input.
RC1/T1OSI/CCP2	RC1	ST	CMOS	General purpose I/O.
	T1OSI	XTAL	—	Timer1 oscillator input.
	CCP2	ST	CMOS	Capture/Compare/PWM2.
RC2/P1A/CCP1	RC2	ST	CMOS	General purpose I/O.
	P1A	ST	CMOS	PWM output.
	CCP1	—	CMOS	Capture/Compare/PWM1.
RC3/SCK/SCL	RC3	ST	CMOS	General purpose I/O.
	SCK	ST	CMOS	SPI clock.
	SCL	ST	OD	I^2C™ clock.
RC4/SDI/SDA	RC4	ST	CMOS	General purpose I/O.
	SDI	ST	—	SPI data input.
	SDA	ST	OD	I^2C data input/output.
RC5/SDO	RC5	ST	CMOS	General purpose I/O.
	SDO	—	CMOS	SPI data output.
RC6/TX/CK	RC6	ST	CMOS	General purpose I/O.
	TX	—	CMOS	EUSART asynchronous transmit.
	CK	ST	CMOS	EUSART synchronous clock.
RC7/RX/DT	RC7	ST	CMOS	General purpose I/O.
	RX	ST	—	EUSART asynchronous input.
	DT	ST	CMOS	EUSART synchronous data.
RD0	RD0	TTL	CMOS	General purpose I/O.
RD1	RD1	TTL	CMOS	General purpose I/O.
RD2	RD2	TTL	CMOS	General purpose I/O.
RD3	RD3	TTL	CMOS	General purpose I/O.
RD4	RD4	TTL	CMOS	General purpose I/O.
RD5/P1B	RD5	TTL	CMOS	General purpose I/O.
	P1B	—	CMOS	PWM output.
RD6/P1C	RD6	TTL	CMOS	General purpose I/O.
	P1C	—	CMOS	PWM output.

Legend: AN = Analog input or output; CMOS = CMOS compatible input or output; OD = Open Drain
TTL = TTL compatible input; ST = Schmitt Trigger input with CMOS levels
HV = High Voltage; XTAL = Crystal

PIC16F882/883/884/886/887

TABLE 1-2: PIC16F884/887 PINOUT DESCRIPTION (CONTINUED)

Name	Function	Input Type	Output Type	Description
RD7/P1D	RD7	TTL	CMOS	General purpose I/O.
	P1D	AN	—	PWM output.
RE0/AN5	RE0	TTL	CMOS	General purpose I/O.
	AN5	AN	—	A/D Channel 5.
RE1/AN6	RE1	TTL	CMOS	General purpose I/O.
	AN6	AN	—	A/D Channel 6.
RE2/AN7	RE2	TTL	CMOS	General purpose I/O.
	AN7	AN	—	A/D Channel 7.
RE3/$\overline{MCLR}$/VPP	RE3	TTL	—	General purpose input.
	$\overline{MCLR}$	ST	—	Master Clear with internal pull-up.
	VPP	HV	—	Programming voltage.
VSS	VSS	Power	—	Ground reference.
VDD	VDD	Power	—	Positive supply.

Legend: AN = Analog input or output CMOS = CMOS compatible input or output OD = Open Drain
TTL = TTL compatible input ST = Schmitt Trigger input with CMOS levels
HV = High Voltage XTAL = Crystal

2.0 MEMORY ORGANIZATION

2.1 Program Memory Organization

The PIC16F882/883/884/886/887 has a 13-bit program counter capable of addressing a 2K x 14 (0000h-07FFh) for the PIC16F882, 4K x 14 (0000h-0FFFh) for the PIC16F883/PIC16F884, and 8K x 14 (0000h-1FFFh) for the PIC16F886/PIC16F887 program memory space. Accessing a location above these boundaries will cause a wrap-around within the first 8K x 14 space. The Reset vector is at 0000h and the interrupt vector is at 0004h (see Figures 2-2 and 2-3).

FIGURE 2-1: PROGRAM MEMORY MAP AND STACK FOR THE PIC16F882

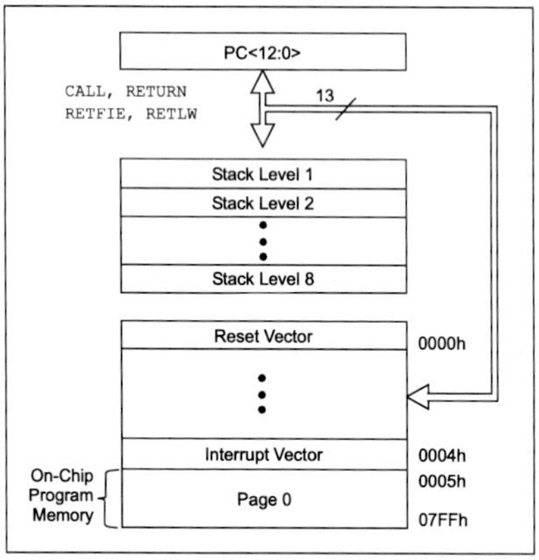

FIGURE 2-2: PROGRAM MEMORY MAP AND STACK FOR THE PIC16F883/PIC16F884

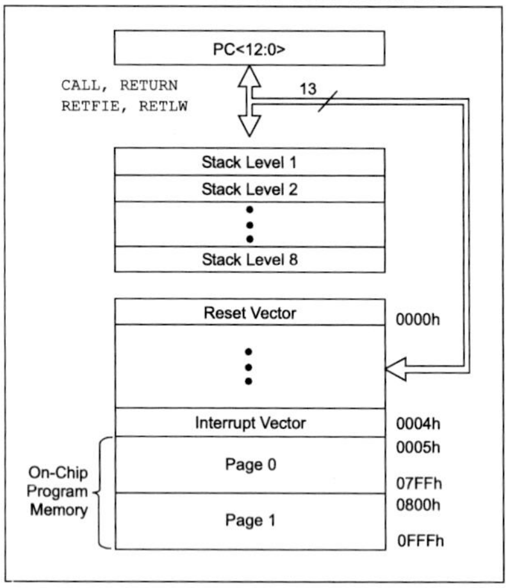

FIGURE 2-3: PROGRAM MEMORY MAP AND STACK FOR THE PIC16F886/PIC16F887

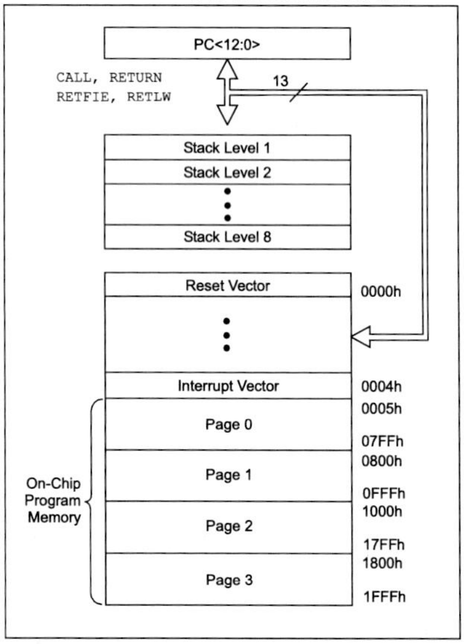

2.2 Data Memory Organization

The data memory (see Figures 2-2 and 2-3) is partitioned into four banks which contain the General Purpose Registers (GPR) and the Special Function Registers (SFR). The Special Function Registers are located in the first 32 locations of each bank. The General Purpose Registers, implemented as static RAM, are located in the last 96 locations of each Bank. Register locations F0h-FFh in Bank 1, 170h-17Fh in Bank 2 and 1F0h-1FFh in Bank 3, point to addresses 70h-7Fh in Bank 0. The actual number of General Purpose Resisters (GPR) implemented in each Bank depends on the device. Details are shown in Figures 2-5 and 2-6. All other RAM is unimplemented and returns '0' when read. RP<1:0> of the STATUS register are the bank select bits:

RP1	RP0	
0	0	→Bank 0 is selected
0	1	→Bank 1 is selected
1	0	→Bank 2 is selected
1	1	→Bank 3 is selected

2.2.1 GENERAL PURPOSE REGISTER FILE

The register file is organized as 128 x 8 in the PIC16F882, 256 x 8 in the PIC16F883/PIC16F884, and 368 x 8 in the PIC16F886/PIC16F887. Each register is accessed, either directly or indirectly, through the File Select Register (FSR) (see **Section 2.4 "Indirect Addressing, INDF and FSR Registers"**).

2.2.2 SPECIAL FUNCTION REGISTERS

The Special Function Registers are registers used by the CPU and peripheral functions for controlling the desired operation of the device (see Table 2-1). These registers are static RAM.

The special registers can be classified into two sets: core and peripheral. The Special Function Registers associated with the "core" are described in this section. Those related to the operation of the peripheral features are described in the section of that peripheral feature.

FIGURE 2-4: PIC16F882 SPECIAL FUNCTION REGISTERS

Bank 0	File Address	Bank 1	File Address	Bank 2	File Address	Bank 3	File Address
Indirect addr. (1)	00h	Indirect addr. (1)	80h	Indirect addr. (1)	100h	Indirect addr. (1)	180h
TMR0	01h	OPTION_REG	81h	TMR0	101h	OPTION_REG	181h
PCL	02h	PCL	82h	PCL	102h	PCL	182h
STATUS	03h	STATUS	83h	STATUS	103h	STATUS	183h
FSR	04h	FSR	84h	FSR	104h	FSR	184h
PORTA	05h	TRISA	85h	WDTCON	105h	SRCON	185h
PORTB	06h	TRISB	86h	PORTB	106h	TRISB	186h
PORTC	07h	TRISC	87h	CM1CON0	107h	BAUDCTL	187h
	08h		88h	CM2CON0	108h	ANSEL	188h
PORTE	09h	TRISE	89h	CM2CON1	109h	ANSELH	189h
PCLATH	0Ah	PCLATH	8Ah	PCLATH	10Ah	PCLATH	18Ah
INTCON	0Bh	INTCON	8Bh	INTCON	10Bh	INTCON	18Bh
PIR1	0Ch	PIE1	8Ch	EEDAT	10Ch	EECON1	18Ch
PIR2	0Dh	PIE2	8Dh	EEADR	10Dh	EECON2(1)	18Dh
TMR1L	0Eh	PCON	8Eh	EEDATH	10Eh	Reserved	18Eh
TMR1H	0Fh	OSCCON	8Fh	EEADRH	10Fh	Reserved	18Fh
T1CON	10h	OSCTUNE	90h		110h		190h
TMR2	11h	SSPCON2	91h		111h		191h
T2CON	12h	PR2	92h		112h		192h
SSPBUF	13h	SSPADD	93h		113h		193h
SSPCON	14h	SSPSTAT	94h		114h		194h
CCPR1L	15h	WPUB	95h		115h		195h
CCPR1H	16h	IOCB	96h		116h		196h
CCP1CON	17h	VRCON	97h		117h		197h
RCSTA	18h	TXSTA	98h		118h		198h
TXREG	19h	SPBRG	99h		119h		199h
RCREG	1Ah	SPBRGH	9Ah		11Ah		19Ah
CCPR2L	1Bh	PWM1CON	9Bh		11Bh		19Bh
CCPR2H	1Ch	ECCPAS	9Ch		11Ch		19Ch
CCP2CON	1Dh	PSTRCON	9Dh		11Dh		19Dh
ADRESH	1Eh	ADRESL	9Eh		11Eh		19Eh
ADCON0	1Fh	ADCON1	9Fh		11Fh		19Fh
General Purpose Registers 96 Bytes	20h	General Purpose Registers 32 Bytes	A0h		120h		1A0h
			BFh				
			C0h				
			EFh		16Fh		1EFh
		accesses 70h-7Fh	F0h	accesses 70h-7Fh	170h	accesses 70h-7Fh	1F0h
	7Fh		FFh		17Fh		1FFh
Bank 0		Bank 1		Bank 2		Bank 3	

☐ Unimplemented data memory locations, read as '0'.

Note 1: Not a physical register.

PIC16F882/883/884/886/887

FIGURE 2-5: PIC16F883/PIC16F884 SPECIAL FUNCTION REGISTERS

Bank 0	File Address	Bank 1	File Address	Bank 2	File Address	Bank 3	File Address
Indirect addr. (1)	00h	Indirect addr. (1)	80h	Indirect addr. (1)	100h	Indirect addr. (1)	180h
TMR0	01h	OPTION_REG	81h	TMR0	101h	OPTION_REG	181h
PCL	02h	PCL	82h	PCL	102h	PCL	182h
STATUS	03h	STATUS	83h	STATUS	103h	STATUS	183h
FSR	04h	FSR	84h	FSR	104h	FSR	184h
PORTA	05h	TRISA	85h	WDTCON	105h	SRCON	185h
PORTB	06h	TRISB	86h	PORTB	106h	TRISB	186h
PORTC	07h	TRISC	87h	CM1CON0	107h	BAUDCTL	187h
PORTD(2)	08h	TRISD(2)	88h	CM2CON0	108h	ANSEL	188h
PORTE	09h	TRISE	89h	CM2CON1	109h	ANSELH	189h
PCLATH	0Ah	PCLATH	8Ah	PCLATH	10Ah	PCLATH	18Ah
INTCON	0Bh	INTCON	8Bh	INTCON	10Bh	INTCON	18Bh
PIR1	0Ch	PIE1	8Ch	EEDAT	10Ch	EECON1	18Ch
PIR2	0Dh	PIE2	8Dh	EEADR	10Dh	EECON2(1)	18Dh
TMR1L	0Eh	PCON	8Eh	EEDATH	10Eh	Reserved	18Eh
TMR1H	0Fh	OSCCON	8Fh	EEADRH	10Fh	Reserved	18Fh
T1CON	10h	OSCTUNE	90h		110h		190h
TMR2	11h	SSPCON2	91h		111h		191h
T2CON	12h	PR2	92h		112h		192h
SSPBUF	13h	SSPADD	93h		113h		193h
SSPCON	14h	SSPSTAT	94h		114h		194h
CCPR1L	15h	WPUB	95h		115h		195h
CCPR1H	16h	IOCB	96h		116h		196h
CCP1CON	17h	VRCON	97h		117h		197h
RCSTA	18h	TXSTA	98h		118h		198h
TXREG	19h	SPBRG	99h		119h		199h
RCREG	1Ah	SPBRGH	9Ah		11Ah		19Ah
CCPR2L	1Bh	PWM1CON	9Bh		11Bh		19Bh
CCPR2H	1Ch	ECCPAS	9Ch		11Ch		19Ch
CCP2CON	1Dh	PSTRCON	9Dh		11Dh		19Dh
ADRESH	1Eh	ADRESL	9Eh		11Eh		19Eh
ADCON0	1Fh	ADCON1	9Fh		11Fh		19Fh
General Purpose Registers 96 Bytes	20h	General Purpose Registers 80 Bytes	A0h	General Purpose Registers 80 Bytes	120h		1A0h
			EFh		16Fh		1EFh
		accesses 70h-7Fh	F0h	accesses 70h-7Fh	170h	accesses 70h-7Fh	1F0h
	7Fh		FFh		17Fh		1FFh

■ Unimplemented data memory locations, read as '0'.

Note 1: Not a physical register.

2: PIC16F884 only.

PIC16F882/883/884/886/887

FIGURE 2-6: PIC16F886/PIC16F887 SPECIAL FUNCTION REGISTERS

Bank 0	File Address	Bank 1	File Address	Bank 2	File Address	Bank 3	File Address
Indirect addr. (1)	00h	Indirect addr. (1)	80h	Indirect addr. (1)	100h	Indirect addr. (1)	180h
TMR0	01h	OPTION_REG	81h	TMR0	101h	OPTION_REG	181h
PCL	02h	PCL	82h	PCL	102h	PCL	182h
STATUS	03h	STATUS	83h	STATUS	103h	STATUS	183h
FSR	04h	FSR	84h	FSR	104h	FSR	184h
PORTA	05h	TRISA	85h	WDTCON	105h	SRCON	185h
PORTB	06h	TRISB	86h	PORTB	106h	TRISB	186h
PORTC	07h	TRISC	87h	CM1CON0	107h	BAUDCTL	187h
PORTD(2)	08h	TRISD(2)	88h	CM2CON0	108h	ANSEL	188h
PORTE	09h	TRISE	89h	CM2CON1	109h	ANSELH	189h
PCLATH	0Ah	PCLATH	8Ah	PCLATH	10Ah	PCLATH	18Ah
INTCON	0Bh	INTCON	8Bh	INTCON	10Bh	INTCON	18Bh
PIR1	0Ch	PIE1	8Ch	EEDAT	10Ch	EECON1	18Ch
PIR2	0Dh	PIE2	8Dh	EEADR	10Dh	EECON2(1)	18Dh
TMR1L	0Eh	PCON	8Eh	EEDATH	10Eh	Reserved	18Eh
TMR1H	0Fh	OSCCON	8Fh	EEADRH	10Fh	Reserved	18Fh
T1CON	10h	OSCTUNE	90h	General Purpose Registers 16 Bytes	110h	General Purpose Registers 16 Bytes	190h
TMR2	11h	SSPCON2	91h		111h		191h
T2CON	12h	PR2	92h		112h		192h
SSPBUF	13h	SSPADD	93h		113h		193h
SSPCON	14h	SSPSTAT	94h		114h		194h
CCPR1L	15h	WPUB	95h		115h		195h
CCPR1H	16h	IOCB	96h		116h		196h
CCP1CON	17h	VRCON	97h		117h		197h
RCSTA	18h	TXSTA	98h		118h		198h
TXREG	19h	SPBRG	99h		119h		199h
RCREG	1Ah	SPBRGH	9Ah		11Ah		19Ah
CCPR2L	1Bh	PWM1CON	9Bh		11Bh		19Bh
CCPR2H	1Ch	ECCPAS	9Ch		11Ch		19Ch
CCP2CON	1Dh	PSTRCON	9Dh		11Dh		19Dh
ADRESH	1Eh	ADRESL	9Eh		11Eh		19Eh
ADCON0	1Fh	ADCON1	9Fh		11Fh		19Fh
General Purpose Registers 96 Bytes	20h	General Purpose Registers 80 Bytes	A0h	General Purpose Registers 80 Bytes	120h	General Purpose Registers 80 Bytes	1A0h
	3Fh						
	40h						
	6Fh		EFh		16Fh		1EFh
	70h	accesses 70h-7Fh	F0h	accesses 70h-7Fh	170h	accesses 70h-7Fh	1F0h
	7Fh		FFh		17Fh		1FFh

☐ Unimplemented data memory locations, read as '0'.

Note 1: Not a physical register.

2: PIC16F887 only.

PIC16F882/883/884/886/887

TABLE 2-1: PIC16F882/883/884/886/887 SPECIAL FUNCTION REGISTERS SUMMARY BANK 0

Addr	Name	Bit 7	Bit 6	Bit 5	Bit 4	Bit 3	Bit 2	Bit 1	Bit 0	Value on POR, BOR	Page
Bank 0											
00h	INDF	Addressing this location uses contents of FSR to address data memory (not a physical register)								xxxx xxxx	37,217
01h	TMR0	Timer0 Module Register								xxxx xxxx	73,217
02h	PCL	Program Counter's (PC) Least Significant Byte								0000 0000	37,217
03h	STATUS	IRP	RP1	RP0	$\overline{TO}$	$\overline{PD}$	Z	DC	C	0001 1xxx	29,217
04h	FSR	Indirect Data Memory Address Pointer								xxxx xxxx	37,217
05h	PORTA[3]	RA7	RA6	RA5	RA4	RA3	RA2	RA1	RA0	xxxx xxxx	39,217
06h	PORTB[3]	RB7	RB6	RB5	RB4	RB3	RB2	RB1	RB0	xxxx xxxx	48,217
07h	PORTC[3]	RC7	RC6	RC5	RC4	RC3	RC2	RC1	RC0	xxxx xxxx	53,217
08h	PORTD[3,4]	RD7	RD6	RD5	RD4	RD3	RD2	RD1	RD0	xxxx xxxx	57,217
09h	PORTE[3]	—	—	—	—	RE3	RE2[4]	RE1[4]	RE0[4]	---- xxxx	59,217
0Ah	PCLATH	—	—	—	Write Buffer for upper 5 bits of Program Counter					---0 0000	37,217
0Bh	INTCON	GIE	PEIE	T0IE	INTE	RBIE	T0IF	INTF	RBIF[1]	0000 000x	31,217
0Ch	PIR1	—	ADIF	RCIF	TXIF	SSPIF	CCP1IF	TMR2IF	TMR1IF	-000 0000	34,217
0Dh	PIR2	OSFIF	C2IF	C1IF	EEIF	BCLIF	ULPWUIF	—	CCP2IF	0000 00-0	35,217
0Eh	TMR1L	Holding Register for the Least Significant Byte of the 16-bit TMR1 Register								xxxx xxxx	76,217
0Fh	TMR1H	Holding Register for the Most Significant Byte of the 16-bit TMR1 Register								xxxx xxxx	76,217
10h	T1CON	T1GINV	TMR1GE	T1CKPS1	T1CKPS0	T1OSCEN	$\overline{T1SYNC}$	TMR1CS	TMR1ON	0000 0000	79,217
11h	TMR2	Timer2 Module Register								0000 0000	81,217
12h	T2CON	—	TOUTPS3	TOUTPS2	TOUTPS1	TOUTPS0	TMR2ON	T2CKPS1	T2CKPS0	-000 0000	82,217
13h	SSPBUF	Synchronous Serial Port Receive Buffer/Transmit Register								xxxx xxxx	183,217
14h	SSPCON[2]	WCOL	SSPOV	SSPEN	CKP	SSPM3	SSPM2	SSPM1	SSPM0	0000 0000	181,217
15h	CCPR1L	Capture/Compare/PWM Register 1 Low Byte (LSB)								xxxx xxxx	126,217
16h	CCPR1H	Capture/Compare/PWM Register 1 High Byte (MSB)								xxxx xxxx	126,217
17h	CCP1CON	P1M1	P1M0	DC1B1	DC1B0	CCP1M3	CCP1M2	CCP1M1	CCP1M0	0000 0000	124,217
18h	RCSTA	SPEN	RX9	SREN	CREN	ADDEN	FERR	OERR	RX9D	0000 000x	161,217
19h	TXREG	EUSART Transmit Data Register								0000 0000	153,217
1Ah	RCREG	EUSART Receive Data Register								0000 0000	158,217
1Bh	CCPR2L	Capture/Compare/PWM Register 2 Low Byte (LSB)								xxxx xxxx	126,217
1Ch	CCPR2H	Capture/Compare/PWM Register 2 High Byte (MSB)								xxxx xxxx	126,218
1Dh	CCP2CON	—	—	DC2B1	DC2B0	CCP2M3	CCP2M2	CCP2M1	CCP2M0	--00 0000	125,218
1Eh	ADRESH	A/D Result Register High Byte								xxxx xxxx	99,218
1Fh	ADCON0	ADCS1	ADCS0	CHS3	CHS2	CHS1	CHS0	GO/$\overline{DONE}$	ADON	0000 0000	104,218

Legend: – = Unimplemented locations read as '0', u = unchanged, x = unknown, q = value depends on condition, shaded = unimplemented

Note 1: $\overline{MCLR}$ and WDT Reset do not affect the previous value data latch. The RBIF bit will be cleared upon Reset but will set again if the mismatch exists.

2: When SSPCON register bits SSPM<3:0> = 1001, any reads or writes to the SSPADD SFR address are accessed through the SSPMSK register. See Registers • and 13-4 for more detail.

3: Port pins with analog functions controlled by the ANSEL and ANSELH registers will read '0' immediately after a Reset even though the data latches are either undefined (POR) or unchanged (other Resets).

4: PIC16F884/PIC16F887 only.

TABLE 2-2: PIC16F882/883/884/886/887 SPECIAL FUNCTION REGISTERS SUMMARY BANK 1

Addr	Name	Bit 7	Bit 6	Bit 5	Bit 4	Bit 3	Bit 2	Bit 1	Bit 0	Value on POR, BOR	Page
Bank 1											
80h	INDF	Addressing this location uses contents of FSR to address data memory (not a physical register)								xxxx xxxx	37,217
81h	OPTION_REG	$\overline{\text{RBPU}}$	INTEDG	T0CS	T0SE	PSA	PS2	PS1	PS0	1111 1111	30,218
82h	PCL	Program Counter's (PC) Least Significant Byte								0000 0000	37,217
83h	STATUS	IRP	RP1	RP0	$\overline{\text{TO}}$	$\overline{\text{PD}}$	Z	DC	C	0001 1xxx	29,217
84h	FSR	Indirect Data Memory Address Pointer								xxxx xxxx	37,217
85h	TRISA	TRISA7	TRISA6	TRISA5	TRISA4	TRISA3	TRISA2	TRISA1	TRISA0	1111 1111	39,218
86h	TRISB	TRISB7	TRISB6	TRISB5	TRISB4	TRISB3	TRISB2	TRISB1	TRISB0	1111 1111	48,218
87h	TRISC	TRISC7	TRISC6	TRISC5	TRISC4	TRISC3	TRISC2	TRISC1	TRISC0	1111 1111	53,218
88h	TRISD(3)	TRISD7	TRISD6	TRISD5	TRISD4	TRISD3	TRISD2	TRISD1	TRISD0	1111 1111	57,218
89h	TRISE	—	—	—	—	TRISE3	TRISE2(3)	TRISE1(3)	TRISE0(3)	---- 1111	59,218
8Ah	PCLATH	—	—	—	Write Buffer for the upper 5 bits of the Program Counter					---0 0000	37,217
8Bh	INTCON	GIE	PEIE	T0IE	INTE	RBIE	T0IF	INTF	RBIF(1)	0000 000x	31,217
8Ch	PIE1	—	ADIE	RCIE	TXIE	SSPIE	CCP1IE	TMR2IE	TMR1IE	-000 0000	32,218
8Dh	PIE2	OSFIE	C2IE	C1IE	EEIE	BCLIE	ULPWUIE	—	CCP2IE	0000 00-0	33,218
8Eh	PCON	—	—	ULPWUE	SBOREN	—	—	$\overline{\text{POR}}$	$\overline{\text{BOR}}$	--01 --qq	36,218
8Fh	OSCCON	—	IRCF2	IRCF1	IRCF0	OSTS	HTS	LTS	SCS	-110 q000	62,218
90h	OSCTUNE	—	—	—	TUN4	TUN3	TUN2	TUN1	TUN0	---0 0000	66,218
91h	SSPCON2	GCEN	ACKSTAT	ACKDT	ACKEN	RCEN	PEN	RSEN	SEN	0000 0000	181,218
92h	PR2	Timer2 Period Register								1111 1111	81,218
93h	SSPADD(2)	Synchronous Serial Port (I²C mode) Address Register								0000 0000	189,218
93h	SSPMSK(2)	MSK7	MSK6	MSK5	MSK4	MSK3	MSK2	MSK1	MSK0	1111 1111	189,218
94h	SSPSTAT	SMP	CKE	D/$\overline{\text{A}}$	P	S	R/$\overline{\text{W}}$	UA	BF	0000 0000	189,218
95h	WPUB	WPUB7	WPUB6	WPUB5	WPUB4	WPUB3	WPUB2	WPUB1	WPUB0	1111 1111	49,218
96h	IOCB	IOCB7	IOCB6	IOCB5	IOCB4	IOCB3	IOCB2	IOCB1	IOCB0	0000 0000	49,218
97h	VRCON	VREN	VROE	VRR	VRSS	VR3	VR2	VR1	VR0	0000 0000	97,218
98h	TXSTA	CSRC	TX9	TXEN	SYNC	SENDB	BRGH	TRMT	TX9D	0000 0010	160,218
99h	SPBRG	BRG7	BRG6	BRG5	BRG4	BRG3	BRG2	BRG1	BRG0	0000 0000	163,218
9Ah	SPBRGH	BRG15	BRG14	BRG13	BRG12	BRG11	BRG10	BRG9	BRG8	0000 0000	163,218
9Bh	PWM1CON	PRSEN	PDC6	PDC5	PDC4	PDC3	PDC2	PDC1	PDC0	0000 0000	145,218
9Ch	ECCPAS	ECCPASE	ECCPAS2	ECCPAS1	ECCPAS0	PSSAC1	PSSAC0	PSSBD1	PSSBD0	0000 0000	142,218
9Dh	PSTRCON	—	—	—	STRSYNC	STRD	STRC	STRB	STRA	---0 0001	146,218
9Eh	ADRESL	A/D Result Register Low Byte								xxxx xxxx	99,218
9Fh	ADCON1	ADFM	—	VCFG1	VCFG0	—	—	—	—	0-00 ----	105,218

Legend: — = Unimplemented locations read as '0', u = unchanged, x = unknown, q = value depends on condition, shaded = unimplemented

Note 1: $\overline{\text{MCLR}}$ and WDT Reset do not affect the previous value data latch. The RBIF bit will be cleared upon Reset but will set again if the mismatch exists.

2: Accessible only when SSPCON register bits SSPM<3:0> = 1001.

3: PIC16F884/PIC16F887 only.

PIC16F882/883/884/886/887

TABLE 2-3: PIC16F882/883/884/886/887 SPECIAL FUNCTION REGISTERS SUMMARY BANK 2

Addr	Name	Bit 7	Bit 6	Bit 5	Bit 4	Bit 3	Bit 2	Bit 1	Bit 0	Value on POR, BOR	Page
Bank 2											
100h	INDF	Addressing this location uses contents of FSR to address data memory (not a physical register)								xxxx xxxx	37,217
101h	TMR0	Timer0 Module Register								xxxx xxxx	73,217
102h	PCL	Program Counter's (PC) Least Significant Byte								0000 0000	37,217
103h	STATUS	IRP	RP1	RP0	$\overline{TO}$	$\overline{PD}$	Z	DC	C	0001 1xxx	29,217
104h	FSR	Indirect Data Memory Address Pointer								xxxx xxxx	37,217
105h	WDTCON	—	—	—	WDTPS3	WDTPS2	WDTPS1	WDTPS0	SWDTEN	---0 1000	225,218
106h	PORTB	RB7	RB6	RB5	RB4	RB3	RB2	RB1	RB0	xxxx xxxx	48,217
107h	CM1CON0	C1ON	C1OUT	C1OE	C1POL	—	C1R	C1CH1	C1CH0	0000 -000	88,218
108h	CM2CON0	C2ON	C2OUT	C2OE	C2POL	—	C2R	C2CH1	C2CH0	0000 -000	89,218
109h	CM2CON1	MC1OUT	MC2OUT	C1RSEL	C2RSEL	—	—	T1GSS	C2SYNC	0000 --10	91,219
10Ah	PCLATH	—	—	—	Write Buffer for the upper 5 bits of the Program Counter					---0 0000	37,217
10Bh	INTCON	GIE	PEIE	T0IE	INTE	RBIE	T0IF	INTF	RBIF[(1)]	0000 000x	31,217
10Ch	EEDAT	EEDAT7	EEDAT6	EEDAT5	EEDAT4	EEDAT3	EEDAT2	EEDAT1	EEDAT0	0000 0000	112,219
10Dh	EEADR	EEADR7	EEADR6	EEADR5	EEADR4	EEADR3	EEADR2	EEADR1	EEADR0	0000 0000	112,219
10Eh	EEDATH	—	—	EEDATH5	EEDATH4	EEDATH3	EEDATH2	EEDATH1	EEDATH0	--00 0000	112,219
10Fh	EEADRH	—	—	—	EEADRH4[(2)]	EEADRH3	EEADRH2	EEADRH1	EEADRH0	---- 0000	112,219

Legend: – = Unimplemented locations read as '0', u = unchanged, x = unknown, q = value depends on condition, shaded = unimplemented

Note 1: $\overline{MCLR}$ and WDT Reset does not affect the previous value data latch. The RBIF bit will be cleared upon Reset but will set again if the mismatch exists.

2: PIC16F886/PIC16F887 only.

TABLE 2-4: PIC16F882/883/884/886/887 SPECIAL FUNCTION REGISTERS SUMMARY BANK 3

Addr	Name	Bit 7	Bit 6	Bit 5	Bit 4	Bit 3	Bit 2	Bit 1	Bit 0	Value on POR, BOR	Page
Bank 3											
180h	INDF	Addressing this location uses contents of FSR to address data memory (not a physical register)								xxxx xxxx	37,217
181h	OPTION_REG	$\overline{RBPU}$	INTEDG	T0CS	T0SE	PSA	PS2	PS1	PS0	1111 1111	30,218
182h	PCL	Program Counter's (PC) Least Significant Byte								0000 0000	37,217
183h	STATUS	IRP	RP1	RP0	$\overline{TO}$	$\overline{PD}$	Z	DC	C	0001 1xxx	29,217
184h	FSR	Indirect Data Memory Address Pointer								xxxx xxxx	37,217
185h	SRCON	SR1	SR0	C1SEN	C2REN	PULSS	PULSR	—	FVREN	0000 00-0	93,219
186h	TRISB	TRISB7	TRISB6	TRISB5	TRISB4	TRISB3	TRISB2	TRISB1	TRISB0	1111 1111	48,218
187h	BAUDCTL	ABDOVF	RCIDL	—	SCKP	BRG16	—	WUE	ABDEN	01-0 0-00	162,219
188h	ANSEL	ANS7[(2)]	ANS6[(2)]	ANS5[(2)]	ANS4	ANS3	ANS2	ANS1	ANS0	1111 1111	40,219
189h	ANSELH	—	—	ANS13	ANS12	ANS11	ANS10	ANS9	ANS8	--11 1111	99,219
18Ah	PCLATH	—	—	—	Write Buffer for the upper 5 bits of the Program Counter					---0 0000	37,217
18Bh	INTCON	GIE	PEIE	T0IE	INTE	RBIE	T0IF	INTF	RBIF[(1)]	0000 000x	31,217
18Ch	EECON1	EEPGD	—	—	—	WRERR	WREN	WR	RD	x--- x000	113,219
18Dh	EECON2	EEPROM Control Register 2 (not a physical register)								---- ----	111,219

Legend: – = Unimplemented locations read as '0', u = unchanged, x = unknown, q = value depends on condition, shaded = unimplemented

Note 1: $\overline{MCLR}$ and WDT Reset does not affect the previous value data latch. The RBIF bit will be cleared upon Reset but will set again if the mismatch exists.

2: PIC16F884/PIC16F887 only.

2.2.2.1 STATUS Register

The STATUS register, shown in Register 2-1, contains:

- the arithmetic status of the ALU
- the Reset status
- the bank select bits for data memory (GPR and SFR)

The STATUS register can be the destination for any instruction, like any other register. If the STATUS register is the destination for an instruction that affects the Z, DC or C bits, then the write to these three bits is disabled. These bits are set or cleared according to the device logic. Furthermore, the $\overline{\text{TO}}$ and $\overline{\text{PD}}$ bits are not writable. Therefore, the result of an instruction with the STATUS register as destination may be different than intended.

For example, `CLRF STATUS`, will clear the upper three bits and set the Z bit. This leaves the STATUS register as `'000u u1uu'` (where `u` = unchanged).

It is recommended, therefore, that only `BCF`, `BSF`, `SWAPF` and `MOVWF` instructions are used to alter the STATUS register, because these instructions do not affect any Status bits. For other instructions not affecting any Status bits, see **Section 15.0 "Instruction Set Summary"**

> **Note 1:** The C and DC bits operate as a $\overline{\text{Borrow}}$ and $\overline{\text{Digit Borrow}}$ out bit, respectively, in subtraction.

REGISTER 2-1: STATUS: STATUS REGISTER

R/W-0	R/W-0	R/W-0	R-1	R-1	R/W-x	R/W-x	R/W-x
IRP	RP1	RP0	$\overline{\text{TO}}$	$\overline{\text{PD}}$	Z	DC(1)	C(1)
bit 7							bit 0

Legend:			
R = Readable bit	W = Writable bit	U = Unimplemented bit, read as '0'	
-n = Value at POR	'1' = Bit is set	'0' = Bit is cleared	x = Bit is unknown

bit 7 **IRP:** Register Bank Select bit (used for indirect addressing)
`1` = Bank 2, 3 (100h-1FFh)
`0` = Bank 0, 1 (00h-FFh)

bit 6-5 **RP<1:0>:** Register Bank Select bits (used for direct addressing)
`00` = Bank 0 (00h-7Fh)
`01` = Bank 1 (80h-FFh)
`10` = Bank 2 (100h-17Fh)
`11` = Bank 3 (180h-1FFh)

bit 4 **$\overline{\text{TO}}$:** Time-out bit
`1` = After power-up, `CLRWDT` instruction or `SLEEP` instruction
`0` = A WDT time-out occurred

bit 3 **$\overline{\text{PD}}$:** Power-down bit
`1` = After power-up or by the `CLRWDT` instruction
`0` = By execution of the `SLEEP` instruction

bit 2 **Z:** Zero bit
`1` = The result of an arithmetic or logic operation is zero
`0` = The result of an arithmetic or logic operation is not zero

bit 1 **DC:** Digit Carry/$\overline{\text{Borrow}}$ bit (`ADDWF`, `ADDLW`, `SUBLW`, `SUBWF` instructions)(1)
`1` = A carry-out from the 4th low-order bit of the result occurred
`0` = No carry-out from the 4th low-order bit of the result

bit 0 **C:** Carry/$\overline{\text{Borrow}}$ bit (`ADDWF`, `ADDLW`, `SUBLW`, `SUBWF` instructions)(1)
`1` = A carry-out from the Most Significant bit of the result occurred
`0` = No carry-out from the Most Significant bit of the result occurred

Note 1: For $\overline{\text{Borrow}}$, the polarity is reversed. A subtraction is executed by adding the two's complement of the second operand. For rotate (`RRF`, `RLF`) instructions, this bit is loaded with either the high-order or low-order bit of the source register.

PIC16F882/883/884/886/887

2.2.2.2 OPTION Register

The OPTION register, shown in Register 2-2, is a readable and writable register, which contains various control bits to configure:

- Timer0/WDT prescaler
- External INT interrupt
- Timer0
- Weak pull-ups on PORTB

Note:	To achieve a 1:1 prescaler assignment for Timer0, assign the prescaler to the WDT by setting PSA bit of the OPTION register to '1'. See **Section 6.3 "Timer1 Prescaler"**.

REGISTER 2-2: OPTION_REG: OPTION REGISTER

R/W-1	R/W-1	R/W-1	R/W-1	R/W-1	R/W-1	R/W-1	R/W-1
$\overline{RBPU}$	INTEDG	T0CS	T0SE	PSA	PS2	PS1	PS0
bit 7							bit 0

Legend:			
R = Readable bit	W = Writable bit	U = Unimplemented bit, read as '0'	
-n = Value at POR	'1' = Bit is set	'0' = Bit is cleared	x = Bit is unknown

bit 7 **$\overline{RBPU}$:** PORTB Pull-up Enable bit
1 = PORTB pull-ups are disabled
0 = PORTB pull-ups are enabled by individual PORT latch values

bit 6 **INTEDG:** Interrupt Edge Select bit
1 = Interrupt on rising edge of INT pin
0 = Interrupt on falling edge of INT pin

bit 5 **T0CS:** Timer0 Clock Source Select bit
1 = Transition on T0CKI pin
0 = Internal instruction cycle clock (FOSC/4)

bit 4 **T0SE:** Timer0 Source Edge Select bit
1 = Increment on high-to-low transition on T0CKI pin
0 = Increment on low-to-high transition on T0CKI pin

bit 3 **PSA:** Prescaler Assignment bit
1 = Prescaler is assigned to the WDT
0 = Prescaler is assigned to the Timer0 module

bit 2-0 **PS<2:0>:** Prescaler Rate Select bits

Bit Value	Timer0 Rate	WDT Rate
000	1 : 2	1 : 1
001	1 : 4	1 : 2
010	1 : 8	1 : 4
011	1 : 16	1 : 8
100	1 : 32	1 : 16
101	1 : 64	1 : 32
110	1 : 128	1 : 64
111	1 : 256	1 : 128

2.2.2.3 INTCON Register

The INTCON register, shown in Register 2-3, is a readable and writable register, which contains the various enable and flag bits for TMR0 register overflow, PORTB change and external INT pin interrupts.

> **Note:** Interrupt flag bits are set when an interrupt condition occurs, regardless of the state of its corresponding enable bit or the Global Enable bit, GIE of the INTCON register. User software should ensure the appropriate interrupt flag bits are clear prior to enabling an interrupt.

REGISTER 2-3: INTCON: INTERRUPT CONTROL REGISTER

R/W-0	R/W-0	R/W-0	R/W-0	R/W-0	R/W-0	R/W-0	R/W-x
GIE	PEIE	T0IE	INTE	RBIE(1)	T0IF(2)	INTF	RBIF
bit 7							bit 0

Legend:			
R = Readable bit	W = Writable bit	U = Unimplemented bit, read as '0'	
-n = Value at POR	'1' = Bit is set	'0' = Bit is cleared	x = Bit is unknown

bit 7 **GIE:** Global Interrupt Enable bit
1 = Enables all unmasked interrupts
0 = Disables all interrupts

bit 6 **PEIE:** Peripheral Interrupt Enable bit
1 = Enables all unmasked peripheral interrupts
0 = Disables all peripheral interrupts

bit 5 **T0IE:** Timer0 Overflow Interrupt Enable bit
1 = Enables the Timer0 interrupt
0 = Disables the Timer0 interrupt

bit 4 **INTE:** INT External Interrupt Enable bit
1 = Enables the INT external interrupt
0 = Disables the INT external interrupt

bit 3 **RBIE:** PORTB Change Interrupt Enable bit(1)
1 = Enables the PORTB change interrupt
0 = Disables the PORTB change interrupt

bit 2 **T0IF:** Timer0 Overflow Interrupt Flag bit(2)
1 = TMR0 register has overflowed (must be cleared in software)
0 = TMR0 register did not overflow

bit 1 **INTF:** INT External Interrupt Flag bit
1 = The INT external interrupt occurred (must be cleared in software)
0 = The INT external interrupt did not occur

bit 0 **RBIF:** PORTB Change Interrupt Flag bit
1 = When at least one of the PORTB general purpose I/O pins changed state (must be cleared in software)
0 = None of the PORTB general purpose I/O pins have changed state

Note 1: IOCB register must also be enabled.
2: T0IF bit is set when Timer0 rolls over. Timer0 is unchanged on Reset and should be initialized before clearing T0IF bit.

PIC16F882/883/884/886/887

2.2.2.4 PIE1 Register

The PIE1 register contains the interrupt enable bits, as shown in Register 2-4.

Note:	Bit PEIE of the INTCON register must be set to enable any peripheral interrupt.

REGISTER 2-4: PIE1: PERIPHERAL INTERRUPT ENABLE REGISTER 1

U-0	R/W-0	R/W-0	R/W-0	R/W-0	R/W-0	R/W-0	R/W-0
—	ADIE	RCIE	TXIE	SSPIE	CCP1IE	TMR2IE	TMR1IE
bit 7							bit 0

Legend:			
R = Readable bit	W = Writable bit	U = Unimplemented bit, read as ‘0’	
-n = Value at POR	‘1’ = Bit is set	‘0’ = Bit is cleared	x = Bit is unknown

bit 7 **Unimplemented:** Read as ‘0’

bit 6 **ADIE:** A/D Converter (ADC) Interrupt Enable bit
1 = Enables the ADC interrupt
0 = Disables the ADC interrupt

bit 5 **RCIE:** EUSART Receive Interrupt Enable bit
1 = Enables the EUSART receive interrupt
0 = Disables the EUSART receive interrupt

bit 4 **TXIE:** EUSART Transmit Interrupt Enable bit
1 = Enables the EUSART transmit interrupt
0 = Disables the EUSART transmit interrupt

bit 3 **SSPIE:** Master Synchronous Serial Port (MSSP) Interrupt Enable bit
1 = Enables the MSSP interrupt
0 = Disables the MSSP interrupt

bit 2 **CCP1IE:** CCP1 Interrupt Enable bit
1 = Enables the CCP1 interrupt
0 = Disables the CCP1 interrupt

bit 1 **TMR2IE:** Timer2 to PR2 Match Interrupt Enable bit
1 = Enables the Timer2 to PR2 match interrupt
0 = Disables the Timer2 to PR2 match interrupt

bit 0 **TMR1IE:** Timer1 Overflow Interrupt Enable bit
1 = Enables the Timer1 overflow interrupt
0 = Disables the Timer1 overflow interrupt

2.2.2.5 PIE2 Register

The PIE2 register contains the interrupt enable bits, as shown in Register 2-5.

Note:	Bit PEIE of the INTCON register must be set to enable any peripheral interrupt.

REGISTER 2-5: PIE2: PERIPHERAL INTERRUPT ENABLE REGISTER 2

R/W-0	R/W-0	R/W-0	R/W-0	R/W-0	R/W-0	U-0	R/W-0
OSFIE	C2IE	C1IE	EEIE	BCLIE	ULPWUIE	—	CCP2IE
bit 7							bit 0

Legend:			
R = Readable bit	W = Writable bit	U = Unimplemented bit, read as '0'	
-n = Value at POR	'1' = Bit is set	'0' = Bit is cleared	x = Bit is unknown

bit 7 **OSFIE:** Oscillator Fail Interrupt Enable bit
1 = Enables oscillator fail interrupt
0 = Disables oscillator fail interrupt

bit 6 **C2IE:** Comparator C2 Interrupt Enable bit
1 = Enables Comparator C2 interrupt
0 = Disables Comparator C2 interrupt

bit 5 **C1IE:** Comparator C1 Interrupt Enable bit
1 = Enables Comparator C1 interrupt
0 = Disables Comparator C1 interrupt

bit 4 **EEIE:** EEPROM Write Operation Interrupt Enable bit
1 = Enables EEPROM write operation interrupt
0 = Disables EEPROM write operation interrupt

bit 3 **BCLIE:** Bus Collision Interrupt Enable bit
1 = Enables Bus Collision interrupt
0 = Disables Bus Collision interrupt

bit 2 **ULPWUIE:** Ultra Low-Power Wake-up Interrupt Enable bit
1 = Enables Ultra Low-Power Wake-up interrupt
0 = Disables Ultra Low-Power Wake-up interrupt

bit 1 **Unimplemented:** Read as '0'

bit 0 **CCP2IE:** CCP2 Interrupt Enable bit
1 = Enables CCP2 interrupt
0 = Disables CCP2 interrupt

PIC16F882/883/884/886/887

2.2.2.6 PIR1 Register

The PIR1 register contains the interrupt flag bits, as shown in Register 2-6.

Note:	Interrupt flag bits are set when an interrupt condition occurs, regardless of the state of its corresponding enable bit or the Global Enable bit, GIE of the INTCON register. User software should ensure the appropriate interrupt flag bits are clear prior to enabling an interrupt.

REGISTER 2-6: PIR1: PERIPHERAL INTERRUPT REQUEST REGISTER 1

U-0	R/W-0	R-0	R-0	R/W-0	R/W-0	R/W-0	R/W-0
—	ADIF	RCIF	TXIF	SSPIF	CCP1IF	TMR2IF	TMR1IF
bit 7							bit 0

Legend:			
R = Readable bit	W = Writable bit	U = Unimplemented bit, read as '0'	
-n = Value at POR	'1' = Bit is set	'0' = Bit is cleared	x = Bit is unknown

bit 7 **Unimplemented:** Read as '0'

bit 6 **ADIF:** A/D Converter Interrupt Flag bit
1 = A/D conversion complete (must be cleared in software)
0 = A/D conversion has not completed or has not been started

bit 5 **RCIF:** EUSART Receive Interrupt Flag bit
1 = The EUSART receive buffer is full (cleared by reading RCREG)
0 = The EUSART receive buffer is not full

bit 4 **TXIF:** EUSART Transmit Interrupt Flag bit
1 = The EUSART transmit buffer is empty (cleared by writing to TXREG)
0 = The EUSART transmit buffer is full

bit 3 **SSPIF:** Master Synchronous Serial Port (MSSP) Interrupt Flag bit
1 = The MSSP interrupt condition has occurred, and must be cleared in software before returning from the Interrupt Service Routine. The conditions that will set this bit are:
SPI
A transmission/reception has taken place
I^2C Slave/Master
A transmission/reception has taken place
I^2C Master
The initiated Start condition was completed by the MSSP module
The initiated Stop condition was completed by the MSSP module
The initiated restart condition was completed by the MSSP module
The initiated Acknowledge condition was completed by the MSSP module
A Start condition occurred while the MSSP module was idle (Multi-master system)
A Stop condition occurred while the MSSP module was idle (Multi-master system)
0 = No MSSP interrupt condition has occurred

bit 2 **CCP1IF:** CCP1 Interrupt Flag bit
Capture mode:
1 = A TMR1 register capture occurred (must be cleared in software)
0 = No TMR1 register capture occurred
Compare mode:
1 = A TMR1 register compare match occurred (must be cleared in software)
0 = No TMR1 register compare match occurred
PWM mode:
Unused in this mode

bit 1 **TMR2IF:** Timer2 to PR2 Interrupt Flag bit
1 = A Timer2 to PR2 match occurred (must be cleared in software)
0 = No Timer2 to PR2 match occurred

bit 0 **TMR1IF:** Timer1 Overflow Interrupt Flag bit
1 = The TMR1 register overflowed (must be cleared in software)
0 = The TMR1 register did not overflow

2.2.2.7 PIR2 Register

The PIR2 register contains the interrupt flag bits, as shown in Register 2-7.

Note: Interrupt flag bits are set when an interrupt condition occurs, regardless of the state of its corresponding enable bit or the Global Enable bit, GIE of the INTCON register. User software should ensure the appropriate interrupt flag bits are clear prior to enabling an interrupt.

REGISTER 2-7: PIR2: PERIPHERAL INTERRUPT REQUEST REGISTER 2

R/W-0	R/W-0	R/W-0	R/W-0	R/W-0	R/W-0	U-0	R/W-0
OSFIF	C2IF	C1IF	EEIF	BCLIF	ULPWUIF	—	CCP2IF
bit 7							bit 0

Legend:			
R = Readable bit	W = Writable bit	U = Unimplemented bit, read as '0'	
-n = Value at POR	'1' = Bit is set	'0' = Bit is cleared	x = Bit is unknown

bit 7 **OSFIF:** Oscillator Fail Interrupt Flag bit
1 = System oscillator failed, clock input has changed to INTOSC (must be cleared in software)
0 = System clock operating

bit 6 **C2IF:** Comparator C2 Interrupt Flag bit
1 = Comparator output (C2OUT bit) has changed (must be cleared in software)
0 = Comparator output (C2OUT bit) has not changed

bit 5 **C1IF:** Comparator C1 Interrupt Flag bit
1 = Comparator output (C1OUT bit) has changed (must be cleared in software)
0 = Comparator output (C1OUT bit) has not changed

bit 4 **EEIF:** EE Write Operation Interrupt Flag bit
1 = Write operation completed (must be cleared in software)
0 = Write operation has not completed or has not started

bit 3 **BCLIF:** Bus Collision Interrupt Flag bit
1 = A bus collision has occurred in the MSSP when configured for I^2C Master mode
0 = No bus collision has occurred

bit 2 **ULPWUIF:** Ultra Low-Power Wake-up Interrupt Flag bit
1 = Wake-up condition has occurred (must be cleared in software)
0 = No Wake-up condition has occurred

bit 1 **Unimplemented:** Read as '0'

bit 0 **CCP2IF:** CCP2 Interrupt Flag bit
Capture mode:
1 = A TMR1 register capture occurred (must be cleared in software)
0 = No TMR1 register capture occurred
Compare mode:
1 = A TMR1 register compare match occurred (must be cleared in software)
0 = No TMR1 register compare match occurred
PWM mode:
Unused in this mode

PIC16F882/883/884/886/887

2.2.2.8 PCON Register

The Power Control (PCON) register (see Register 2-8) contains flag bits to differentiate between a:

- Power-on Reset ($\overline{POR}$)
- Brown-out Reset ($\overline{BOR}$)
- Watchdog Timer Reset (WDT)
- External $\overline{MCLR}$ Reset

The PCON register also controls the Ultra Low-Power Wake-up and software enable of the $\overline{BOR}$.

REGISTER 2-8: PCON: POWER CONTROL REGISTER

U-0	U-0	R/W-0	R/W-1	U-0	U-0	R/W-0	R/W-x
—	—	ULPWUE	SBOREN[1]	—	—	$\overline{POR}$	$\overline{BOR}$
bit 7							bit 0

Legend:			
R = Readable bit	W = Writable bit	U = Unimplemented bit, read as '0'	
-n = Value at POR	'1' = Bit is set	'0' = Bit is cleared	x = Bit is unknown

bit 7-6	**Unimplemented:** Read as '0'
bit 5	**ULPWUE:** Ultra Low-Power Wake-up Enable bit 1 = Ultra Low-Power Wake-up enabled 0 = Ultra Low-Power Wake-up disabled
bit 4	**SBOREN:** Software BOR Enable bit[1] 1 = BOR enabled 0 = BOR disabled
bit 3-2	**Unimplemented:** Read as '0'
bit 1	**$\overline{POR}$:** Power-on Reset Status bit 1 = No Power-on Reset occurred 0 = A Power-on Reset occurred (must be set in software after a Power-on Reset occurs)
bit 0	**$\overline{BOR}$:** Brown-out Reset Status bit 1 = No Brown-out Reset occurred 0 = A Brown-out Reset occurred (must be set in software after a Brown-out Reset occurs)

Note 1: BOREN<1:0> = 01 in the Configuration Word Register 1 for this bit to control the $\overline{BOR}$.

2.3 PCL and PCLATH

The Program Counter (PC) is 13 bits wide. The low byte comes from the PCL register, which is a readable and writable register. The high byte (PC<12:8>) is not directly readable or writable and comes from PCLATH. On any Reset, the PC is cleared. Figure 2-7 shows the two situations for the loading of the PC. The upper example in Figure 2-7 shows how the PC is loaded on a write to PCL (PCLATH<4:0> → PCH). The lower example in Figure 2-7 shows how the PC is loaded during a `CALL` or `GOTO` instruction (PCLATH<4:3> → PCH).

FIGURE 2-7: LOADING OF PC IN DIFFERENT SITUATIONS

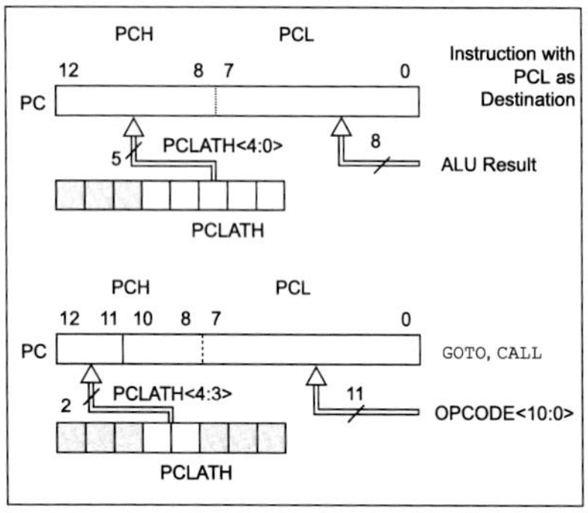

2.3.1 MODIFYING PCL

Executing any instruction with the PCL register as the destination simultaneously causes the Program Counter PC<12:8> bits (PCH) to be replaced by the contents of the PCLATH register. This allows the entire contents of the program counter to be changed by writing the desired upper 5 bits to the PCLATH register. When the lower 8 bits are written to the PCL register, all 13 bits of the program counter will change to the values contained in the PCLATH register and those being written to the PCL register.

A computed `GOTO` is accomplished by adding an offset to the program counter (`ADDWF PCL`). Care should be exercised when jumping into a look-up table or program branch table (computed `GOTO`) by modifying the PCL register. Assuming that PCLATH is set to the table start address, if the table length is greater than 255 instructions or if the lower 8 bits of the memory address rolls over from 0xFF to 0x00 in the middle of the table, then PCLATH must be incremented for each address rollover that occurs between the table beginning and the target location within the table.

For more information refer to Application Note AN556, "*Implementing a Table Read*" (DS00556).

2.3.2 STACK

The PIC16F882/883/884/886/887 devices have an 8-level x 13-bit wide hardware stack (see Figures 2-2 and 2-3). The stack space is not part of either program or data space and the Stack Pointer is not readable or writable. The PC is PUSHed onto the stack when a `CALL` instruction is executed or an interrupt causes a branch. The stack is POPed in the event of a `RETURN`, `RETLW` or a `RETFIE` instruction execution. PCLATH is not affected by a PUSH or POP operation.

The stack operates as a circular buffer. This means that after the stack has been PUSHed eight times, the ninth push overwrites the value that was stored from the first push. The tenth push overwrites the second push (and so on).

> **Note 1:** There are no Status bits to indicate stack overflow or stack underflow conditions.
>
> **2:** There are no instructions/mnemonics called PUSH or POP. These are actions that occur from the execution of the `CALL`, `RETURN`, `RETLW` and `RETFIE` instructions or the vectoring to an interrupt address.

2.4 Indirect Addressing, INDF and FSR Registers

The INDF register is not a physical register. Addressing the INDF register will cause indirect addressing.

Indirect addressing is possible by using the INDF register. Any instruction using the INDF register actually accesses data pointed to by the File Select Register (FSR). Reading INDF itself indirectly will produce 00h. Writing to the INDF register indirectly results in a no operation (although Status bits may be affected). An effective 9-bit address is obtained by concatenating the 8-bit FSR and the IRP bit of the STATUS register, as shown in Figure 2-8.

A simple program to clear RAM location 20h-2Fh using indirect addressing is shown in Example 2-1.

EXAMPLE 2-1: INDIRECT ADDRESSING

```
         MOVLW   0x20    ;initialize pointer
         MOVWF   FSR     ;to RAM
NEXT     CLRF    INDF    ;clear INDF register
         INCF    FSR     ;inc pointer
         BTFSS   FSR,4   ;all done?
         GOTO    NEXT    ;no clear next
CONTINUE                 ;yes continue
```


PIC16F882/883/884/886/887

FIGURE 2-8: DIRECT/INDIRECT ADDRESSING PIC16F882/883/884/886/887

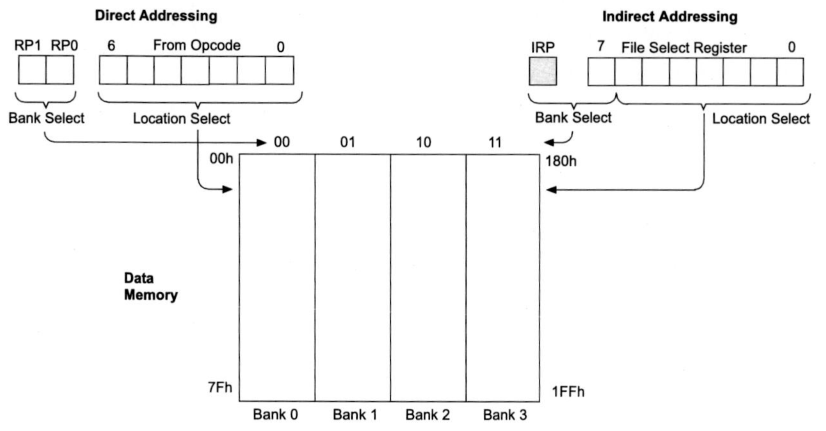

Note: For memory map detail, see Figures 2-2 and 2-3.

3.0 I/O PORTS

There are as many as thirty-five general purpose I/O pins available. Depending on which peripherals are enabled, some or all of the pins may not be available as general purpose I/O. In general, when a peripheral is enabled, the associated pin may not be used as a general purpose I/O pin.

3.1 PORTA and the TRISA Registers

PORTA is a 8-bit wide, bidirectional port. The corresponding data direction register is TRISA (Register 3-2). Setting a TRISA bit (= 1) will make the corresponding PORTA pin an input (i.e., disable the output driver). Clearing a TRISA bit (= 0) will make the corresponding PORTA pin an output (i.e., enables output driver and puts the contents of the output latch on the selected pin). Example 3-1 shows how to initialize PORTA.

Reading the PORTA register (Register 3-1) reads the status of the pins, whereas writing to it will write to the PORT latch. All write operations are read-modify-write operations. Therefore, a write to a port implies that the port pins are read, this value is modified and then written to the PORT data latch.

The TRISA register (Register 3-2) controls the PORTA pin output drivers, even when they are being used as analog inputs. The user should ensure the bits in the TRISA register are maintained set when using them as analog inputs. I/O pins configured as analog input always read '0'.

> **Note:** The ANSEL register must be initialized to configure an analog channel as a digital input. Pins configured as analog inputs will read '0'.

EXAMPLE 3-1: INITIALIZING PORTA

```
BANKSEL PORTA       ;
CLRF    PORTA       ;Init PORTA
BANKSEL ANSEL       ;
CLRF    ANSEL       ;digital I/O
BANKSEL TRISA       ;
MOVLW   0Ch         ;Set RA<3:2> as inputs
MOVWF   TRISA       ;and set RA<5:4,1:0>
                    ;as outputs
```

REGISTER 3-1: PORTA: PORTA REGISTER

R/W-x	R/W-x	R/W-x	R/W-x	R/W-x	R/W-x	R/W-x	R/W-x
RA7	RA6	RA5	RA4	RA3	RA2	RA1	RA0
bit 7							bit 0

Legend:			
R = Readable bit	W = Writable bit	U = Unimplemented bit, read as '0'	
-n = Value at POR	'1' = Bit is set	'0' = Bit is cleared	x = Bit is unknown

bit 7-0 **RA<7:0>**: PORTA I/O Pin bit
1 = Port pin is > V_{IH}
0 = Port pin is < V_{IL}

REGISTER 3-2: TRISA: PORTA TRI-STATE REGISTER

R/W-1(1)	R/W-1(1)	R/W-1	R/W-1	R/W-1	R/W-1	R/W-1	R/W-1
TRISA7	TRISA6	TRISA5	TRISA4	TRISA3	TRISA2	TRISA1	TRISA0
bit 7							bit 0

Legend:			
R = Readable bit	W = Writable bit	U = Unimplemented bit, read as '0'	
-n = Value at POR	'1' = Bit is set	'0' = Bit is cleared	x = Bit is unknown

bit 7-0 **TRISA<7:0>**: PORTA Tri-State Control bit
1 = PORTA pin configured as an input (tri-stated)
0 = PORTA pin configured as an output

Note 1: TRISA<7:6> always reads '1' in XT, HS and LP Oscillator modes.

PIC16F882/883/884/886/887

3.2 Additional Pin Functions

RA0 also has an Ultra Low-Power Wake-up option. The next three sections describe these functions.

3.2.1 ANSEL REGISTER

The ANSEL register (Register 3-3) is used to configure the Input mode of an I/O pin to analog. Setting the appropriate ANSEL bit high will cause all digital reads on the pin to be read as '0' and allow analog functions on the pin to operate correctly.

The state of the ANSEL bits has no affect on digital output functions. A pin with TRIS clear and ANSEL set will still operate as a digital output, but the Input mode will be analog. This can cause unexpected behavior when executing read-modify-write instructions on the affected port.

REGISTER 3-3: ANSEL: ANALOG SELECT REGISTER

R/W-1	R/W-1	R/W-1	R/W-1	R/W-1	R/W-1	R/W-1	R/W-1
ANS7[2]	ANS6[2]	ANS5[2]	ANS4	ANS3	ANS2	ANS1	ANS0
bit 7							bit 0

Legend:			
R = Readable bit	W = Writable bit	U = Unimplemented bit, read as '0'	
-n = Value at POR	'1' = Bit is set	'0' = Bit is cleared	x = Bit is unknown

bit 7-0 **ANS<7:0>**: Analog Select bits
Analog select between analog or digital function on pins AN<7:0>, respectively.
1 = Analog input. Pin is assigned as analog input[1].
0 = Digital I/O. Pin is assigned to port or special function.

Note 1: Setting a pin to an analog input automatically disables the digital input circuitry, weak pull-ups, and interrupt-on-change if available. The corresponding TRIS bit must be set to Input mode in order to allow external control of the voltage on the pin.

2: Not implemented on PIC16F883/886.

3.2.2 ULTRA LOW-POWER WAKE-UP

The Ultra Low-Power Wake-up (ULPWU) on RA0 allows a slow falling voltage to generate an interrupt-on-change on RA0 without excess current consumption. The mode is selected by setting the ULPWUE bit of the PCON register. This enables a small current sink, which can be used to discharge a capacitor on RA0.

Follow these steps to use this feature:

a) Charge the capacitor on RA0 by configuring the RA0 pin to output (= `1`).
b) Configure RA0 as an input.
c) Set the ULPWUIE bit of the PIE2 register to enable interrupt.
d) Set the ULPWUE bit of the PCON register to begin the capacitor discharge.
e) Execute a `SLEEP` instruction.

When the voltage on RA0 drops below V_{IL}, an interrupt will be generated which will cause the device to wake-up and execute the next instruction. If the GIE bit of the INTCON register is set, the device will then call the interrupt vector (0004h).

This feature provides a low-power technique for periodically waking up the device from Sleep. The time-out is dependent on the discharge time of the RC circuit on RA0. See Example 3-2 for initializing the Ultra Low-Power Wake-up module.

A series resistor between RA0 and the external capacitor provides overcurrent protection for the RA0/AN0/ULPWU/C12IN0- pin and can allow for software calibration of the time-out (see Figure 3-1). A timer can be used to measure the charge time and discharge time of the capacitor. The charge time can then be adjusted to provide the desired interrupt delay. This technique will compensate for the affects of temperature, voltage and component accuracy. The Ultra Low-Power Wake-up peripheral can also be configured as a simple Programmable Low Voltage Detect or temperature sensor.

> **Note:** For more information, refer to AN879, "*Using the Microchip Ultra Low-Power Wake-up Module*" Application Note (DS00879).

EXAMPLE 3-2: ULTRA LOW-POWER WAKE-UP INITIALIZATION

```
BANKSEL PORTA           ;
BSF     PORTA,0         ;Set RA0 data latch
BANKSEL ANSEL           ;
BCF     ANSEL,0         ;RA0 to digital I/O
BANKSEL TRISA           ;
BCF     TRISA,0         ;Output high to
CALL    CapDelay        ;charge capacitor
BANKSEL PIR2            ;
BCF     PIR2,ULPWUIF    ;Clear flag
BANKSEL PCON
BSF     PCON,ULPWUE     ;Enable ULP Wake-up
BSF     TRISA,0         ;RA0 to input
BSF     PIE2, ULPWUIE   ;Enable interrupt
MOVLW   B'11000000'     ;Enable peripheral
MOVWF   INTCON          ;interrupt
SLEEP                   ;Wait for IOC
NOP                     ;
```

3.2.3 PIN DESCRIPTIONS AND DIAGRAMS

Each PORTA pin is multiplexed with other functions. The pins and their combined functions are briefly described here. For specific information about individual functions such as the comparator or the A/D Converter (ADC), refer to the appropriate section in this data sheet.

3.2.3.1 RA0/AN0/ULPWU/C12IN0-

Figure 3-1 shows the diagram for this pin. This pin is configurable to function as one of the following:

- a general purpose I/O
- an analog input for the ADC
- a negative analog input to Comparator C1 or C2
- an analog input for the Ultra Low-Power Wake-up

FIGURE 3-1: BLOCK DIAGRAM OF RA0

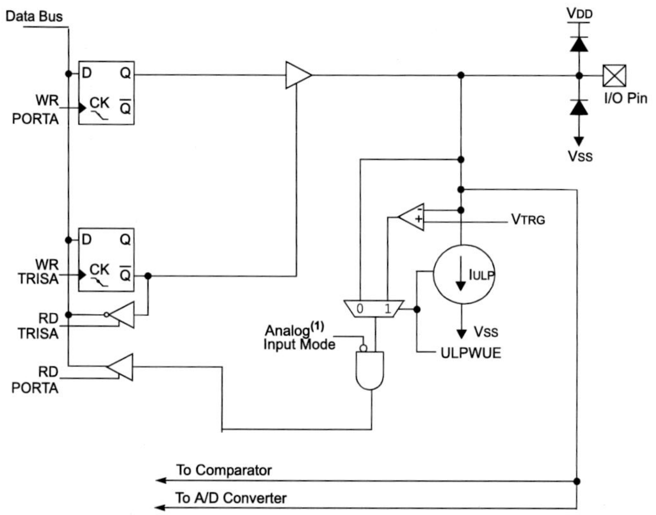

Note 1: ANSEL determines Analog Input mode.

3.2.3.2 RA1/AN1/C12IN1-

Figure 3-2 shows the diagram for this pin. This pin is configurable to function as one of the following:

- a general purpose I/O
- an analog input for the ADC
- a negative analog input to Comparator C1 or C2

FIGURE 3-2: BLOCK DIAGRAM OF RA1

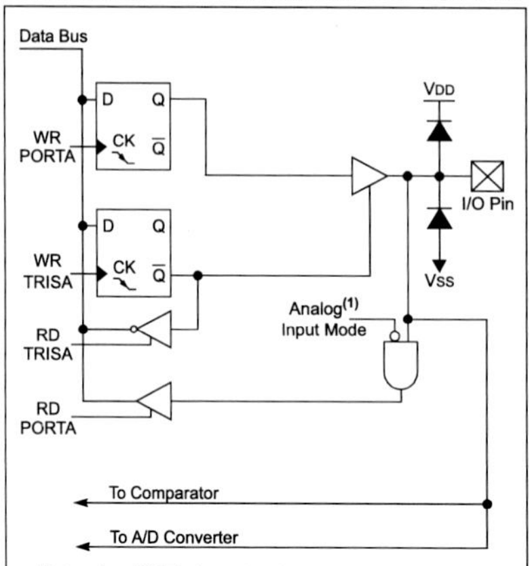

Note 1: ANSEL determines Analog Input mode.

3.2.3.3 RA2/AN2/VREF-/CVREF/C2IN+

Figure 3-3 shows the diagram for this pin. This pin is configurable to function as one of the following:

- a general purpose I/O
- an analog input for the ADC
- a negative voltage reference input for the ADC and CVREF
- a comparator voltage reference output
- a positive analog input to Comparator C2

FIGURE 3-3: BLOCK DIAGRAM OF RA2

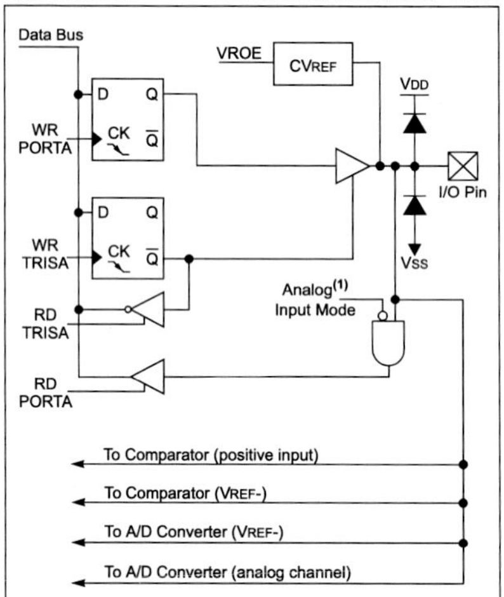

Note 1: ANSEL determines Analog Input mode.

PIC16F882/883/884/886/887

3.2.3.4 RA3/AN3/VREF+/C1IN+

Figure 3-4 shows the diagram for this pin. This pin is configurable to function as one of the following:

- a general purpose input
- an analog input for the ADC
- a positive voltage reference input for the ADC and CVREF
- a positive analog input to Comparator C1

FIGURE 3-4: BLOCK DIAGRAM OF RA3

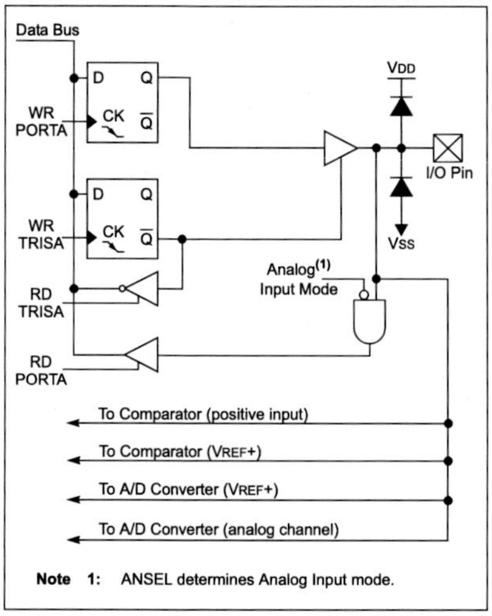

3.2.3.5 RA4/T0CKI/C1OUT

Figure 3-5 shows the diagram for this pin. This pin is configurable to function as one of the following:

- a general purpose I/O
- a clock input for Timer0
- a digital output from Comparator C1

FIGURE 3-5: BLOCK DIAGRAM OF RA4

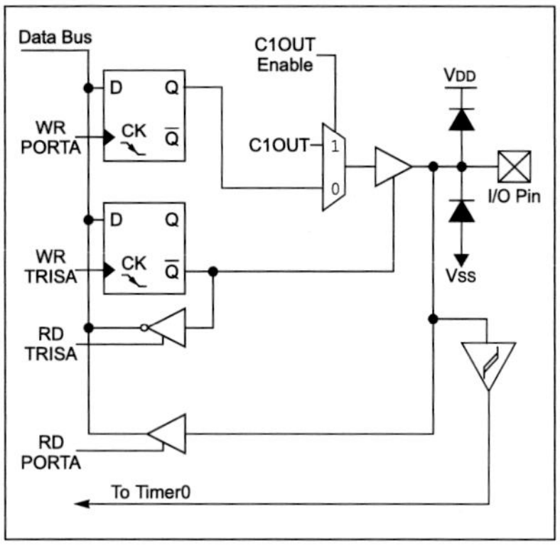

3.2.3.6 RA5/AN4/$\overline{SS}$/C2OUT

Figure 3-6 shows the diagram for this pin. This pin is configurable to function as one of the following:

- a general purpose I/O
- an analog input for the ADC
- a slave select input
- a digital output from Comparator C2

FIGURE 3-6: BLOCK DIAGRAM OF RA5

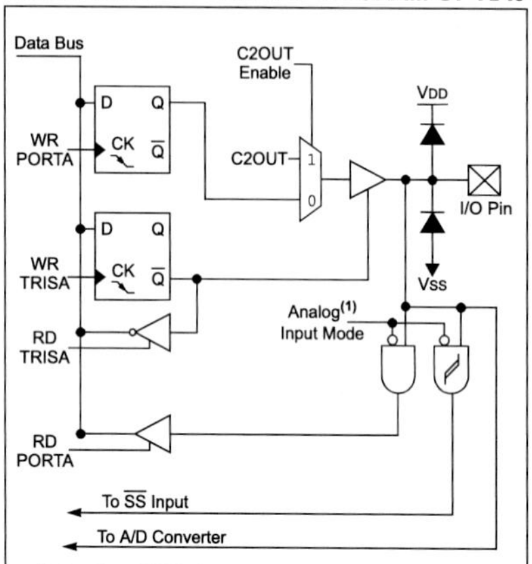

Note 1: ANSEL determines Analog Input mode.

3.2.3.7 RA6/OSC2/CLKOUT

Figure 3-7 shows the diagram for this pin. This pin is configurable to function as one of the following:

- a general purpose I/O
- a crystal/resonator connection
- a clock output

FIGURE 3-7: BLOCK DIAGRAM OF RA6

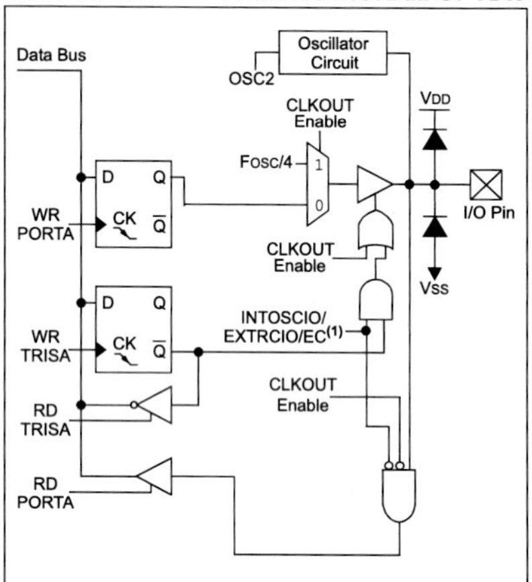

Note 1: With I/O option.

PIC16F882/883/884/886/887

3.2.3.8 RA7/OSC1/CLKIN

Figure 3-8 shows the diagram for this pin. This pin is configurable to function as one of the following:

- a general purpose I/O
- a crystal/resonator connection
- a clock input

FIGURE 3-8: BLOCK DIAGRAM OF RA7

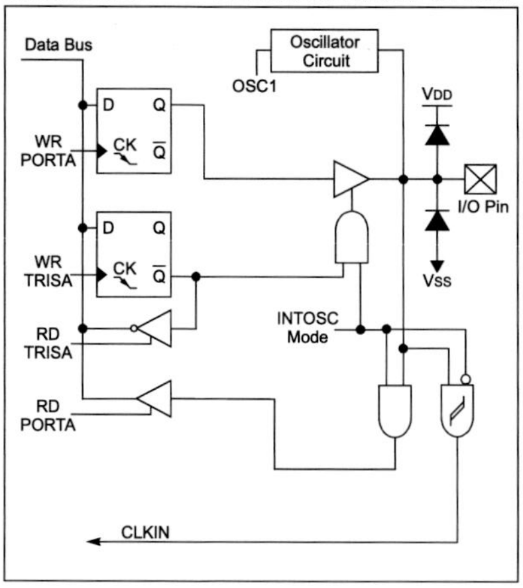

TABLE 3-1: SUMMARY OF REGISTERS ASSOCIATED WITH PORTA

Name	Bit 7	Bit 6	Bit 5	Bit 4	Bit 3	Bit 2	Bit 1	Bit 0	Value on POR, BOR	Value on all other Resets
ADCON0	ADCS1	ADCS0	CHS3	CHS2	CHS1	CHS0	GO/$\overline{\text{DONE}}$	ADON	0000 0000	0000 0000
ANSEL	ANS7	ANS6	ANS5	ANS4	ANS3	ANS2	ANS1	ANS0	1111 1111	1111 1111
CM1CON0	C1ON	C1OUT	C1OE	C1POL	—	C1R	C1CH1	C1CH0	0000 -000	0000 -000
CM2CON0	C2ON	C2OUT	C2OE	C2POL	—	C2R	C2CH1	C2CH0	0000 -000	0000 -000
CM2CON1	MC1OUT	MC2OUT	C1RSEL	C2RSEL	—	—	T1GSS	C2SYNC	0000 --10	0000 --10
PCON	—	—	ULPWUE	SBOREN	—	—	$\overline{\text{POR}}$	$\overline{\text{BOR}}$	--01 --qq	--0u --uu
OPTION_REG	$\overline{\text{RBPU}}$	INTEDG	T0CS	T0SE	PSA	PS2	PS1	PS0	1111 1111	1111 1111
PORTA	RA7	RA6	RA5	RA4	RA3	RA2	RA1	RA0	xxxx xxxx	uuuu uuuu
SSPCON	WCOL	SSPOV	SSPEN	CKP	SSPM3	SSPM2	SSPM1	SSPM0	0000 0000	0000 0000
TRISA	TRISA7	TRISA6	TRISA5	TRISA4	TRISA3	TRISA2	TRISA1	TRISA0	1111 1111	1111 1111

Legend: x = unknown, u = unchanged, – = unimplemented locations read as '0'. Shaded cells are not used by PORTA.

3.3 PORTB and TRISB Registers

PORTB is an 8-bit wide, bidirectional port. The corresponding data direction register is TRISB (Register 3-6). Setting a TRISB bit (= `1`) will make the corresponding PORTB pin an input (i.e., put the corresponding output driver in a High-Impedance mode). Clearing a TRISB bit (= `0`) will make the corresponding PORTB pin an output (i.e., enable the output driver and put the contents of the output latch on the selected pin). Example 3-3 shows how to initialize PORTB.

Reading the PORTB register (Register 3-5) reads the status of the pins, whereas writing to it will write to the PORT latch. All write operations are read-modify-write operations. Therefore, a write to a port implies that the port pins are read, this value is modified and then written to the PORT data latch.

The TRISB register (Register 3-6) controls the PORTB pin output drivers, even when they are being used as analog inputs. The user should ensure the bits in the TRISB register are maintained set when using them as analog inputs. I/O pins configured as analog input always read '0'. Example 3-3 shows how to initialize PORTB.

EXAMPLE 3-3: INITIALIZING PORTB

```
BANKSEL PORTB        ;
CLRF    PORTB        ;Init PORTB
BANKSEL TRISB        ;
MOVLW   B'11110000'  ;Set RB<7:4> as inputs
                     ;and RB<3:0> as outputs
MOVWF   TRISB        ;
```

> **Note:** The ANSELH register must be initialized to configure an analog channel as a digital input. Pins configured as analog inputs will read '0'.

3.4 Additional PORTB Pin Functions

PORTB pins RB<7:0> on the device family device have an interrupt-on-change option and a weak pull-up option. The following three sections describe these PORTB pin functions.

Every PORTB pin on this device family has an interrupt-on-change option and a weak pull-up option.

3.4.1 ANSELH REGISTER

The ANSELH register (Register 3-4) is used to configure the Input mode of an I/O pin to analog. Setting the appropriate ANSELH bit high will cause all digital reads on the pin to be read as '0' and allow analog functions on the pin to operate correctly.

The state of the ANSELH bits has no affect on digital output functions. A pin with TRIS clear and ANSELH set will still operate as a digital output, but the Input mode will be analog. This can cause unexpected behavior when executing read-modify-write instructions on the affected port.

3.4.2 WEAK PULL-UPS

Each of the PORTB pins has an individually configurable internal weak pull-up. Control bits WPUB<7:0> enable or disable each pull-up (see Register 3-7). Each weak pull-up is automatically turned off when the port pin is configured as an output. All pull-ups are disabled on a Power-on Reset by the $\overline{\text{RBPU}}$ bit of the OPTION register.

3.4.3 INTERRUPT-ON-CHANGE

All of the PORTB pins are individually configurable as an interrupt-on-change pin. Control bits IOCB<7:0> enable or disable the interrupt function for each pin. Refer to Register 3-8. The interrupt-on-change feature is disabled on a Power-on Reset.

For enabled interrupt-on-change pins, the present value is compared with the old value latched on the last read of PORTB to determine which bits have changed or mismatched the old value. The 'mismatch' outputs of the last read are OR'd together to set the PORTB Change Interrupt flag bit (RBIF) in the INTCON register.

This interrupt can wake the device from Sleep. The user, in the Interrupt Service Routine, clears the interrupt by:

a) Any read or write of PORTB. This will end the mismatch condition.

b) Clear the flag bit RBIF.

A mismatch condition will continue to set flag bit RBIF. Reading or writing PORTB will end the mismatch condition and allow flag bit RBIF to be cleared. The latch holding the last read value is not affected by a $\overline{\text{MCLR}}$ nor Brown-out Reset. After these Resets, the RBIF flag will continue to be set if a mismatch is present.

> **Note:** If a change on the I/O pin should occur when the read operation is being executed (start of the Q2 cycle), then the RBIF interrupt flag may not get set. Furthermore, since a read or write on a port affects all bits of that port, care must be taken when using multiple pins in Interrupt-on-Change mode. Changes on one pin may not be seen while servicing changes on another pin.

PIC16F882/883/884/886/887

REGISTER 3-4: ANSELH: ANALOG SELECT HIGH REGISTER

U-0	U-0	R/W-1	R/W-1	R/W-1	R/W-1	R/W-1	R/W-1
—	—	ANS13	ANS12	ANS11	ANS10	ANS9	ANS8
bit 7							bit 0

Legend:			
R = Readable bit	W = Writable bit	U = Unimplemented bit, read as '0'	
-n = Value at POR	'1' = Bit is set	'0' = Bit is cleared	x = Bit is unknown

bit 7-6 **Unimplemented:** Read as '0'

bit 5-0 **ANS<13:8>**: Analog Select bits
Analog select between analog or digital function on pins AN<13:8>, respectively.
1 = Analog input. Pin is assigned as analog input(1).
0 = Digital I/O. Pin is assigned to port or special function.

Note 1: Setting a pin to an analog input automatically disables the digital input circuitry, weak pull-ups, and interrupt-on-change if available. The corresponding TRIS bit must be set to Input mode in order to allow external control of the voltage on the pin.

REGISTER 3-5: PORTB: PORTB REGISTER

R/W-x	R/W-x	R/W-x	R/W-x	R/W-x	R/W-x	R/W-x	R/W-x
RB7	RB6	RB5	RB4	RB3	RB2	RB1	RB0
bit 7							bit 0

Legend:			
R = Readable bit	W = Writable bit	U = Unimplemented bit, read as '0'	
-n = Value at POR	'1' = Bit is set	'0' = Bit is cleared	x = Bit is unknown

bit 7-0 **RB<7:0>**: PORTB I/O Pin bit
1 = Port pin is > V_{IH}
0 = Port pin is < V_{IL}

REGISTER 3-6: TRISB: PORTB TRI-STATE REGISTER

R/W-1	R/W-1	R/W-1	R/W-1	R/W-1	R/W-1	R/W-1	R/W-1
TRISB7	TRISB6	TRISB5	TRISB4	TRISB3	TRISB2	TRISB1	TRISB0
bit 7							bit 0

Legend:			
R = Readable bit	W = Writable bit	U = Unimplemented bit, read as '0'	
-n = Value at POR	'1' = Bit is set	'0' = Bit is cleared	x = Bit is unknown

bit 7-0 **TRISB<7:0>:** PORTB Tri-State Control bit
1 = PORTB pin configured as an input (tri-stated)
0 = PORTB pin configured as an output

REGISTER 3-7: WPUB: WEAK PULL-UP PORTB REGISTER

R/W-1	R/W-1	R/W-1	R/W-1	R/W-1	R/W-1	R/W-1	R/W-1
WPUB7	WPUB6	WPUB5	WPUB4	WPUB3	WPUB2	WPUB1	WPUB0
bit 7							bit 0

Legend:			
R = Readable bit	W = Writable bit	U = Unimplemented bit, read as ‘0’	
-n = Value at POR	‘1’ = Bit is set	‘0’ = Bit is cleared	x = Bit is unknown

bit 7-0 **WPUB<7:0>**: Weak Pull-up Register bit
1 = Pull-up enabled
0 = Pull-up disabled

Note 1: Global $\overline{RBPU}$ bit of the OPTION register must be cleared for individual pull-ups to be enabled.
2: The weak pull-up device is automatically disabled if the pin is in configured as an output.

REGISTER 3-8: IOCB: INTERRUPT-ON-CHANGE PORTB REGISTER

R/W-0	R/W-0	R/W-0	R/W-0	R/W-0	R/W-0	R/W-0	R/W-0
IOCB7	IOCB6	IOCB5	IOCB4	IOCB3	IOCB2	IOCB1	IOCB0
bit 7							bit 0

Legend:			
R = Readable bit	W = Writable bit	U = Unimplemented bit, read as ‘0’	
-n = Value at POR	‘1’ = Bit is set	‘0’ = Bit is cleared	x = Bit is unknown

bit 7-0 **IOCB<7:0>:** Interrupt-on-Change PORTB Control bit
1 = Interrupt-on-change enabled
0 = Interrupt-on-change disabled

3.4.4 PIN DESCRIPTIONS AND DIAGRAMS

Each PORTB pin is multiplexed with other functions. The pins and their combined functions are briefly described here. For specific information about individual functions such as the SSP, I^{2}C or interrupts, refer to the appropriate section in this data sheet.

3.4.4.1 RB0/AN12/INT

Figure 3-9 shows the diagram for this pin. This pin is configurable to function as one of the following:

- a general purpose I/O
- an analog input for the ADC
- an external edge triggered interrupt

3.4.4.2 RB1/AN10/P1C[1]/C12IN3-

Figure 3-9 shows the diagram for this pin. This pin is configurable to function as one of the following:

- a general purpose I/O
- an analog input for the ADC
- a PWM output[1]
- an analog input to Comparator C1 or C2

Note 1: P1C is available on PIC16F882/883/886 only.

3.4.4.3 RB2/AN8/P1B[1]

Figure 3-9 shows the diagram for this pin. This pin is configurable to function as one of the following:

- a general purpose I/O
- an analog input for the ADC
- a PWM output[1]

Note 1: P1B is available on PIC16F882/883/886 only.

3.4.4.4 RB3/AN9/PGM/C12IN2-

Figure 3-9 shows the diagram for this pin. This pin is configurable to function as one of the following:

- a general purpose I/O
- an analog input for the ADC
- Low-voltage In-Circuit Serial Programming enable pin
- an analog input to Comparator C1 or C2

FIGURE 3-9: BLOCK DIAGRAM OF RB<3:0>

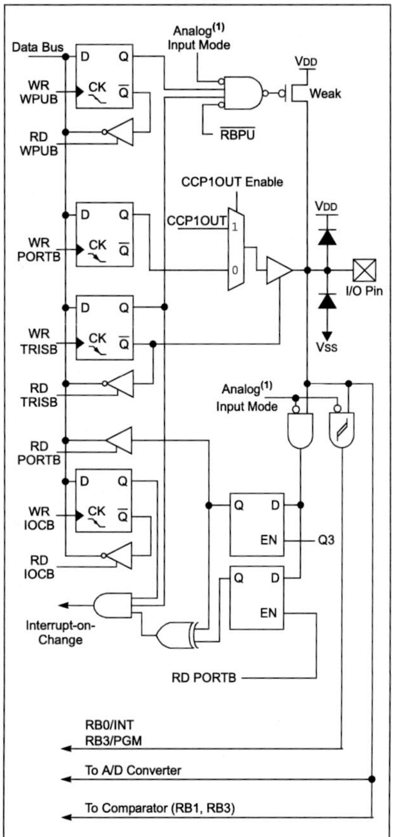

Note 1: ANSELH determines Analog Input mode.

3.4.4.5 RB4/AN11/P1D(1)

Figure 3-10 shows the diagram for this pin. This pin is configurable to function as one of the following:

- a general purpose I/O
- an analog input for the ADC
- a PWM output(1)

> **Note 1:** P1D is available on PIC16F882/883/886 only.

3.4.4.6 RB5/AN13/$\overline{\text{T1G}}$

Figure 3-10 shows the diagram for this pin. This pin is configurable to function as one of the following:

- a general purpose I/O
- an analog input for the ADC
- a Timer1 gate input

3.4.4.7 RB6/ICSPCLK

Figure 3-10 shows the diagram for this pin. This pin is configurable to function as one of the following:

- a general purpose I/O
- In-Circuit Serial Programming clock

3.4.4.8 RB7/ICSPDAT

Figure 3-10 shows the diagram for this pin. This pin is configurable to function as one of the following:

- a general purpose I/O
- In-Circuit Serial Programming data

FIGURE 3-10: BLOCK DIAGRAM OF RB<7:4>

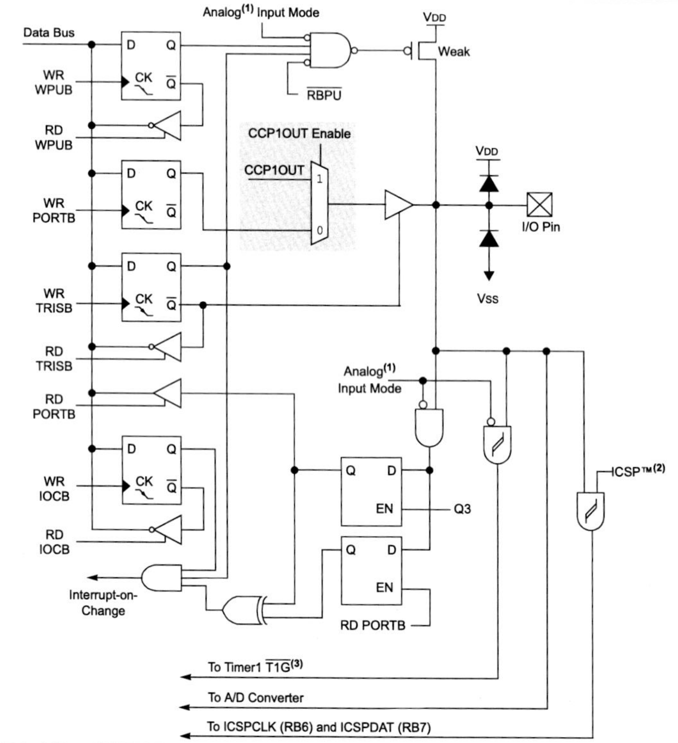

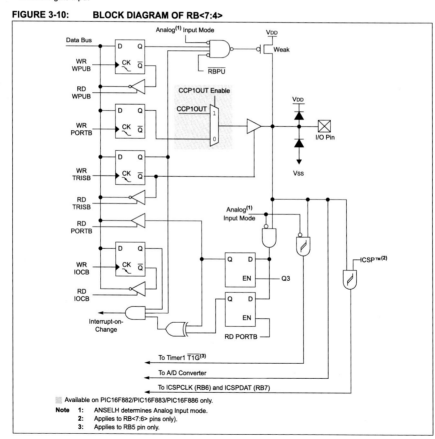

Available on PIC16F882/PIC16F883/PIC16F886 only.

Note 1: ANSELH determines Analog Input mode.
2: Applies to RB<7:6> pins only).
3: Applies to RB5 pin only.

PIC16F882/883/884/886/887

TABLE 3-2: SUMMARY OF REGISTERS ASSOCIATED WITH PORTB

Name	Bit 7	Bit 6	Bit 5	Bit 4	Bit 3	Bit 2	Bit 1	Bit 0	Value on POR, BOR	Value on all other Resets
ANSELH	—	—	ANS13	ANS12	ANS11	ANS10	ANS9	ANS8	--11 1111	--11 1111
CCP1CON	P1M1	P1M0	DC1B1	DC1B0	CCP1M3	CCP1M2	CCP1M1	CCP1M0	0000 0000	0000 0000
CM2CON1	MC1OUT	MC2OUT	C1RSEL	C2RSEL	—	—	T1GSS	C2SYNC	0000 --10	0000 --10
IOCB	IOCB7	IOCB6	IOCB5	IOCB4	IOCB3	IOCB2	IOCB1	IOCB0	0000 0000	0000 0000
INTCON	GIE	PEIE	T0IE	INTE	RBIE	T0IF	INTF	RBIF	0000 000x	0000 000x
OPTION_REG	$\overline{\text{RBPU}}$	INTEDG	T0CS	T0SE	PSA	PS2	PS1	PS0	1111 1111	1111 1111
PORTB	RB7	RB6	RB5	RB4	RB3	RB2	RB1	RB0	xxxx xxxx	uuuu uuuu
TRISB	TRISB7	TRISB6	TRISB5	TRISB4	TRISB3	TRISB2	TRISB1	TRISB0	1111 1111	1111 1111
WPUB	WPUB7	WPUB6	WPUB5	WPUB4	WPUB3	WPUB2	WPUB1	WPUB0	1111 1111	1111 1111

Legend: x = unknown, u = unchanged, – = unimplemented read as '0'. Shaded cells are not used by PORTB.

3.5 PORTC and TRISC Registers

PORTC is a 8-bit wide, bidirectional port. The corresponding data direction register is TRISC (Register 3-10). Setting a TRISC bit (= 1) will make the corresponding PORTC pin an input (i.e., put the corresponding output driver in a High-Impedance mode). Clearing a TRISC bit (= 0) will make the corresponding PORTC pin an output (i.e., enable the output driver and put the contents of the output latch on the selected pin). Example 3-4 shows how to initialize PORTC.

Reading the PORTC register (Register 3-9) reads the status of the pins, whereas writing to it will write to the PORT latch. All write operations are read-modify-write operations. Therefore, a write to a port implies that the port pins are read, this value is modified and then written to the PORT data latch.

The TRISC register (Register 3-10) controls the PORTC pin output drivers, even when they are being used as analog inputs. The user should ensure the bits in the TRISC register are maintained set when using them as analog inputs. I/O pins configured as analog input always read '0'.

EXAMPLE 3-4: INITIALIZING PORTC

```
BANKSEL PORTC          ;
CLRF    PORTC          ;Init PORTC
BANKSEL TRISC          ;
MOVLW   B'00001100'    ;Set RC<3:2> as inputs
MOVWF   TRISC          ;and set RC<7:4,1:0>
                       ;as outputs
```

REGISTER 3-9: PORTC: PORTC REGISTER

R/W-x	R/W-x	R/W-x	R/W-x	R/W-x	R/W-x	R/W-x	R/W-x
RC7	RC6	RC5	RC4	RC3	RC2	RC1	RC0
bit 7							bit 0

Legend:			
R = Readable bit	W = Writable bit	U = Unimplemented bit, read as '0'	
-n = Value at POR	'1' = Bit is set	'0' = Bit is cleared	x = Bit is unknown

bit 7-0 **RC<7:0>**: PORTC General Purpose I/O Pin bit
1 = Port pin is > V_{IH}
0 = Port pin is < V_{IL}

REGISTER 3-10: TRISC: PORTC TRI-STATE REGISTER

R/W-1	R/W-1	R/W-1	R/W-1	R/W-1	R/W-1	R/W-1[(1)]	R/W-1[(1)]
TRISC7	TRISC6	TRISC5	TRISC4	TRISC3	TRISC2	TRISC1	TRISC0
bit 7							bit 0

Legend:			
R = Readable bit	W = Writable bit	U = Unimplemented bit, read as '0'	
-n = Value at POR	'1' = Bit is set	'0' = Bit is cleared	x = Bit is unknown

bit 7-0 **TRISC<7:0>**: PORTC Tri-State Control bit
1 = PORTC pin configured as an input (tri-stated)
0 = PORTC pin configured as an output

Note 1: TRISC<1:0> always reads '1' in LP Oscillator mode.

PIC16F882/883/884/886/887

3.5.1 RC0/T1OSO/T1CKI

Figure 3-11 shows the diagram for this pin. This pin is configurable to function as one of the following:

- a general purpose I/O
- a Timer1 oscillator output
- a Timer1 clock input

FIGURE 3-11: BLOCK DIAGRAM OF RC0

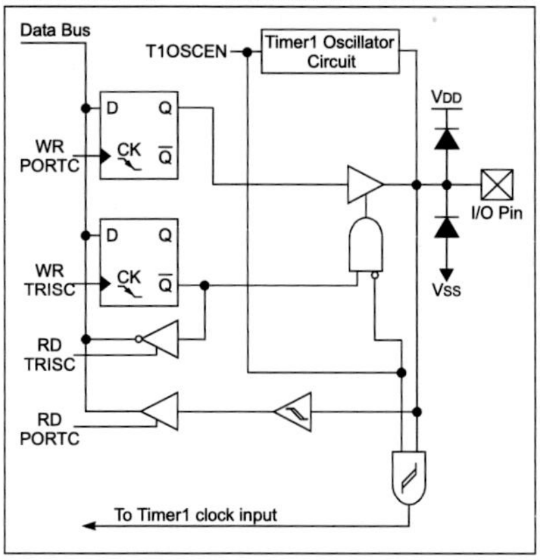

3.5.2 RC1/T1OSI/CCP2

Figure 3-12 shows the diagram for this pin. This pin is configurable to function as one of the following:

- a general purpose I/O
- a Timer1 oscillator input
- a Capture input and Compare/PWM output for Comparator C2

FIGURE 3-12: BLOCK DIAGRAM OF RC1

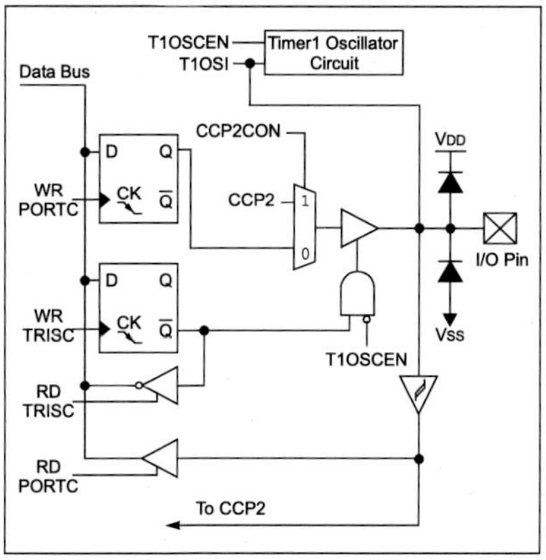

3.5.3 RC2/P1A/CCP1

Figure 3-13 shows the diagram for this pin. This pin is configurable to function as one of the following:

- a general purpose I/O
- a PWM output
- a Capture input and Compare output for Comparator C1

FIGURE 3-13: BLOCK DIAGRAM OF RC2

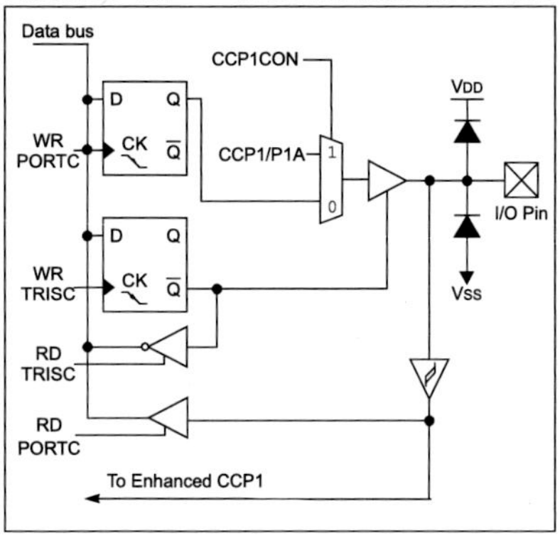

PIC16F882/883/884/886/887

3.5.4 RC3/SCK/SCL

Figure 3-14 shows the diagram for this pin. This pin is configurable to function as one of the following:

- a general purpose I/O
- a SPI clock
- an I²C™ clock

FIGURE 3-14: BLOCK DIAGRAM OF RC3

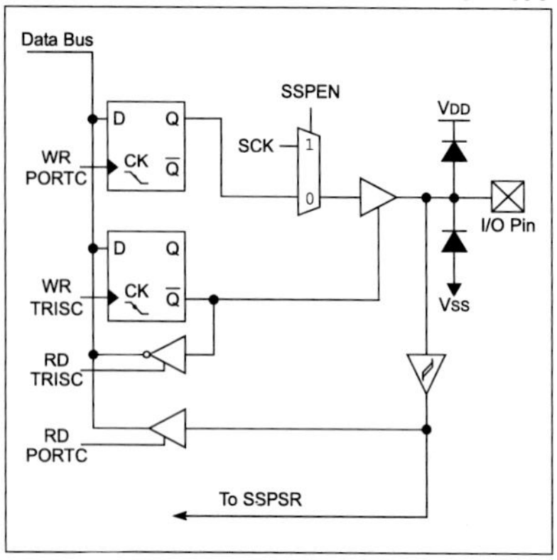

3.5.5 RC4/SDI/SDA

Figure 3-15 shows the diagram for this pin. This pin is configurable to function as one of the following:

- a general purpose I/O
- a SPI data I/O
- an I²C data I/O

FIGURE 3-15: BLOCK DIAGRAM OF RC4

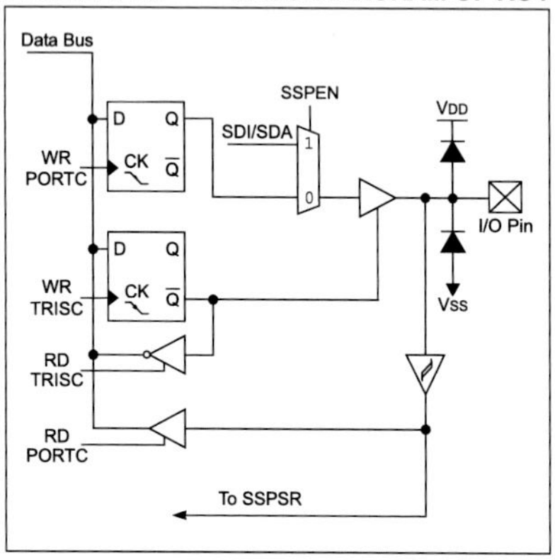

3.5.6 RC5/SDO

Figure 3-16 shows the diagram for this pin. This pin is configurable to function as one of the following:

- a general purpose I/O
- a serial data output

FIGURE 3-16: BLOCK DIAGRAM OF RC5

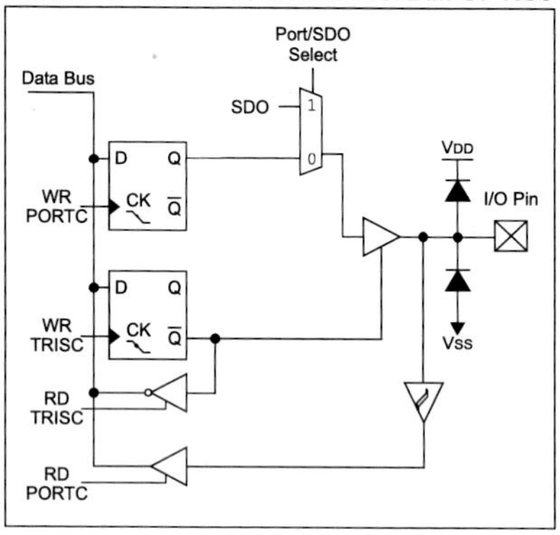

3.5.7 RC6/TX/CK

Figure 3-17 shows the diagram for this pin. This pin is configurable to function as one of the following:

- a general purpose I/O
- an asynchronous serial output
- a synchronous clock I/O

FIGURE 3-17: BLOCK DIAGRAM OF RC6

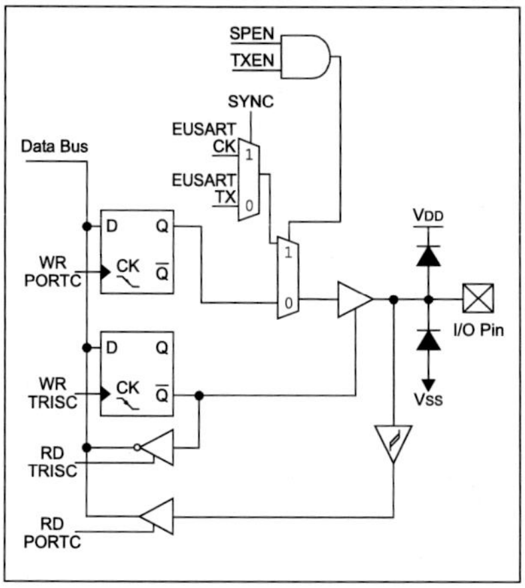

3.5.8 RC7/RX/DT

Figure 3-18 shows the diagram for this pin. This pin is configurable to function as one of the following:

- a general purpose I/O
- an asynchronous serial input
- a synchronous serial data I/O

FIGURE 3-18: BLOCK DIAGRAM OF RC7

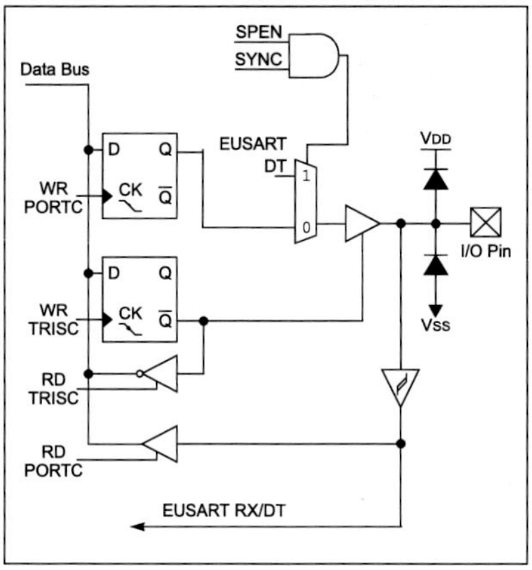

TABLE 3-3: SUMMARY OF REGISTERS ASSOCIATED WITH PORTC

Name	Bit 7	Bit 6	Bit 5	Bit 4	Bit 3	Bit 2	Bit 1	Bit 0	Value on POR, BOR	Value on all other Resets
CCP1CON	P1M1	P1M0	DC1B1	DC1B0	CCP1M3	CCP1M2	CCP1M1	CCP1M0	0000 0000	0000 0000
CCP2CON	—	—	DC2B1	DC2B0	CCP2M3	CCP2M2	CCP2M1	CCP2M0	--00 0000	--00 0000
PORTC	RC7	RC6	RC5	RC4	RC3	RC2	RC1	RC0	xxxx xxxx	uuuu uuuu
PSTRCON	—	—	—	STRSYNC	STRD	STRC	STRB	STRA	---0 0001	---0 0001
RCSTA	SPEN	RX9	SREN	CREN	ADDEN	FERR	OERR	RX9D	0000 000x	0000 000x
SSPCON	WCOL	SSPOV	SSPEN	CKP	SSPM3	SSPM2	SSPM1	SSPM0	0000 0000	0000 0000
T1CON	T1GINV	TMR1GE	T1CKPS1	T1CKPS0	T1OSCEN	$\overline{\text{T1SYNC}}$	TMR1CS	TMR1ON	0000 0000	0000 0000
TRISC	TRISC7	TRISC6	TRISC5	TRISC4	TRISC3	TRISC2	TRISC1	TRISC0	1111 1111	1111 1111

Legend: x = unknown, u = unchanged, – = unimplemented locations read as '0'. Shaded cells are not used by PORTC.

3.6 PORTD and TRISD Registers

PORTD[(1)] is a 8-bit wide, bidirectional port. The corresponding data direction register is TRISD (Register 3-12). Setting a TRISD bit (= 1) will make the corresponding PORTD pin an input (i.e., put the corresponding output driver in a High-Impedance mode). Clearing a TRISD bit (= 0) will make the corresponding PORTD pin an output (i.e., enable the output driver and put the contents of the output latch on the selected pin). Example 3-5 shows how to initialize PORTD.

Reading the PORTD register (Register 3-11) reads the status of the pins, whereas writing to it will write to the PORT latch. All write operations are read-modify-write operations. Therefore, a write to a port implies that the port pins are read, this value is modified and then written to the PORT data latch.

Note 1: PORTD is available on PIC16F884/887 only.

The TRISD register (Register 3-12) controls the PORTD pin output drivers, even when they are being used as analog inputs. The user should ensure the bits in the TRISD register are maintained set when using them as analog inputs. I/O pins configured as analog input always read '0'.

EXAMPLE 3-5: INITIALIZING PORTD

```
BANKSEL PORTD          ;
CLRF    PORTD          ;Init PORTD
BANKSEL TRISD          ;
MOVLW   B'00001100'    ;Set RD<3:2> as inputs
MOVWF   TRISD          ;and set RD<7:4,1:0>
                       ;as outputs
```

REGISTER 3-11: PORTD: PORTD REGISTER

R/W-x	R/W-x	R/W-x	R/W-x	R/W-x	R/W-x	R/W-x	R/W-x
RD7	RD6	RD5	RD4	RD3	RD2	RD1	RD0
bit 7							bit 0

Legend:			
R = Readable bit	W = Writable bit	U = Unimplemented bit, read as '0'	
-n = Value at POR	'1' = Bit is set	'0' = Bit is cleared	x = Bit is unknown

bit 7-0 **RD<7:0>**: PORTD General Purpose I/O Pin bit
1 = Port pin is > V_{IH}
0 = Port pin is < V_{IL}

REGISTER 3-12: TRISD: PORTD TRI-STATE REGISTER

R/W-1	R/W-1	R/W-1	R/W-1	R/W-1	R/W-1	R/W-1	R/W-1
TRISD7	TRISD6	TRISD5	TRISD4	TRISD3	TRISD2	TRISD1	TRISD0
bit 7							bit 0

Legend:			
R = Readable bit	W = Writable bit	U = Unimplemented bit, read as '0'	
-n = Value at POR	'1' = Bit is set	'0' = Bit is cleared	x = Bit is unknown

bit 7-0 **TRISD<7:0>**: PORTD Tri-State Control bit
1 = PORTD pin configured as an input (tri-stated)
0 = PORTD pin configured as an output

PIC16F882/883/884/886/887

3.6.1 RD<4:0>

Figure 3-19 shows the diagram for these pins. These pins are configured to function as general purpose I/O's.

> **Note:** RD<4:0> is available on PIC16F884/887 only.

FIGURE 3-19: BLOCK DIAGRAM OF RD<4:0>

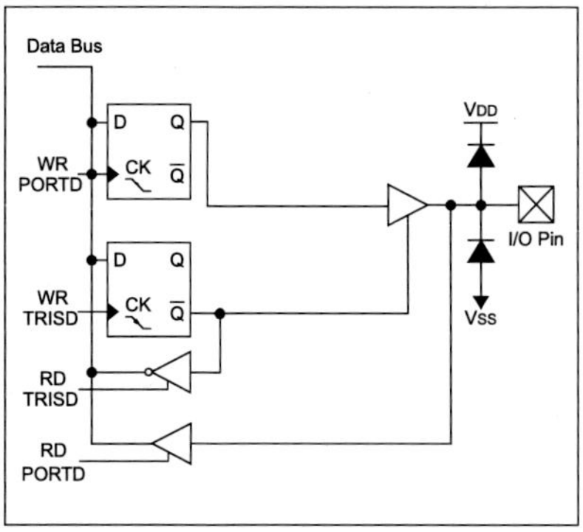

3.6.2 RD5/P1B(1)

Figure 3-20 shows the diagram for this pin. This pin is configurable to function as one of the following:

- a general purpose I/O
- a PWM output

> **Note 1:** RD5/P1B is available on PIC16F884/887 only. See RB2/AN8/P1B for this function on PIC16F882/883/886.

3.6.3 RD6/P1C(1)

Figure 3-20 shows the diagram for this pin. This pin is configurable to function as one of the following:

- a general purpose I/O
- a PWM output

> **Note 1:** RD6/P1C is available on PIC16F884/887 only. See RB1/AN10/P1C/C12IN3- for this function on PIC16F882/883/886.

3.6.4 RD7/P1D(1)

Figure 3-20 shows the diagram for this pin. This pin is configurable to function as one of the following:

- a general purpose I/O
- a PWM output

> **Note 1:** RD7/P1D is available on PIC16F884/887 only. See RB4/AN11/P1D for this function on PIC16F882/883/886.

FIGURE 3-20: BLOCK DIAGRAM OF RD<7:5>

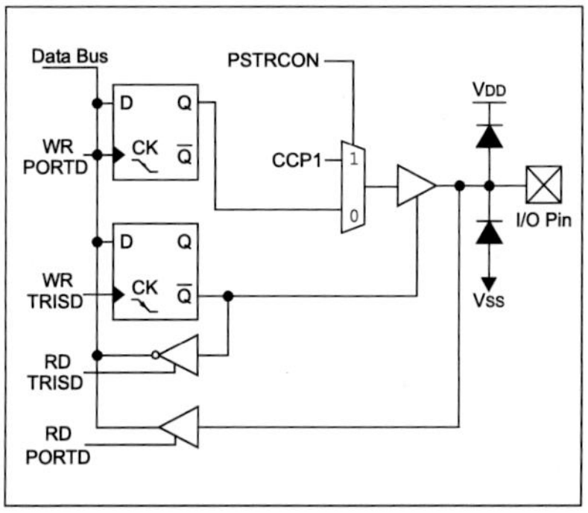

TABLE 3-4: SUMMARY OF REGISTERS ASSOCIATED WITH PORTD

Name	Bit 7	Bit 6	Bit 5	Bit 4	Bit 3	Bit 2	Bit 1	Bit 0	Value on POR, BOR	Value on all other Resets
PORTD	RD7	RD6	RD5	RD4	RD3	RD2	RD1	RD0	xxxx xxxx	uuuu uuuu
PSTRCON	—	—	—	STRSYNC	STRD	STRC	STRB	STRA	---0 0001	---0 0001
TRISD	TRISD7	TRISD6	TRISD5	TRISD4	TRISD3	TRISD2	TRISD1	TRISD0	1111 1111	1111 1111

Legend: x = unknown, u = unchanged, – = unimplemented locations read as '0'. Shaded cells are not used by PORTD.

3.7 PORTE and TRISE Registers

PORTE[1] is a 4-bit wide, bidirectional port. The corresponding data direction register is TRISE. Setting a TRISE bit (= 1) will make the corresponding PORTE pin an input (i.e., put the corresponding output driver in a High-Impedance mode). Clearing a TRISE bit (= 0) will make the corresponding PORTE pin an output (i.e., enable the output driver and put the contents of the output latch on the selected pin). The exception is RE3, which is input only and its TRIS bit will always read as '1'. Example 3-6 shows how to initialize PORTE.

Reading the PORTE register (Register 3-13) reads the status of the pins, whereas writing to it will write to the PORT latch. All write operations are read-modify-write operations. Therefore, a write to a port implies that the port pins are read, this value is modified and then written to the PORT data latch. RE3 reads '0' when MCLRE = 1.

> **Note 1:** RE<2:0> pins are available on PIC16F884/887 only.

The TRISE register (Register 3-14) controls the PORTE pin output drivers, even when they are being used as analog inputs. The user should ensure the bits in the TRISE register are maintained set when using them as analog inputs. I/O pins configured as analog input always read '0'.

> **Note:** The ANSEL register must be initialized to configure an analog channel as a digital input. Pins configured as analog inputs will read '0'.

EXAMPLE 3-6: INITIALIZING PORTE

```
BANKSEL PORTE          ;
CLRF    PORTE          ;Init PORTE
BANKSEL ANSEL          ;
CLRF    ANSEL          ;digital I/O
BCF     STATUS,RP1     ;Bank 1
BANKSEL TRISE          ;
MOVLW   B'00001100'    ;Set RE<3:2> as inputs
MOVWF   TRISE          ;and set RE<1:0>
                       ;as outputs
```

REGISTER 3-13: PORTE: PORTE REGISTER

U-0	U-0	U-0	U-0	R-x	R/W-x	R/W-x	R/W-x
—	—	—	—	RE3	RE2	RE1	RE0
bit 7							bit 0

Legend:			
R = Readable bit	W = Writable bit	U = Unimplemented bit, read as '0'	
-n = Value at POR	'1' = Bit is set	'0' = Bit is cleared	x = Bit is unknown

bit 7-4 **Unimplemented**: Read as '0'

bit 3-0 **RD<3:0>**: PORTE General Purpose I/O Pin bit
1 = Port pin is > V_{IH}
0 = Port pin is < V_{IL}

REGISTER 3-14: TRISE: PORTE TRI-STATE REGISTER

U-0	U-0	U-0	U-0	R-1[1]	R/W-1	R/W-1	R/W-1
—	—	—	—	TRISE3	TRISE2	TRISE1	TRISE0
bit 7							bit 0

Legend:			
R = Readable bit	W = Writable bit	U = Unimplemented bit, read as '0'	
-n = Value at POR	'1' = Bit is set	'0' = Bit is cleared	x = Bit is unknown

bit 7-4 **Unimplemented**: Read as '0'

bit 3-0 **TRISE<3:0>**: PORTE Tri-State Control bit
1 = PORTE pin configured as an input (tri-stated)
0 = PORTE pin configured as an output

Note 1: TRISE<3> always reads '1'.

3.7.1 RE0/AN5[(1)]

This pin is configurable to function as one of the following:

- a general purpose I/O
- an analog input for the ADC

> **Note 1:** RE0/AN5 is available on PIC16F884/887 only.

3.7.2 RE1/AN6[(1)]

This pin is configurable to function as one of the following:

- a general purpose I/O
- an analog input for the ADC

> **Note 1:** RE1/AN6 is available on PIC16F884/887 only.

3.7.3 RE2/AN7[(1)]

This pin is configurable to function as one of the following:

- a general purpose I/O
- an analog input for the ADC

> **Note 1:** RE2/AN7 is available on PIC16F884/887 only.

FIGURE 3-21: BLOCK DIAGRAM OF RE<2:0>

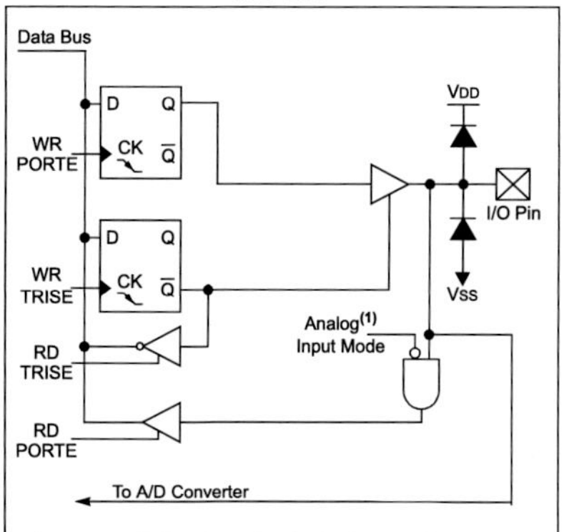

Note 1: ANSEL determines Analog Input mode.

3.7.4 RE3/$\overline{MCLR}$/VPP

Figure 3-22 shows the diagram for this pin. This pin is configurable to function as one of the following:

- a general purpose input
- as Master Clear Reset with weak pull-up

FIGURE 3-22: BLOCK DIAGRAM OF RE3

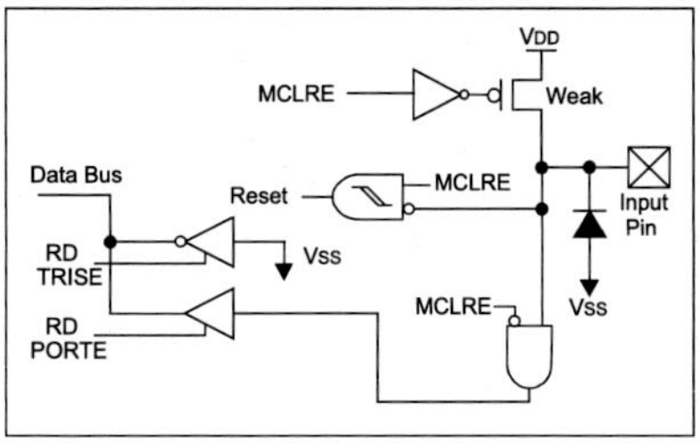

TABLE 3-5: SUMMARY OF REGISTERS ASSOCIATED WITH PORTE

Name	Bit 7	Bit 6	Bit 5	Bit 4	Bit 3	Bit 2	Bit 1	Bit 0	Value on POR, BOR	Value on all other Resets
ANSEL	ANS7	ANS6	ANS5	ANS4	ANS3	ANS2	ANS1	ANS0	1111 1111	1111 1111
PORTE	—	—	—	—	RE3	RE2	RE1	RE0	---- xxxx	---- uuuu
TRISE	—	—	—	—	TRISE3	TRISE2	TRISE1	TRISE0	---- 1111	---- 1111

Legend: x = unknown, u = unchanged, – = unimplemented locations read as '0'. Shaded cells are not used by PORTE

4.0 OSCILLATOR MODULE (WITH FAIL-SAFE CLOCK MONITOR)

4.1 Overview

The oscillator module has a wide variety of clock sources and selection features that allow it to be used in a wide range of applications while maximizing performance and minimizing power consumption. Figure 4-1 illustrates a block diagram of the oscillator module.

Clock sources can be configured from external oscillators, quartz crystal resonators, ceramic resonators and Resistor-Capacitor (RC) circuits. In addition, the system clock source can be configured from one of two internal oscillators, with a choice of speeds selectable via software. Additional clock features include:

- Selectable system clock source between external or internal via software.
- Two-Speed Start-up mode, which minimizes latency between external oscillator start-up and code execution.
- Fail-Safe Clock Monitor (FSCM) designed to detect a failure of the external clock source (LP, XT, HS, EC or RC modes) and switch automatically to the internal oscillator.

The oscillator module can be configured in one of eight clock modes.

1. EC – External clock with I/O on OSC2/CLKOUT.
2. LP – 32 kHz Low-Power Crystal mode.
3. XT – Medium Gain Crystal or Ceramic Resonator Oscillator mode.
4. HS – High Gain Crystal or Ceramic Resonator mode.
5. RC – External Resistor-Capacitor (RC) with FOSC/4 output on OSC2/CLKOUT.
6. RCIO – External Resistor-Capacitor (RC) with I/O on OSC2/CLKOUT.
7. INTOSC – Internal oscillator with FOSC/4 output on OSC2 and I/O on OSC1/CLKIN.
8. INTOSCIO – Internal oscillator with I/O on OSC1/CLKIN and OSC2/CLKOUT.

Clock Source modes are configured by the FOSC<2:0> bits in the Configuration Word Register 1 (CONFIG1). The internal clock can be generated from two internal oscillators. The HFINTOSC is a calibrated high-frequency oscillator. The LFINTOSC is an uncalibrated low-frequency oscillator.

FIGURE 4-1: SIMPLIFIED PIC® MCU CLOCK SOURCE BLOCK DIAGRAM

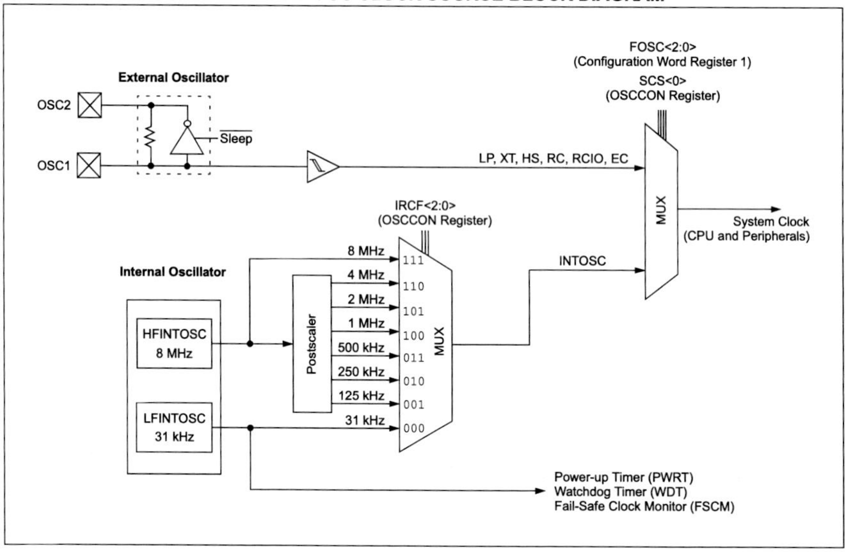

PIC16F882/883/884/886/887

5.0 TIMER0 MODULE

The Timer0 module is an 8-bit timer/counter with the following features:

- 8-bit timer/counter register (TMR0)
- 8-bit prescaler (shared with Watchdog Timer)
- Programmable internal or external clock source
- Programmable external clock edge selection
- Interrupt on overflow

Figure 5-1 is a block diagram of the Timer0 module.

5.1 Timer0 Operation

When used as a timer, the Timer0 module can be used as either an 8-bit timer or an 8-bit counter.

5.1.1 8-BIT TIMER MODE

When used as a timer, the Timer0 module will increment every instruction cycle (without prescaler). Timer mode is selected by clearing the T0CS bit of the OPTION register to '0'.

When TMR0 is written, the increment is inhibited for two instruction cycles immediately following the write.

Note:	The value written to the TMR0 register can be adjusted, in order to account for the two instruction cycle delay when TMR0 is written.

5.1.2 8-BIT COUNTER MODE

When used as a counter, the Timer0 module will increment on every rising or falling edge of the T0CKI pin. The incrementing edge is determined by the T0SE bit of the OPTION register. Counter mode is selected by setting the T0CS bit of the OPTION register to '1'.

FIGURE 5-1: TIMER0/WDT PRESCALER BLOCK DIAGRAM

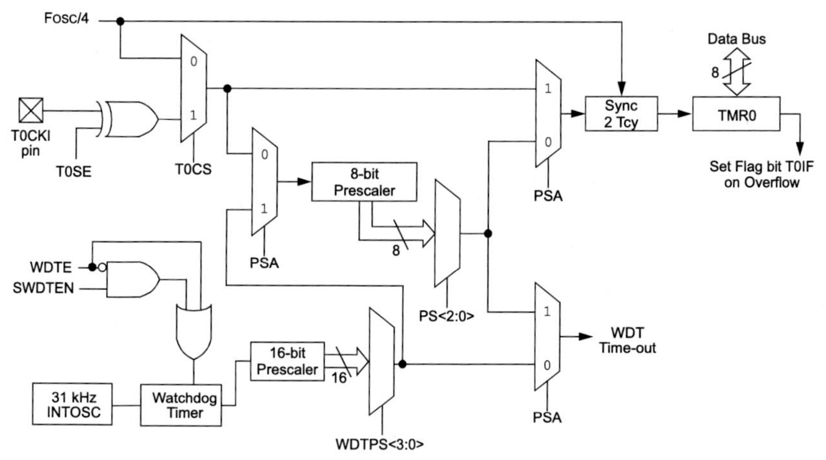

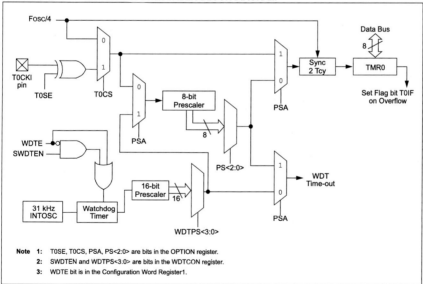

Note 1: T0SE, T0CS, PSA, PS<2:0> are bits in the OPTION register.
2: SWDTEN and WDTPS<3:0> are bits in the WDTCON register.
3: WDTE bit is in the Configuration Word Register1.

PIC16F882/883/884/886/887

5.1.3 SOFTWARE PROGRAMMABLE PRESCALER

A single software programmable prescaler is available for use with either Timer0 or the Watchdog Timer (WDT), but not both simultaneously. The prescaler assignment is controlled by the PSA bit of the OPTION register. To assign the prescaler to Timer0, the PSA bit must be cleared to a '0'.

There are 8 prescaler options for the Timer0 module ranging from 1:2 to 1:256. The prescale values are selectable via the PS<2:0> bits of the OPTION register. In order to have a 1:1 prescaler value for the Timer0 module, the prescaler must be assigned to the WDT module.

The prescaler is not readable or writable. When assigned to the Timer0 module, all instructions writing to the TMR0 register will clear the prescaler.

When the prescaler is assigned to WDT, a `CLRWDT` instruction will clear the prescaler along with the WDT.

5.1.3.1 Switching Prescaler Between Timer0 and WDT Modules

As a result of having the prescaler assigned to either Timer0 or the WDT, it is possible to generate an unintended device Reset when switching prescaler values. When changing the prescaler assignment from Timer0 to the WDT module, the instruction sequence shown in Example 5-1, must be executed.

EXAMPLE 5-1: CHANGING PRESCALER (TIMER0 → WDT)

```
BANKSEL  TMR0             ;
CLRWDT                    ;Clear WDT
CLRF     TMR0             ;Clear TMR0 and
                          ;prescaler
BANKSEL  OPTION_REG       ;
BSF      OPTION_REG,PSA   ;Select WDT
CLRWDT                    ;
                          ;
MOVLW    b'11111000'      ;Mask prescaler
ANDWF    OPTION_REG,W     ;bits
IORLW    b'00000101'      ;Set WDT prescaler
MOVWF    OPTION_REG       ;to 1:32
```

When changing the prescaler assignment from the WDT to the Timer0 module, the following instruction sequence must be executed (see Example 5-2).

EXAMPLE 5-2: CHANGING PRESCALER (WDT → TIMER0)

```
CLRWDT                  ;Clear WDT and
                        ;prescaler
BANKSEL OPTION_REG      ;
MOVLW   b'11110000'     ;Mask TMR0 select and
ANDWF   OPTION_REG,W    ;prescaler bits
IORLW   b'00000011'     ;Set prescale to 1:16
MOVWF   OPTION_REG      ;
```

5.1.4 TIMER0 INTERRUPT

Timer0 will generate an interrupt when the TMR0 register overflows from FFh to 00h. The T0IF interrupt flag bit of the INTCON register is set every time the TMR0 register overflows, regardless of whether or not the Timer0 interrupt is enabled. The T0IF bit must be cleared in software. The Timer0 interrupt enable is the T0IE bit of the INTCON register.

Note: The Timer0 interrupt cannot wake the processor from Sleep since the timer is frozen during Sleep.

5.1.5 USING TIMER0 WITH AN EXTERNAL CLOCK

When Timer0 is in Counter mode, the synchronization of the T0CKI input and the Timer0 register is accomplished by sampling the prescaler output on the Q2 and Q4 cycles of the internal phase clocks. Therefore, the high and low periods of the external clock source must meet the timing requirements as shown in the **Section 17.0 "Electrical Specifications"**.

REGISTER 5-1: OPTION_REG: OPTION REGISTER

R/W-1	R/W-1	R/W-1	R/W-1	R/W-1	R/W-1	R/W-1	R/W-1
$\overline{RBPU}$	INTEDG	T0CS	T0SE	PSA	PS2	PS1	PS0
bit 7							bit 0

Legend:			
R = Readable bit	W = Writable bit	U = Unimplemented bit, read as '0'	
-n = Value at POR	'1' = Bit is set	'0' = Bit is cleared	x = Bit is unknown

bit 7 **$\overline{RBPU}$:** PORTB Pull-up Enable bit
1 = PORTB pull-ups are disabled
0 = PORTB pull-ups are enabled by individual PORT latch values

bit 6 **INTEDG:** Interrupt Edge Select bit
1 = Interrupt on rising edge of INT pin
0 = Interrupt on falling edge of INT pin

bit 5 **T0CS:** TMR0 Clock Source Select bit
1 = Transition on T0CKI pin
0 = Internal instruction cycle clock (FOSC/4)

bit 4 **T0SE:** TMR0 Source Edge Select bit
1 = Increment on high-to-low transition on T0CKI pin
0 = Increment on low-to-high transition on T0CKI pin

bit 3 **PSA:** Prescaler Assignment bit
1 = Prescaler is assigned to the WDT
0 = Prescaler is assigned to the Timer0 module

bit 2-0 **PS<2:0>:** Prescaler Rate Select bits

BIT VALUE	TMR0 RATE	WDT RATE
000	1 : 2	1 : 1
001	1 : 4	1 : 2
010	1 : 8	1 : 4
011	1 : 16	1 : 8
100	1 : 32	1 : 16
101	1 : 64	1 : 32
110	1 : 128	1 : 64
111	1 : 256	1 : 128

Note 1: A dedicated 16-bit WDT postscaler is available. See **Section 14.5 "Watchdog Timer (WDT)"** for more information.

TABLE 5-1: SUMMARY OF REGISTERS ASSOCIATED WITH TIMER0

Name	Bit 7	Bit 6	Bit 5	Bit 4	Bit 3	Bit 2	Bit 1	Bit 0	Value on POR, BOR	Value on all other Resets
TMR0	Timer0 Module Register								xxxx xxxx	uuuu uuuu
INTCON	GIE	PEIE	T0IE	INTE	RBIE	T0IF	INTF	RBIF	0000 000x	0000 000x
OPTION_REG	$\overline{RBPU}$	INTEDG	T0CS	T0SE	PSA	PS2	PS1	PS0	1111 1111	1111 1111
TRISA	TRISA7	TRISA6	TRISA5	TRISA4	TRISA3	TRISA2	TRISA1	TRISA0	1111 1111	1111 1111

Legend: — = Unimplemented locations, read as '0', u = unchanged, x = unknown. Shaded cells are not used by the Timer0 module.

6.0 TIMER1 MODULE WITH GATE CONTROL

The Timer1 module is a 16-bit timer/counter with the following features:

- 16-bit timer/counter register pair (TMR1H:TMR1L)
- Programmable internal or external clock source
- 3-bit prescaler
- Optional LP oscillator
- Synchronous or asynchronous operation
- Timer1 gate (count enable) via comparator or $\overline{\text{T1G}}$ pin
- Interrupt on overflow
- Wake-up on overflow (external clock, Asynchronous mode only)
- Time base for the Capture/Compare function
- Special Event Trigger (with ECCP)
- Comparator output synchronization to Timer1 clock

Figure 6-1 is a block diagram of the Timer1 module.

6.1 Timer1 Operation

The Timer1 module is a 16-bit incrementing counter which is accessed through the TMR1H:TMR1L register pair. Writes to TMR1H or TMR1L directly update the counter.

When used with an internal clock source, the module is a timer. When used with an external clock source, the module can be used as either a timer or counter.

6.2 Clock Source Selection

The TMR1CS bit of the T1CON register is used to select the clock source. When TMR1CS = 0, the clock source is FOSC/4. When TMR1CS = 1, the clock source is supplied externally.

Clock Source	TMR1CS
FOSC/4	0
T1CKI pin	1

FIGURE 6-1: TIMER1 BLOCK DIAGRAM

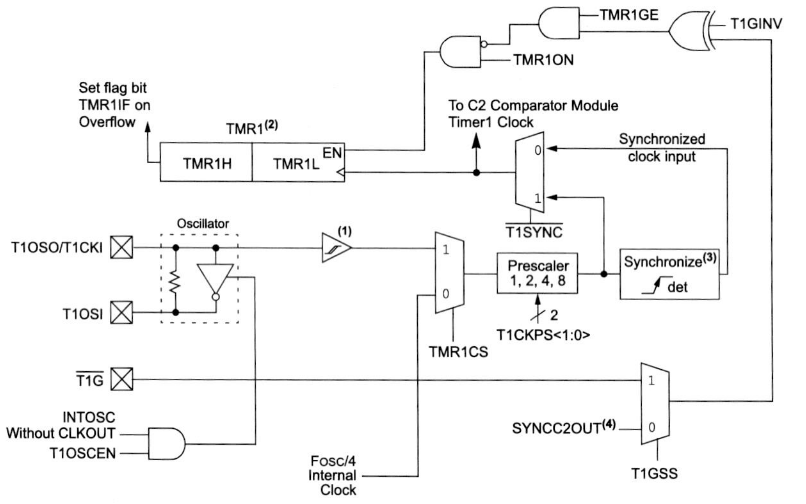

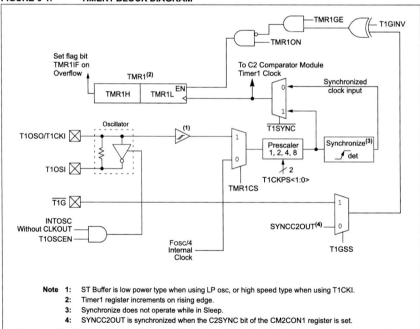

Note 1: ST Buffer is low power type when using LP osc, or high speed type when using T1CKI.
2: Timer1 register increments on rising edge.
3: Synchronize does not operate while in Sleep.
4: SYNCC2OUT is synchronized when the C2SYNC bit of the CM2CON1 register is set.

6.2.1 INTERNAL CLOCK SOURCE

When the internal clock source is selected the TMR1H:TMR1L register pair will increment on multiples of FOSC as determined by the Timer1 prescaler.

6.2.2 EXTERNAL CLOCK SOURCE

When the external clock source is selected, the Timer1 module may work as a timer or a counter.

When counting, Timer1 is incremented on the rising edge of the external clock input T1CKI. In addition, the Counter mode clock can be synchronized to the microcontroller system clock or run asynchronously.

If an external clock oscillator is needed (and the microcontroller is using the INTOSC without CLKOUT), Timer1 can use the LP oscillator as a clock source.

In Counter mode, a falling edge must be registered by the counter prior to the first incrementing rising edge after one or more of the following conditions (see Figure 6-2):

- Timer1 is enabled after POR or BOR Reset
- A write to TMR1H or TMR1L
- T1CKI is high when Timer1 is disabled and when Timer1 is reenabled T1CKI is low.

6.3 Timer1 Prescaler

Timer1 has four prescaler options allowing 1, 2, 4 or 8 divisions of the clock input. The T1CKPS bits of the T1CON register control the prescale counter. The prescale counter is not directly readable or writable; however, the prescaler counter is cleared upon a write to TMR1H or TMR1L.

6.4 Timer1 Oscillator

A low-power 32.768 kHz oscillator is built-in between pins T1OSI (input) and T1OSO (amplifier output). The oscillator is enabled by setting the T1OSCEN control bit of the T1CON register. The oscillator will continue to run during Sleep.

The Timer1 oscillator is identical to the LP oscillator. The user must provide a software time delay to ensure proper oscillator start-up.

TRISC0 and TRISC1 bits are set when the Timer1 oscillator is enabled. RC0 and RC1 bits read as '0' and TRISC0 and TRISC1 bits read as '1'.

> **Note:** The oscillator requires a start-up and stabilization time before use. Thus, T1OSCEN should be set and a suitable delay observed prior to enabling Timer1.

6.5 Timer1 Operation in Asynchronous Counter Mode

If control bit $\overline{\text{T1SYNC}}$ of the T1CON register is set, the external clock input is not synchronized. The timer continues to increment asynchronous to the internal phase clocks. The timer will continue to run during Sleep and can generate an interrupt on overflow, which will wake-up the processor. However, special precautions in software are needed to read/write the timer (see **Section 6.5.1 "Reading and Writing Timer1 in Asynchronous Counter Mode"**).

> **Note:** When switching from synchronous to asynchronous operation, it is possible to skip an increment. When switching from asynchronous to synchronous operation, it is possible to produce a single spurious increment.

6.5.1 READING AND WRITING TIMER1 IN ASYNCHRONOUS COUNTER MODE

Reading TMR1H or TMR1L while the timer is running from an external asynchronous clock will ensure a valid read (taken care of in hardware). However, the user should keep in mind that reading the 16-bit timer in two 8-bit values itself, poses certain problems, since the timer may overflow between the reads.

For writes, it is recommended that the user simply stop the timer and write the desired values. A write contention may occur by writing to the timer registers, while the register is incrementing. This may produce an unpredictable value in the TMR1H:TTMR1L register pair.

6.6 Timer1 Gate

Timer1 gate source is software configurable to be the $\overline{\text{T1G}}$ pin or the output of Comparator C2. This allows the device to directly time external events using $\overline{\text{T1G}}$ or analog events using Comparator C2. See the CM2CON1 register (Register 8-3) for selecting the Timer1 gate source. This feature can simplify the software for a Delta-Sigma A/D converter and many other applications. For more information on Delta-Sigma A/D converters, see the Microchip web site (www.microchip.com).

> **Note:** TMR1GE bit of the T1CON register must be set to use the Timer1 gate.

Timer1 gate can be inverted using the T1GINV bit of the T1CON register, whether it originates from the $\overline{\text{T1G}}$ pin or Comparator C2 output. This configures Timer1 to measure either the active-high or active-low time between events.

PIC16F882/883/884/886/887

6.7 Timer1 Interrupt

The Timer1 register pair (TMR1H:TMR1L) increments to FFFFh and rolls over to 0000h. When Timer1 rolls over, the Timer1 interrupt flag bit of the PIR1 register is set. To enable the interrupt on rollover, you must set these bits:

- Timer1 interrupt enable bit of the PIE1 register
- PEIE bit of the INTCON register
- GIE bit of the INTCON register

The interrupt is cleared by clearing the TMR1IF bit in the Interrupt Service Routine.

Note:	The TMR1H:TTMR1L register pair and the TMR1IF bit should be cleared before enabling interrupts.

6.8 Timer1 Operation During Sleep

Timer1 can only operate during Sleep when setup in Asynchronous Counter mode. In this mode, an external crystal or clock source can be used to increment the counter. To set up the timer to wake the device:

- TMR1ON bit of the T1CON register must be set
- TMR1IE bit of the PIE1 register must be set
- PEIE bit of the INTCON register must be set

The device will wake-up on an overflow and execute the next instruction. If the GIE bit of the INTCON register is set, the device will call the Interrupt Service Routine (0004h).

6.9 ECCP Capture/Compare Time Base

The ECCP module uses the TMR1H:TMR1L register pair as the time base when operating in Capture or Compare mode.

In Capture mode, the value in the TMR1H:TMR1L register pair is copied into the CCPRxH:CCPRxL register pair on a configured event.

In Compare mode, an event is triggered when the value CCPRxH:CCPRxL register pair matches the value in the TMR1H:TMR1L register pair. This event can be a Special Event Trigger.

See **Section 11.0 "Capture/Compare/PWM Modules (CCP1 and CCP2)"** for more information.

6.10 ECCP Special Event Trigger

If an ECCP is configured to trigger a special event, the trigger will clear the TMR1H:TMR1L register pair. This special event does not cause a Timer1 interrupt. The ECCP module may still be configured to generate a ECCP interrupt.

In this mode of operation, the CCPRxH:CCPRxL register pair effectively becomes the period register for Timer1.

Timer1 should be synchronized to the FOSC to utilize the Special Event Trigger. Asynchronous operation of Timer1 can cause a Special Event Trigger to be missed.

In the event that a write to TMR1H or TMR1L coincides with a Special Event Trigger from the ECCP, the write will take precedence.

For more information, see **Section 11.0 "Capture/Compare/PWM Modules (CCP1 and CCP2)"**.

6.11 Comparator Synchronization

The same clock used to increment Timer1 can also be used to synchronize the comparator output. This feature is enabled in the Comparator module.

When using the comparator for Timer1 gate, the comparator output should be synchronized to Timer1. This ensures Timer1 does not miss an increment if the comparator changes.

For more information, see **Section 8.0 "Comparator Module"**.

FIGURE 6-2: TIMER1 INCREMENTING EDGE

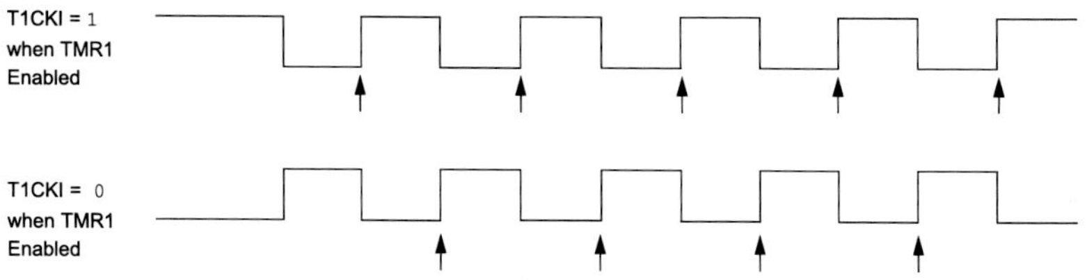

Note 1: Arrows indicate counter increments.

2: In Counter mode, a falling edge must be registered by the counter prior to the first incrementing rising edge of the clock.

6.12 Timer1 Control Register

The Timer1 Control register (T1CON), shown in Register 6-1, is used to control Timer1 and select the various features of the Timer1 module.

REGISTER 6-1: T1CON: TIMER1 CONTROL REGISTER

R/W-0	R/W-0	R/W-0	R/W-0	R/W-0	R/W-0	R/W-0	R/W-0
T1GINV(1)	TMR1GE(2)	T1CKPS1	T1CKPS0	T1OSCEN	$\overline{\text{T1SYNC}}$	TMR1CS	TMR1ON
bit 7							bit 0

Legend:			
R = Readable bit	W = Writable bit	U = Unimplemented bit, read as '0'	
-n = Value at POR	'1' = Bit is set	'0' = Bit is cleared	x = Bit is unknown

bit 7 **T1GINV:** Timer1 Gate Invert bit(1)
1 = Timer1 gate is active-high (Timer1 counts when gate is high)
0 = Timer1 gate is active-low (Timer1 counts when gate is low)

bit 6 **TMR1GE:** Timer1 Gate Enable bit(2)
If TMR1ON = 0:
This bit is ignored
If TMR1ON = 1:
1 = Timer1 counting is controlled by the Timer1 Gate function
0 = Timer1 is always counting

bit 5-4 **T1CKPS<1:0>:** Timer1 Input Clock Prescale Select bits
11 = 1:8 Prescale Value
10 = 1:4 Prescale Value
01 = 1:2 Prescale Value
00 = 1:1 Prescale Value

bit 3 **T1OSCEN:** LP Oscillator Enable Control bit
1 = LP oscillator is enabled for Timer1 clock
0 = LP oscillator is off

bit 2 **$\overline{\text{T1SYNC}}$:** Timer1 External Clock Input Synchronization Control bit
TMR1CS = 1:
1 = Do not synchronize external clock input
0 = Synchronize external clock input
TMR1CS = 0:
This bit is ignored. Timer1 uses the internal clock

bit 1 **TMR1CS:** Timer1 Clock Source Select bit
1 = External clock from T1CKI pin (on the rising edge)
0 = Internal clock (FOSC/4)

bit 0 **TMR1ON:** Timer1 On bit
1 = Enables Timer1
0 = Stops Timer1

Note 1: T1GINV bit inverts the Timer1 gate logic, regardless of source.
2: TMR1GE bit must be set to use either $\overline{\text{T1G}}$ pin or C2OUT, as selected by the T1GSS bit of the CM2CON1 register, as a Timer1 gate source.

PIC16F882/883/884/886/887

TABLE 6-1: SUMMARY OF REGISTERS ASSOCIATED WITH TIMER1

Name	Bit 7	Bit 6	Bit 5	Bit 4	Bit 3	Bit 2	Bit 1	Bit 0	Value on POR, BOR	Value on all other Resets
CM2CON1	MC1OUT	MC2OUT	C1RSEL	C2RSEL	—	—	T1GSS	C2SYNC	0000 --10	0000 --10
INTCON	GIE	PEIE	T0IE	INTE	RBIE	T0IF	INTF	RBIF	0000 000x	0000 000x
PIE1	—	ADIE	RCIE	TXIE	SSPIE	CCP1IE	TMR2IE	TMR1IE	-000 0000	-000 0000
PIR1	—	ADIF	RCIF	TXIF	SSPIF	CCP1IF	TMR2IF	TMR1IF	-000 0000	-000 0000
TMR1H	Holding Register for the Most Significant Byte of the 16-bit TMR1 Register								xxxx xxxx	uuuu uuuu
TMR1L	Holding Register for the Least Significant Byte of the 16-bit TMR1 Register								xxxx xxxx	uuuu uuuu
T1CON	T1GINV	TMR1GE	T1CKPS1	T1CKPS0	T1OSCEN	$\overline{\text{T1SYNC}}$	TMR1CS	TMR1ON	0000 0000	uuuu uuuu

Legend: x = unknown, u = unchanged, – = unimplemented, read as '0'. Shaded cells are not used by the Timer1 module.

7.0 TIMER2 MODULE

The Timer2 module is an eight-bit timer with the following features:

- 8-bit timer register (TMR2)
- 8-bit period register (PR2)
- Interrupt on TMR2 match with PR2
- Software programmable prescaler (1:1, 1:4, 1:16)
- Software programmable postscaler (1:1 to 1:16)

See Figure 7-1 for a block diagram of Timer2.

7.1 Timer2 Operation

The clock input to the Timer2 module is the system instruction clock (FOSC/4). The clock is fed into the Timer2 prescaler, which has prescale options of 1:1, 1:4 or 1:16. The output of the prescaler is then used to increment the TMR2 register.

The values of TMR2 and PR2 are constantly compared to determine when they match. TMR2 will increment from 00h until it matches the value in PR2. When a match occurs, two things happen:

- TMR2 is reset to 00h on the next increment cycle
- The Timer2 postscaler is incremented

The match output of the Timer2/PR2 comparator is then fed into the Timer2 postscaler. The postscaler has postscale options of 1:1 to 1:16 inclusive. The output of the Timer2 postscaler is used to set the TMR2IF interrupt flag bit in the PIR1 register.

The TMR2 and PR2 registers are both fully readable and writable. On any Reset, the TMR2 register is set to 00h and the PR2 register is set to FFh.

Timer2 is turned on by setting the TMR2ON bit in the T2CON register to a '1'. Timer2 is turned off by clearing the TMR2ON bit to a '0'.

The Timer2 prescaler is controlled by the T2CKPS bits in the T2CON register. The Timer2 postscaler is controlled by the TOUTPS bits in the T2CON register. The prescaler and postscaler counters are cleared when:

- A write to TMR2 occurs.
- A write to T2CON occurs.
- Any device Reset occurs (Power-on Reset, $\overline{MCLR}$ Reset, Watchdog Timer Reset, or Brown-out Reset).

Note:	TMR2 is not cleared when T2CON is written.

FIGURE 7-1: TIMER2 BLOCK DIAGRAM

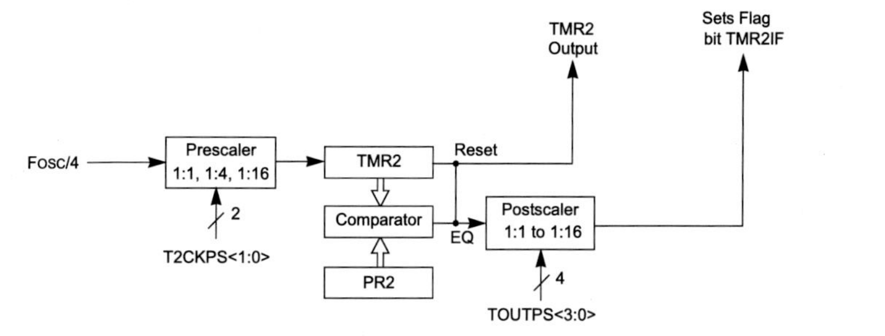

PIC16F882/883/884/886/887

REGISTER 7-1: T2CON: TIMER2 CONTROL REGISTER

U-0	R/W-0	R/W-0	R/W-0	R/W-0	R/W-0	R/W-0	R/W-0
—	TOUTPS3	TOUTPS2	TOUTPS1	TOUTPS0	TMR2ON	T2CKPS1	T2CKPS0
bit 7							bit 0

Legend:			
R = Readable bit	W = Writable bit	U = Unimplemented bit, read as '0'	
-n = Value at POR	'1' = Bit is set	'0' = Bit is cleared	x = Bit is unknown

bit 7 **Unimplemented:** Read as '0'

bit 6-3 **TOUTPS<3:0>:** Timer2 Output Postscaler Select bits
0000 = 1:1 Postscaler
0001 = 1:2 Postscaler
0010 = 1:3 Postscaler
0011 = 1:4 Postscaler
0100 = 1:5 Postscaler
0101 = 1:6 Postscaler
0110 = 1:7 Postscaler
0111 = 1:8 Postscaler
1000 = 1:9 Postscaler
1001 = 1:10 Postscaler
1010 = 1:11 Postscaler
1011 = 1:12 Postscaler
1100 = 1:13 Postscaler
1101 = 1:14 Postscaler
1110 = 1:15 Postscaler
1111 = 1:16 Postscaler

bit 2 **TMR2ON:** Timer2 On bit
1 = Timer2 is on
0 = Timer2 is off

bit 1-0 **T2CKPS<1:0>:** Timer2 Clock Prescale Select bits
00 = Prescaler is 1
01 = Prescaler is 4
1x = Prescaler is 16

TABLE 7-1: SUMMARY OF ASSOCIATED TIMER2 REGISTERS

Name	Bit 7	Bit 6	Bit 5	Bit 4	Bit 3	Bit 2	Bit 1	Bit 0	Value on POR, BOR	Value on all other Resets
INTCON	GIE	PEIE	T0IE	INTE	RBIE	T0IF	INTF	RBIF	0000 000x	0000 000x
PIE1	—	ADIE	RCIE	TXIE	SSPIE	CCP1IE	TMR2IE	TMR1IE	-000 0000	-000 0000
PIR1	—	ADIF	RCIF	TXIF	SSPIF	CCP1IF	TMR2IF	TMR1IF	-000 0000	-000 0000
PR2	Timer2 Module Period Register								1111 1111	1111 1111
TMR2	Holding Register for the 8-bit TMR2 Register								0000 0000	0000 0000
T2CON	—	TOUTPS3	TOUTPS2	TOUTPS1	TOUTPS0	TMR2ON	T2CKPS1	T2CKPS0	-000 0000	-000 0000

Legend: x = unknown, u = unchanged, — = unimplemented read as '0'. Shaded cells are not used for Timer2 module.

8.0 COMPARATOR MODULE

Comparators are used to interface analog circuits to a digital circuit by comparing two analog voltages and providing a digital indication of their relative magnitudes. The comparators are very useful mixed signal building blocks because they provide analog functionality independent of the program execution. The analog comparator module includes the following features:

- Independent comparator control
- Programmable input selection
- Comparator output is available internally/externally
- Programmable output polarity
- Interrupt-on-change
- Wake-up from Sleep
- PWM shutdown
- Timer1 gate (count enable)
- Output synchronization to Timer1 clock input
- SR Latch
- Programmable and fixed voltage reference

Note: Only Comparator C2 can be linked to Timer1.

8.1 Comparator Overview

A single comparator is shown in Figure 8-1 along with the relationship between the analog input levels and the digital output. When the analog voltage at VIN+ is less than the analog voltage at VIN-, the output of the comparator is a digital low level. When the analog voltage at VIN+ is greater than the analog voltage at VIN-, the output of the comparator is a digital high level.

FIGURE 8-1: SINGLE COMPARATOR

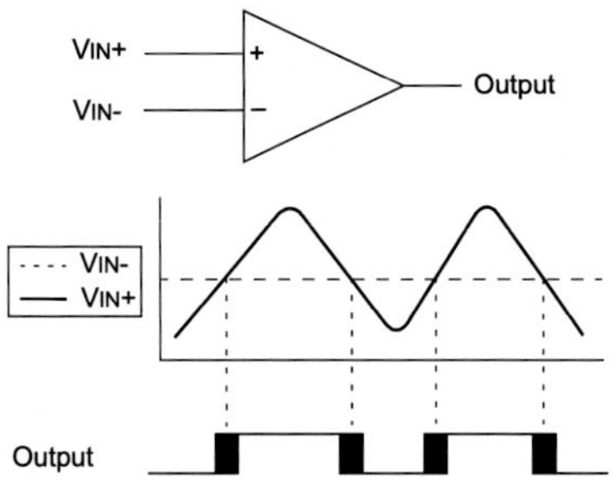

Note: The black areas of the output of the comparator represents the uncertainty due to input offsets and response time.

FIGURE 8-2: COMPARATOR C1 SIMPLIFIED BLOCK DIAGRAM

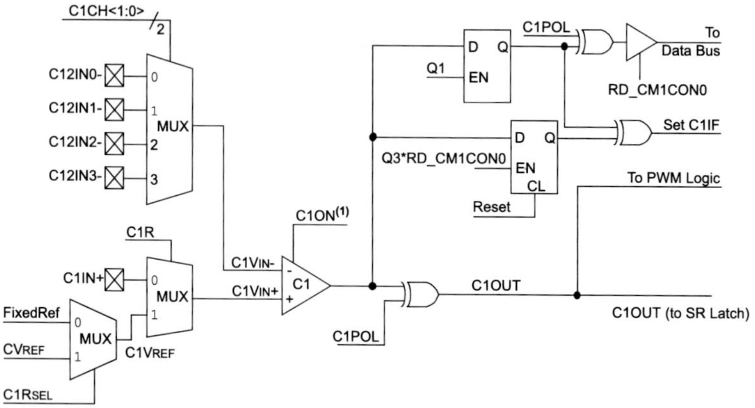

Note 1: When C1ON = 0, the C1 comparator will produce a '0' output to the XOR Gate.
2: Q1 and Q3 are phases of the four phase system clock (FOSC).
3: Q1 is held high during Sleep mode.

FIGURE 8-3: COMPARATOR C2 SIMPLIFIED BLOCK DIAGRAM

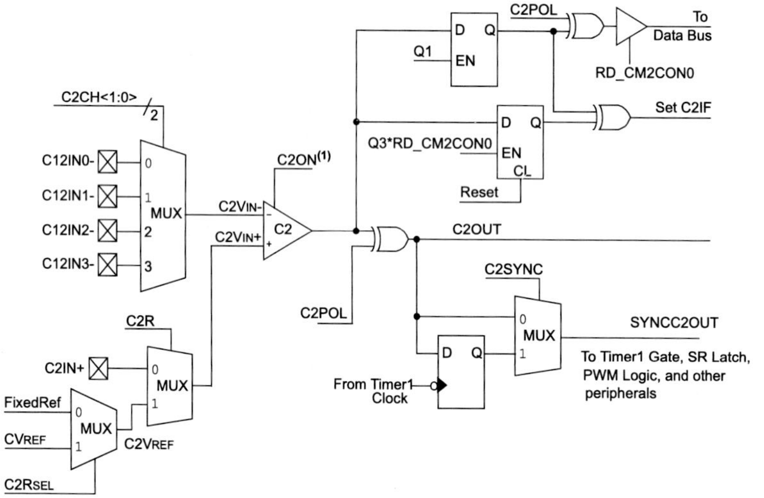

Note 1: When C2ON = 0, the C2 comparator will produce a '0' output to the XOR Gate.
2: Q1 and Q3 are phases of the four-phase system clock (FOSC).
3: Q1 is held high during Sleep mode.

8.2 Comparator Control

Each comparator has a separate control and Configuration register: CM1CON0 for Comparator C1 and CM2CON0 for Comparator C2. In addition, Comparator C2 has a second control register, CM2CON1, for controlling the interaction with Timer1 and simultaneous reading of both comparator outputs.

The CM1CON0 and CM2CON0 registers (see Registers 8-1 and 8-2, respectively) contain the control and Status bits for the following:

- Enable
- Input selection
- Reference selection
- Output selection
- Output polarity

8.2.1 COMPARATOR ENABLE

Setting the CxON bit of the CMxCON0 register enables the comparator for operation. Clearing the CxON bit disables the comparator resulting in minimum current consumption.

8.2.2 COMPARATOR INPUT SELECTION

The CxCH<1:0> bits of the CMxCON0 register direct one of four analog input pins to the comparator inverting input.

> **Note:** To use CxIN+ and CxIN- pins as analog inputs, the appropriate bits must be set in the ANSEL and ANSELH registers and the corresponding TRIS bits must also be set to disable the output drivers.

8.2.3 COMPARATOR REFERENCE SELECTION

Setting the CxR bit of the CMxCON0 register directs an internal voltage reference or an analog input pin to the non-inverting input of the comparator. See **Section 8.10 "Comparator Voltage Reference"** for more information on the internal voltage reference module.

8.2.4 COMPARATOR OUTPUT SELECTION

The output of the comparator can be monitored by reading either the CxOUT bit of the CMxCON0 register or the MCxOUT bit of the CM2CON1 register. In order to make the output available for an external connection, the following conditions must be true:

- CxOE bit of the CMxCON0 register must be set
- Corresponding TRIS bit must be cleared
- CxON bit of the CMxCON0 register must be set

> **Note 1:** The CxOE bit overrides the PORT data latch. Setting the CxON has no impact on the port override.
>
> **2:** The internal output of the comparator is latched with each instruction cycle. Unless otherwise specified, external outputs are not latched.

8.2.5 COMPARATOR OUTPUT POLARITY

Inverting the output of the comparator is functionally equivalent to swapping the comparator inputs. The polarity of the comparator output can be inverted by setting the CxPOL bit of the CMxCON0 register. Clearing the CxPOL bit results in a non-inverted output.

Table 8-1 shows the output state versus input conditions, including polarity control.

TABLE 8-1: COMPARATOR OUTPUT STATE VS. INPUT CONDITIONS

Input Condition	CxPOL	CxOUT
CxVIN- > CxVIN+	0	0
CxVIN- < CxVIN+	0	1
CxVIN- > CxVIN+	1	1
CxVIN- < CxVIN+	1	0

8.3 Comparator Response Time

The comparator output is indeterminate for a period of time after the change of an input source or the selection of a new reference voltage. This period is referred to as the response time. The response time of the comparator differs from the settling time of the voltage reference. Therefore, both of these times must be considered when determining the total response time to a comparator input change. See the Comparator and Voltage Reference specifications in **Section 17.0 "Electrical Specifications"** for more details.

8.4 Comparator Interrupt Operation

The comparator interrupt flag can be set whenever there is a change in the output value of the comparator. Changes are recognized by means of a mismatch circuit which consists of two latches and an exclusive-or gate (see Figures 8-2 and 8-3). One latch is updated with the comparator output level when the CMxCON0 register is read. This latch retains the value until the next read of the CMxCON0 register or the occurrence of a Reset. The other latch of the mismatch circuit is updated on every Q1 system clock. A mismatch condition will occur when a comparator output change is clocked through the second latch on the Q1 clock cycle. At this point the two mismatch latches have opposite output levels which is detected by the exclusive-or gate and fed to the interrupt circuitry. The mismatch condition persists until either the CMxCON0 register is read or the comparator output returns to the previous state.

> **Note 1:** A write operation to the CMxCON0 register will also clear the mismatch condition because all writes include a read operation at the beginning of the write cycle.
>
> **2:** Comparator interrupts will operate correctly regardless of the state of CxOE.

The comparator interrupt is set by the mismatch edge and not the mismatch level. This means that the interrupt flag can be reset without the additional step of reading or writing the CMxCON0 register to clear the mismatch registers. When the mismatch registers are cleared, an interrupt will occur upon the comparator's return to the previous state, otherwise no interrupt will be generated.

Software will need to maintain information about the status of the comparator output, as read from the CMxCON0 register, or CM2CON1 register, to determine the actual change that has occurred.

The CxIF bit of the PIR2 register is the comparator interrupt flag. This bit must be reset in software by clearing it to '0'. Since it is also possible to write a '1' to this register, an interrupt can be generated.

The CxIE bit of the PIE2 register and the PEIE and GIE bits of the INTCON register must all be set to enable comparator interrupts. If any of these bits are cleared, the interrupt is not enabled, although the CxIF bit of the PIR2 register will still be set if an interrupt condition occurs.

FIGURE 8-4: COMPARATOR INTERRUPT TIMING W/O CMxCON0 READ

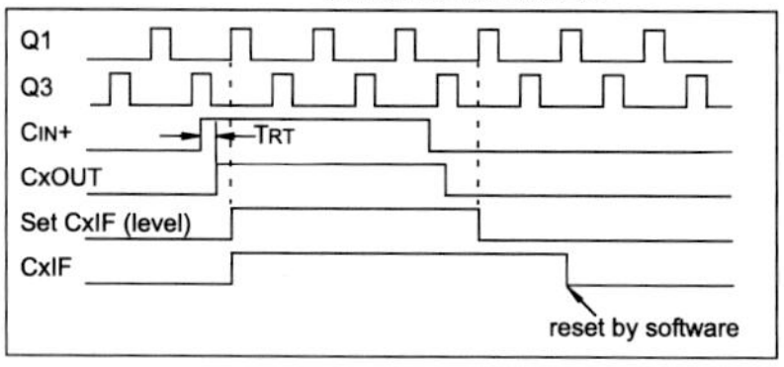

FIGURE 8-5: COMPARATOR INTERRUPT TIMING WITH CMxCON0 READ

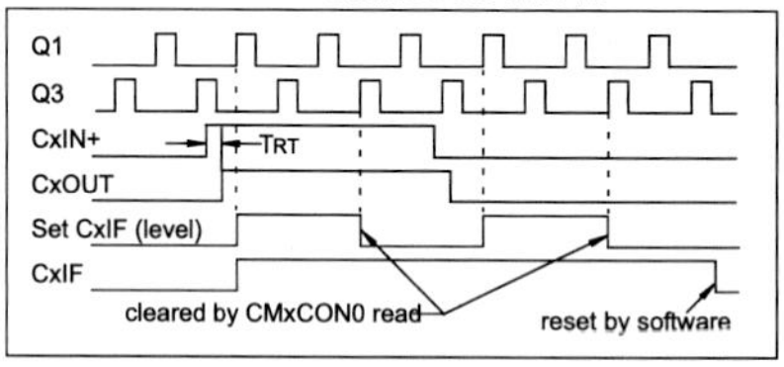

> **Note 1:** If a change in the CMxCON0 register (CxOUT) should occur when a read operation is being executed (start of the Q2 cycle), then the CxIF of the PIR2 register interrupt flag may not get set.
>
> **2:** When either comparator is first enabled, bias circuitry in the comparator module may cause an invalid output from the comparator until the bias circuitry is stable. Allow about 1 μs for bias settling then clear the mismatch condition and interrupt flags before enabling comparator interrupts.

8.5 Operation During Sleep

The comparator, if enabled before entering Sleep mode, remains active during Sleep. The additional current consumed by the comparator is shown separately in the **Section 17.0 "Electrical Specifications"**. If the comparator is not used to wake the device, power consumption can be minimized while in Sleep mode by turning off the comparator. Each comparator is turned off by clearing the CxON bit of the CMxCON0 register.

A change to the comparator output can wake-up the device from Sleep. To enable the comparator to wake the device from Sleep, the CxIE bit of the PIE2 register and the PEIE bit of the INTCON register must be set. The instruction following the Sleep instruction always executes following a wake from Sleep. If the GIE bit of the INTCON register is also set, the device will then execute the Interrupt Service Routine.

8.6 Effects of a Reset

A device Reset forces the CMxCON0 and CM2CON1 registers to their Reset states. This forces both comparators and the voltage references to their Off states.

PIC16F882/883/884/886/887

REGISTER 8-1: CM1CON0: COMPARATOR C1 CONTROL REGISTER 0

R/W-0	R-0	R/W-0	R/W-0	U-0	R/W-0	R/W-0	R/W-0
C1ON	C1OUT	C1OE	C1POL	—	C1R	C1CH1	C1CH0
bit 7							bit 0

Legend:			
R = Readable bit	W = Writable bit	U = Unimplemented bit, read as '0'	
-n = Value at POR	'1' = Bit is set	'0' = Bit is cleared	x = Bit is unknown

bit 7 **C1ON:** Comparator C1 Enable bit
1 = Comparator C1 is enabled
0 = Comparator C1 is disabled

bit 6 **C1OUT:** Comparator C1 Output bit
<u>If C1POL = 1 (inverted polarity):</u>
C1OUT = 0 when C1VIN+ > C1VIN-
C1OUT = 1 when C1VIN+ < C1VIN-
<u>If C1POL = 0 (non-inverted polarity):</u>
C1OUT = 1 when C1VIN+ > C1VIN-
C1OUT = 0 when C1VIN+ < C1VIN-

bit 5 **C1OE:** Comparator C1 Output Enable bit
1 = C1OUT is present on the C1OUT pin(1)
0 = C1OUT is internal only

bit 4 **C1POL:** Comparator C1 Output Polarity Select bit
1 = C1OUT logic is inverted
0 = C1OUT logic is not inverted

bit 3 **Unimplemented:** Read as '0'

bit 2 **C1R:** Comparator C1 Reference Select bit (non-inverting input)
1 = C1VIN+ connects to C1VREF output
0 = C1VIN+ connects to C1IN+ pin

bit 1-0 **C1CH<1:0>:** Comparator C1 Channel Select bit
00 = C12IN0- pin of C1 connects to C1VIN-
01 = C12IN1- pin of C1 connects to C1VIN-
10 = C12IN2- pin of C1 connects to C1VIN-
11 = C12IN3- pin of C1 connects to C1VIN-

Note 1: Comparator output requires the following three conditions: C1OE = 1, C1ON = 1 and corresponding port TRIS bit = 0.

PIC16F882/883/884/886/887

REGISTER 8-2: CM2CON0: COMPARATOR C2 CONTROL REGISTER 0

R/W-0	R-0	R/W-0	R/W-0	U-0	R/W-0	R/W-0	R/W-0
C2ON	C2OUT	C2OE	C2POL	—	C2R	C2CH1	C2CH0
bit 7							bit 0

Legend:			
R = Readable bit	W = Writable bit	U = Unimplemented bit, read as '0'	
-n = Value at POR	'1' = Bit is set	'0' = Bit is cleared	x = Bit is unknown

bit 7 **C2ON:** Comparator C2 Enable bit
1 = Comparator C2 is enabled
0 = Comparator C2 is disabled

bit 6 **C2OUT:** Comparator C2 Output bit
<u>If C2POL = 1 (inverted polarity):</u>
C2OUT = 0 when C2VIN+ > C2VIN-
C2OUT = 1 when C2VIN+ < C2VIN-
<u>If C2POL = 0 (non-inverted polarity):</u>
C2OUT = 1 when C2VIN+ > C2VIN-
C2OUT = 0 when C2VIN+ < C2VIN-

bit 5 **C2OE:** Comparator C2 Output Enable bit
1 = C2OUT is present on C2OUT pin(1)
0 = C2OUT is internal only

bit 4 **C2POL:** Comparator C2 Output Polarity Select bit
1 = C2OUT logic is inverted
0 = C2OUT logic is not inverted

bit 3 **Unimplemented:** Read as '0'

bit 2 **C2R:** Comparator C2 Reference Select bits (non-inverting input)
1 = C2VIN+ connects to C2VREF
0 = C2VIN+ connects to C2IN+ pin

bit 1-0 **C2CH<1:0>:** Comparator C2 Channel Select bits
00 = C12IN0- pin of C2 connects to C2VIN-
01 = C12IN1- pin of C2 connects to C2VIN-
10 = C12IN2- pin of C2 connects to C2VIN-
11 = C12IN3- pin of C2 connects to C2VIN-

Note 1: Comparator output requires the following three conditions: C2OE = 1, C2ON = 1 and corresponding port TRIS bit = 0.

PIC16F882/883/884/886/887

8.7 Analog Input Connection Considerations

A simplified circuit for an analog input is shown in Figure 8-6. Since the analog input pins share their connection with a digital input, they have reverse biased ESD protection diodes to VDD and VSS. The analog input, therefore, must be between VSS and VDD. If the input voltage deviates from this range by more than 0.6V in either direction, one of the diodes is forward biased and a latch-up may occur.

A maximum source impedance of 10 kΩ is recommended for the analog sources. Also, any external component connected to an analog input pin, such as a capacitor or a Zener diode, should have very little leakage current to minimize inaccuracies introduced.

> **Note 1:** When reading a PORT register, all pins configured as analog inputs will read as a '0'. Pins configured as digital inputs will convert as an analog input, according to the input specification.
>
> **2:** Analog levels on any pin defined as a digital input, may cause the input buffer to consume more current than is specified.

FIGURE 8-6: ANALOG INPUT MODEL

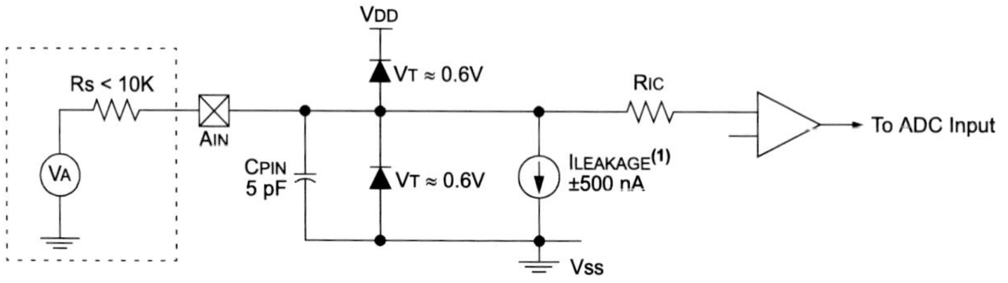

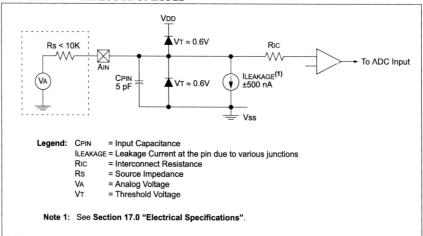

Legend: CPIN = Input Capacitance
ILEAKAGE = Leakage Current at the pin due to various junctions
RIC = Interconnect Resistance
RS = Source Impedance
VA = Analog Voltage
VT = Threshold Voltage

Note 1: See **Section 17.0 "Electrical Specifications"**.

8.8 Additional Comparator Features

There are three additional comparator features:

- Timer1 count enable (gate)
- Synchronizing output with Timer1
- Simultaneous read of comparator outputs

8.8.1 COMPARATOR C2 GATING TIMER1

This feature can be used to time the duration or interval of analog events. Clearing the T1GSS bit of the CM2CON1 register will enable Timer1 to increment based on the output of Comparator C2. This requires that Timer1 is on and gating is enabled. See **Section 6.0 "Timer1 Module with Gate Control"** for details.

It is recommended to synchronize the comparator with Timer1 by setting the C2SYNC bit when the comparator is used as the Timer1 gate source. This ensures Timer1 does not miss an increment if the comparator changes during an increment.

8.8.2 SYNCHRONIZING COMPARATOR C2 OUTPUT TO TIMER1

The Comparator C2 output can be synchronized with Timer1 by setting the C2SYNC bit of the CM2CON1 register. When enabled, the C2 output is latched on the falling edge of the Timer1 clock source. If a prescaler is used with Timer1, the comparator output is latched after the prescaling function. To prevent a race condition, the comparator output is latched on the falling edge of the Timer1 clock source and Timer1 increments on the rising edge of its clock source. See the Comparator Block Diagram (Figures 8-2 and 8-3) and the Timer1 Block Diagram (Figure 6-1) for more information.

8.8.3 SIMULTANEOUS COMPARATOR OUTPUT READ

The MC1OUT and MC2OUT bits of the CM2CON1 register are mirror copies of both comparator outputs. The ability to read both outputs simultaneously from a single register eliminates the timing skew of reading separate registers.

Note 1:	Obtaining the status of C1OUT or C2OUT by reading CM2CON1 does not affect the comparator interrupt mismatch registers.

REGISTER 8-3: CM2CON1: COMPARATOR C2 CONTROL REGISTER 1

R-0	R-0	R/W-0	R/W-0	U-0	U-0	R/W-1	R/W-0
MC1OUT	MC2OUT	C1RSEL	C2RSEL	—	—	T1GSS	C2SYNC
bit 7							bit 0

Legend:			
R = Readable bit	W = Writable bit	U = Unimplemented bit, read as '0'	
-n = Value at POR	'1' = Bit is set	'0' = Bit is cleared	x = Bit is unknown

bit 7 **MC1OUT:** Mirror Copy of C1OUT bit

bit 6 **MC2OUT:** Mirror Copy of C2OUT bit

bit 5 **C1RSEL:** Comparator C1 Reference Select bit
1 = CVREF routed to C1VREF input of Comparator C1
0 = Absolute voltage reference (0.6) routed to C1VREF input of Comparator C1 (or 1.2V precision reference on parts so equipped)

bit 4 **C2RSEL:** Comparator C2 Reference Select bit
1 = CVREF routed to C2VREF input of Comparator C2
0 = Absolute voltage reference (0.6) routed to C2VREF input of Comparator C2 (or 1.2V precision reference on parts so equipped)

bit 3-2 **Unimplemented:** Read as '0'

bit 1 **T1GSS:** Timer1 Gate Source Select bit
1 = Timer1 gate source is $\overline{T1G}$
0 = Timer1 gate source is SYNCC2OUT.

bit 0 **C2SYNC:** Comparator C2 Output Synchronization bit
1 = Output is synchronous to falling edge of Timer1 clock
0 = Output is asynchronous

PIC16F882/883/884/886/887

8.9 Comparator SR Latch

The SR latch module provides additional control of the comparator outputs. The module consists of a single SR latch and output multiplexers. The SR latch can be set, reset or toggled by the comparator outputs. The SR latch may also be set or reset, independent of comparator output, by control bits in the SRCON control register. The SR latch output multiplexers select whether the latch outputs or the comparator outputs are directed to the I/O port logic for eventual output to a pin.

8.9.1 LATCH OPERATION

The latch is a Set-Reset latch that does not depend on a clock source. Each of the Set and Reset inputs are active-high. Each latch input is connected to a comparator output and a software controlled pulse generator. The latch can be set by C1OUT or the PULSS bit of the SRCON register. The latch can be reset by C2OUT or the PULSR bit of the SRCON register. The latch is reset-dominant, therefore, if both Set and Reset inputs are high the latch will go to the Reset state. Both the PULSS and PULSR bits are self resetting which means that a single write to either of the bits is all that is necessary to complete a latch set or Reset operation.

8.9.2 LATCH OUTPUT

The SR<1:0> bits of the SRCON register control the latch output multiplexers and determine four possible output configurations. In these four configurations, the CxOUT I/O port logic is connected to:

- C1OUT and C2OUT
- C1OUT and SR latch $\overline{Q}$
- C2OUT and SR latch Q
- SR latch Q and $\overline{Q}$

After any Reset, the default output configuration is the unlatched C1OUT and C2OUT mode. This maintains compatibility with devices that do not have the SR latch feature.

The applicable TRIS bits of the corresponding ports must be cleared to enable the port pin output drivers. Additionally, the CxOE comparator output enable bits of the CMxCON0 registers must be set in order to make the comparator or latch outputs available on the output pins. The latch configuration enable states are completely independent of the enable states for the comparators.

FIGURE 8-7: SR LATCH SIMPLIFIED BLOCK DIAGRAM

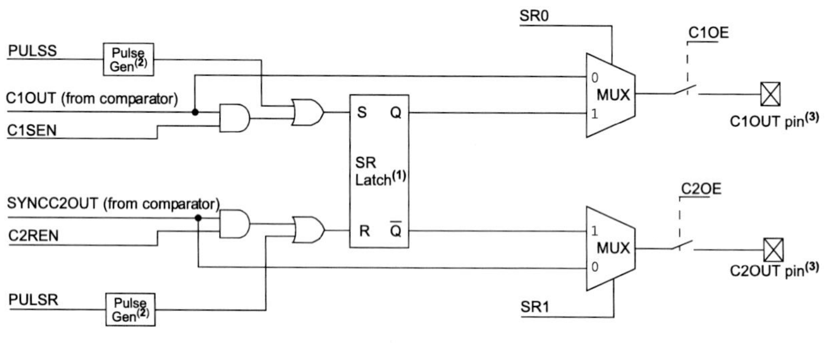

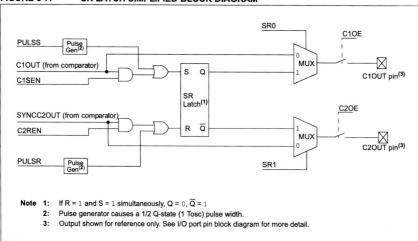

Note 1: If R = 1 and S = 1 simultaneously, Q = 0, $\overline{Q}$ = 1

2: Pulse generator causes a 1/2 Q-state (1 Tosc) pulse width.

3: Output shown for reference only. See I/O port pin block diagram for more detail.

REGISTER 8-4: SRCON: SR LATCH CONTROL REGISTER

R/W-0	R/W-0	R/W-0	R/W-0	R/S-0	R/S-0	U-0	R/W-0
SR1(2)	SR0(2)	C1SEN	C2REN	PULSS	PULSR	—	FVREN
bit 7							bit 0

Legend:		S = Bit is set only -	
R = Readable bit	W = Writable bit	U = Unimplemented bit, read as '0'	
-n = Value at POR	'1' = Bit is set	'0' = Bit is cleared	x = Bit is unknown

bit 7 **SR1:** SR Latch Configuration bit(2)
1 = C2OUT pin is the latch $\overline{Q}$ output
0 = C2OUT pin is the C2 comparator output

bit 6 **SR0:** SR Latch Configuration bits(2)
1 = C1OUT pin is the latch Q output
0 = C1OUT pin is the C1 Comparator output

bit 5 **C1SEN:** C1 Set Enable bit
1 = C1 comparator output sets SR latch
0 = C1 comparator output has no effect on SR latch

bit 4 **C2REN:** C2 Reset Enable bit
1 = C2 comparator output resets SR latch
0 = C2 comparator output has no effect on SR latch

bit 3 **PULSS:** Pulse the SET Input of the SR Latch bit
1 = Triggers pulse generator to set SR latch. Bit is immediately reset by hardware.
0 = Does not trigger pulse generator

bit 2 **PULSR:** Pulse the Reset Input of the SR Latch bit
1 = Triggers pulse generator to reset SR latch. Bit is immediately reset by hardware.
0 = Does not trigger pulse generator

bit 1 **Unimplemented:** Read as '0'

bit 0 **FVREN:** Fixed Voltage Reference Enable bit
1 = 0.6V Reference FROM INTOSC LDO is enabled
0 = 0.6V Reference FROM INTOSC LDO is disabled

Note 1: The CxOUT bit in the CMxCON0 register will always reflect the actual comparator output (not the level on the pin), regardless of the SR latch operation.

2: To enable an SR Latch output to the pin, the appropriate CxOE and TRIS bits must be properly configured.

8.10 Comparator Voltage Reference

The comparator voltage reference module provides an internally generated voltage reference for the comparators. The following features are available:

- Independent from Comparator operation
- Two 16-level voltage ranges
- Output clamped to VSS
- Ratiometric with VDD
- Fixed Reference (0.6V)

The VRCON register (Register 8-5) controls the voltage reference module shown in Figure 8-8.

The voltage source is selectable through both ends of the 16 connection resistor ladder network. Bit VRSS of the VRCON register selects either the internal or external voltage source.

The PIC16F882/883/884/886/887 allows the CVREF signal to be output to the RA2 pin of PORTA under certain configurations only. For more details, see Figure 8-9.

8.10.1 INDEPENDENT OPERATION

The comparator voltage reference is independent of the comparator configuration. Setting the VREN bit of the VRCON register will enable the voltage reference.

8.10.2 OUTPUT VOLTAGE SELECTION

The CVREF voltage reference has 2 ranges with 16 voltage levels in each range. Range selection is controlled by the VRR bit of the VRCON register. The 16 levels are set with the VR<3:0> bits of the VRCON register.

The CVREF output voltage is determined by the following equations:

EQUATION 8-1: CVREF OUTPUT VOLTAGE

$$VRR = 1 \text{ (low range):}$$
$$CVREF = (VR\langle 3:0\rangle / 24) \times VLADDER$$
$$VRR = 0 \text{ (high range):}$$
$$CVREF = (VLADDER/4) + (VR\langle 3:0\rangle \times VLADDER/32)$$
$$VLADDER = VDD \text{ or } ([VREF+] - [VREF-]) \text{ or } VREF+$$

The full range of VSS to VDD cannot be realized due to the construction of the module. See Figure 8-8.

8.10.3 OUTPUT CLAMPED TO VSS

The CVREF output voltage can be set to Vss with no power consumption by clearing the FVREN bit of the VRCON register.

This allows the comparator to detect a zero-crossing while not consuming additional CVREF module current.

> **Note:** Depending on the application, additional components may be required for a zero cross circuit. Reference TB3013, *"Using the ESD Parasitic Diodes on Mixed Signal Microcontrollers"* (DS93013), for more information.

8.10.4 OUTPUT RATIOMETRIC TO VDD

The comparator voltage reference is VDD derived and therefore, the CVREF output changes with fluctuations in VDD. The tested absolute accuracy of the Comparator Voltage Reference can be found in **Section 17.0 "Electrical Specifications"**.

8.10.5 FIXED VOLTAGE REFERENCE

The fixed voltage reference is independent of VDD, with a nominal output voltage of 0.6V. This reference can be enabled by setting the FVREN bit of the SRCON register to '1'. This reference is always enabled when the HFINTOSC oscillator is active.

8.10.6 FIXED VOLTAGE REFERENCE STABILIZATION PERIOD

When the fixed voltage reference module is enabled, it will require some time for the reference and its amplifier circuits to stabilize. The user program must include a small delay routine to allow the module to settle. See **Section 17.0 "Electrical Specifications"** for the minimum delay requirement.

8.10.7 VOLTAGE REFERENCE SELECTION

Multiplexers on the output of the voltage reference module enable selection of either the CVREF or fixed voltage reference for use by the comparators.

Setting the C1RSEL bit of the CM2CON1 register enables current to flow in the CVREF voltage divider and selects the CVREF voltage for use by C1. Clearing the C1RSEL bit selects the fixed voltage for use by C1.

Setting the C2RSEL bit of the CM2CON1 register enables current to flow in the CVREF voltage divider and selects the CVREF voltage for use by C2. Clearing the C2RSEL bit selects the fixed voltage for use by C2.

When both the C1RSEL and C2RSEL bits are cleared, current flow in the CVREF voltage divider is disabled minimizing the power drain of the voltage reference peripheral.

FIGURE 8-8: COMPARATOR VOLTAGE REFERENCE BLOCK DIAGRAM

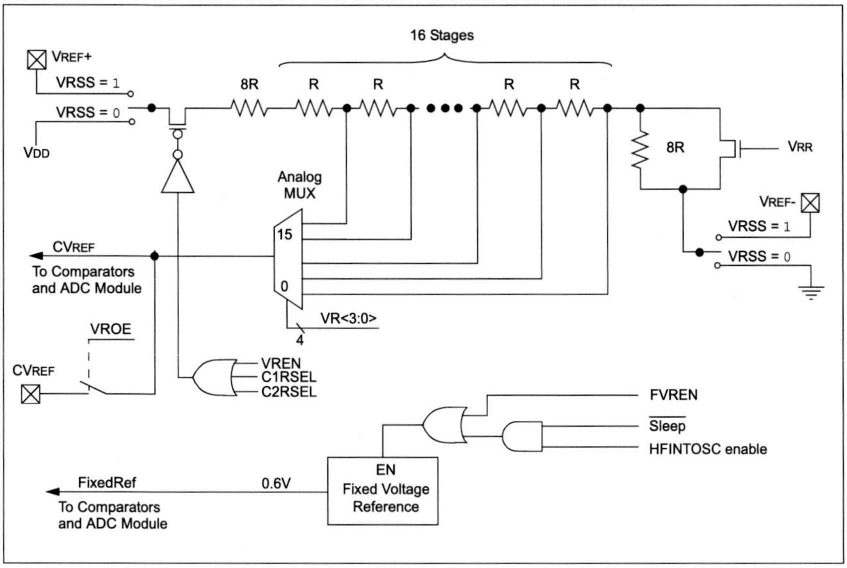

FIGURE 8-9: COMPARATOR AND ADC VOLTAGE REFERENCE BLOCK DIAGRAM

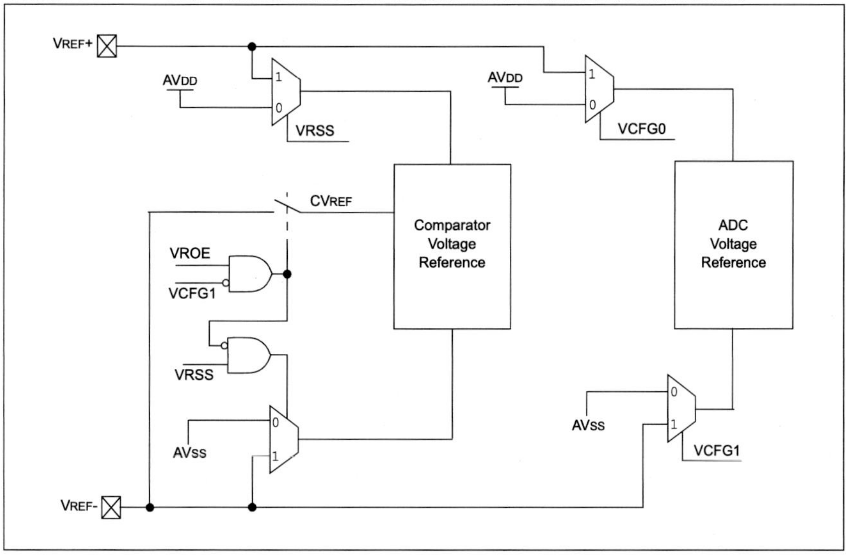

PIC16F882/883/884/886/887

TABLE 8-2: COMPARATOR AND ADC VOLTAGE REFERENCE PRIORITY

RA3	RA2	Comp. Reference (+)	Comp. Reference (-)	ADC Reference (+)	ADC Reference (-)	CFG1	CFG0	VRSS	VROE
I/O	I/O	AVDD	AVSS	AVDD	AVSS	0	0	0	0
I/O	CVREF	AVDD	AVSS	AVDD	AVSS	0	0	0	1
VREF+	VREF-	VREF+	VREF-	AVDD	AVSS	0	0	1	0
VREF+	CVREF	VREF+	AVSS	AVDD	AVSS	0	0	1	1
VREF+	I/O	AVDD	AVSS	VREF+	AVSS	0	1	0	0
VREF+	CVREF	AVDD	AVSS	VREF+	AVSS	0	1	0	1
VREF+	VREF-	VREF+	VREF-	VREF+	AVSS	0	1	1	0
VREF+	CVREF	VREF+	AVSS	VREF+	AVSS	0	1	1	1
I/O	VREF-	AVDD	AVSS	AVDD	VREF-	1	0	0	0
I/O	VREF-	AVDD	AVSS	AVDD	VREF-	1	0	0	1
VREF+	VREF-	VREF+	VREF-	AVDD	VREF-	1	0	1	0
VREF+	VREF-	VREF+	VREF-	AVDD	VREF-	1	0	1	1
VREF+	VREF-	AVDD	AVSS	VREF+	VREF-	1	1	0	0
VREF+	VREF-	AVDD	AVSS	VREF+	VREF-	1	1	0	1
VREF+	VREF-	VREF+	VREF-	VREF+	VREF-	1	1	1	0
VREF+	VREF-	VREF+	VREF-	VREF+	VREF-	1	1	1	1

REGISTER 8-5: VRCON: VOLTAGE REFERENCE CONTROL REGISTER

R/W-0	R/W-0	R/W-0	R/W-0	R/W-0	R/W-0	R/W-0	R/W-0
VREN	VROE	VRR	VRSS	VR3	VR2	VR1	VR0
bit 7							bit 0

Legend:			
R = Readable bit	W = Writable bit	U = Unimplemented bit, read as ‘0’	
-n = Value at POR	‘1’ = Bit is set	‘0’ = Bit is cleared	x = Bit is unknown

bit 7 **VREN:** Comparator C1 Voltage Reference Enable bit
1 = CVREF circuit powered on
0 = CVREF circuit powered down

bit 6 **VROE:** Comparator C2 Voltage Reference Enable bit
1 = CVREF voltage level is also output on the RA2/AN2/VREF-/CVREF/C2IN+ pin
0 = CVREF voltage is disconnected from the RA2/AN2/VREF-/CVREF/C2IN+ pin

bit 5 **VRR:** CVREF Range Selection bit
1 = Low range
0 = High range

bit 4 **VRSS:** Comparator VREF Range Selection bit
1 = Comparator Reference Source, CVRSRC = (VREF+) - (VREF-)
0 = Comparator Reference Source, CVRSRC = VDD - VSS

bit 3-0 **VR<3:0>:** CVREF Value Selection 0 ≤ VR<3:0> ≤ 15
When VRR = 1: CVREF = (VR<3:0>/24) * VDD
When VRR = 0: CVREF = VDD/4 + (VR<3:0>/32) * VDD

TABLE 8-3: SUMMARY OF REGISTERS ASSOCIATED WITH THE COMPARATOR AND VOLTAGE REFERENCE MODULES

Name	Bit 7	Bit 6	Bit 5	Bit 4	Bit 3	Bit 2	Bit 1	Bit 0	Value on POR, BOR	Value on all other Resets
ANSEL	ANS7	ANS6	ANS5	ANS4	ANS3	ANS2	ANS1	ANS0	1111 1111	1111 1111
ANSELH	—	—	ANS13	ANS12	ANS11	ANS10	ANS9	ANS8	--11 1111	--11 1111
CM1CON0	C1ON	C1OUT	C1OE	C1POL	—	C1R	C1CH1	C1CH0	0000 -000	0000 -000
CM2CON0	C2ON	C2OUT	C2OE	C2POL	—	C2R	C2CH1	C2CH0	0000 -000	0000 -000
CM2CON1	MC1OUT	MC2OUT	C1RSEL	C2RSEL	—	—	T1GSS	C2SYNC	0000 --10	0000 --10
INTCON	GIE	PEIE	T0IE	INTE	RBIE	T0IF	INTF	RBIF	0000 000x	0000 000x
PIE2	OSFIE	C2IE	C1IE	EEIE	BCLIE	ULPWUIE	—	CCP2IE	0000 00-0	0000 00-0
PIR2	OSFIF	C2IF	C1IF	EEIF	BCLIF	ULPWUIF	—	CCP2IF	0000 00-0	0000 00-0
PORTA	RA7	RA6	RA5	RA4	RA3	RA2	RA1	RA0	xxxx xxxx	uuuu uuuu
PORTB	RB7	RB6	RB5	RB4	RB3	RB2	RB1	RB0	xxxx xxxx	uuuu uuuu
SRCON	SR1	SR0	C1SEN	C2SEN	PULSS	PULSR	—	FVREN	0000 00-0	0000 00-0
TRISA	TRISA7	TRISA6	TRISA5	TRISA4	TRISA3	TRISA2	TRISA1	TRISA0	1111 1111	1111 1111
TRISB	TRISB7	TRISB6	TRISB5	TRISB4	TRISB3	TRISB2	TRISB1	TRISB0	1111 1111	1111 1111
VRCON	VREN	VROE	VRR	VRSS	VR3	VR2	VR1	VR0	0000 0000	0000 0000

Legend: x = unknown, u = unchanged, – = unimplemented, read as ‘0’. Shaded cells are not used for comparator.

PIC16F882/883/884/886/887

NOTES:

9.0 ANALOG-TO-DIGITAL CONVERTER (ADC) MODULE

The Analog-to-Digital Converter (ADC) allows conversion of an analog input signal to a 10-bit binary representation of that signal. This device uses analog inputs, which are multiplexed into a single sample and hold circuit. The output of the sample and hold is connected to the input of the converter. The converter generates a 10-bit binary result via successive approximation and stores the conversion result into the ADC result registers (ADRESL and ADRESH).

The ADC voltage reference is software selectable to be either internally generated or externally supplied.

The ADC can generate an interrupt upon completion of a conversion. This interrupt can be used to wake-up the device from Sleep.

Figure 9-1 shows the block diagram of the ADC.

FIGURE 9-1: ADC BLOCK DIAGRAM

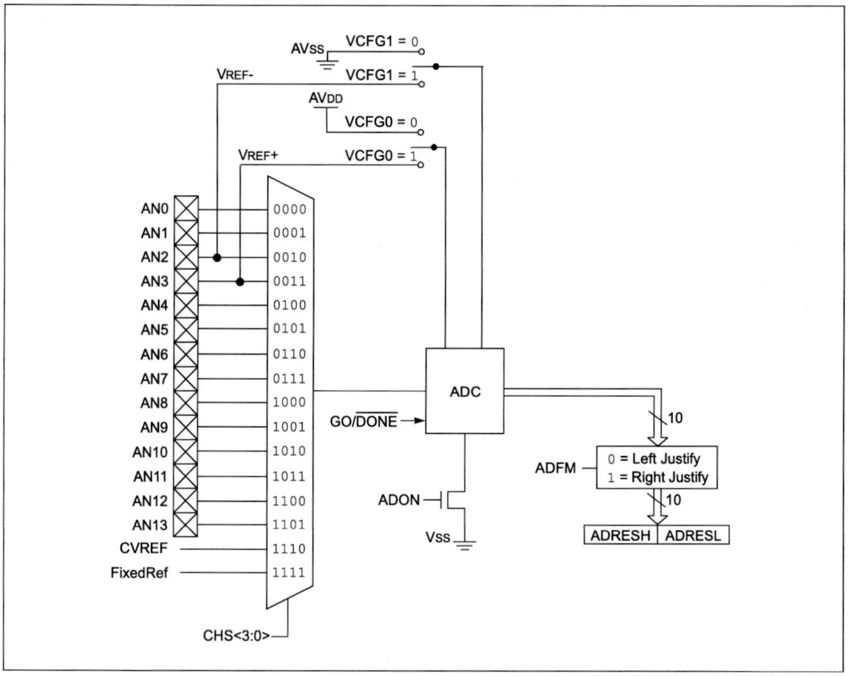

PIC16F882/883/884/886/887

9.1 ADC Configuration

When configuring and using the ADC the following functions must be considered:

- Port configuration
- Channel selection
- ADC voltage reference selection
- ADC conversion clock source
- Interrupt control
- Results formatting

9.1.1 PORT CONFIGURATION

The ADC can be used to convert both analog and digital signals. When converting analog signals, the I/O pin should be configured for analog by setting the associated TRIS and ANSEL bits. See the corresponding Port section for more information.

Note:	Analog voltages on any pin that is defined as a digital input may cause the input buffer to conduct excess current.

9.1.2 CHANNEL SELECTION

The CHS bits of the ADCON0 register determine which channel is connected to the sample and hold circuit.

When changing channels, a delay is required before starting the next conversion. Refer to **Section 9.2 "ADC Operation"** for more information.

9.1.3 ADC VOLTAGE REFERENCE

The VCFG bits of the ADCON0 register provide independent control of the positive and negative voltage references. The positive voltage reference can be either V_{DD} or an external voltage source. Likewise, the negative voltage reference can be either V_{SS} or an external voltage source.

9.1.4 CONVERSION CLOCK

The source of the conversion clock is software selectable via the ADCS bits of the ADCON0 register. There are four possible clock options:

- $F_{OSC}/2$
- $F_{OSC}/8$
- $F_{OSC}/32$
- F_{RC} (dedicated internal oscillator)

The time to complete one bit conversion is defined as T_{AD}. One full 10-bit conversion requires 11 T_{AD} periods as shown in Figure 9-2.

For correct conversion, the appropriate T_{AD} specification must be met. See A/D conversion requirements in **Section 17.0 "Electrical Specifications"** for more information. Table 9-1 gives examples of appropriate ADC clock selections.

Note:	Unless using the F_{RC}, any changes in the system clock frequency will change the ADC clock frequency, which may adversely affect the ADC result.

TABLE 9-1: ADC CLOCK PERIOD (TAD) Vs. DEVICE OPERATING FREQUENCIES (VDD ≥ 3.0V)

ADC Clock Period (TAD)		Device Frequency (FOSC)			
ADC Clock Source	**ADCS<1:0>**	**20 MHz**	**8 MHz**	**4 MHz**	**1 MHz**
FOSC/2	00	100 ns[2]	250 ns[2]	500 ns[2]	2.0 µs
FOSC/8	01	400 ns[2]	1.0 µs[2]	2.0 µs	8.0 µs[3]
FOSC/32	10	1.6 µs	4.0 µs	8.0 µs[3]	32.0 µs[3]
FRC	11	2-6 µs[1,4]	2-6 µs[1,4]	2-6 µs[1,4]	2-6 µs[1,4]

Legend: Shaded cells are outside of recommended range.

Note 1: The FRC source has a typical TAD time of 4 µs for VDD > 3.0V.

2: These values violate the minimum required TAD time.

3: For faster conversion times, the selection of another clock source is recommended.

4: When the device frequency is greater than 1 MHz, the FRC clock source is only recommended if the conversion will be performed during Sleep.

FIGURE 9-2: ANALOG-TO-DIGITAL CONVERSION TAD CYCLES

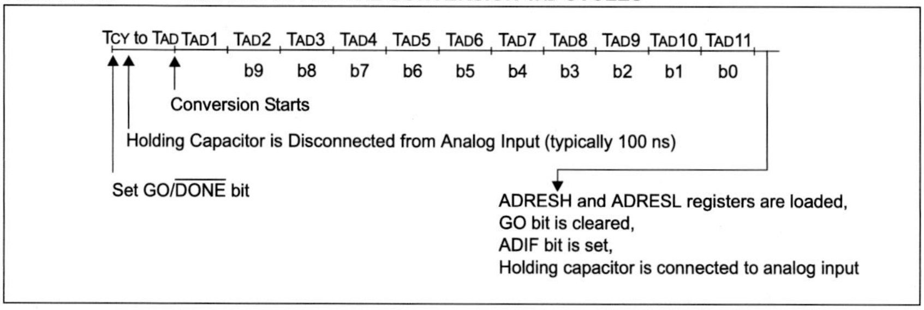

9.1.5 INTERRUPTS

The ADC module allows for the ability to generate an interrupt upon completion of an Analog-to-Digital conversion. The ADC interrupt flag is the ADIF bit in the PIR1 register. The ADC interrupt enable is the ADIE bit in the PIE1 register. The ADIF bit must be cleared in software.

Note:	The ADIF bit is set at the completion of every conversion, regardless of whether or not the ADC interrupt is enabled.

This interrupt can be generated while the device is operating or while in Sleep. If the device is in Sleep, the interrupt will wake-up the device. Upon waking from Sleep, the next instruction following the `SLEEP` instruction is always executed. If the user is attempting to wake-up from Sleep and resume in-line code execution, the global interrupt must be disabled. If the global interrupt is enabled, execution will switch to the Interrupt Service Routine.

Please see **Section 14.3 "Interrupts"** for more information.

9.1.6 RESULT FORMATTING

The 10-bit A/D conversion result can be supplied in two formats, left justified or right justified. The ADFM bit of the ADCON0 register controls the output format.

Figure 9-3 shows the two output formats.

FIGURE 9-3: 10-BIT A/D CONVERSION RESULT FORMAT

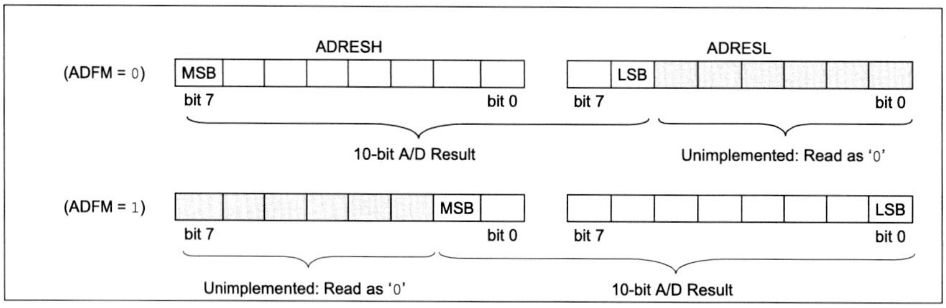

9.2 ADC Operation

9.2.1 STARTING A CONVERSION

To enable the ADC module, the ADON bit of the ADCON0 register must be set to a '1'. Setting the GO/$\overline{\text{DONE}}$ bit of the ADCON0 register to a '1' will start the Analog-to-Digital conversion.

> **Note:** The GO/$\overline{\text{DONE}}$ bit should not be set in the same instruction that turns on the ADC. Refer to **Section 9.2.6 "A/D Conversion Procedure"**.

9.2.2 COMPLETION OF A CONVERSION

When the conversion is complete, the ADC module will:

- Clear the GO/$\overline{\text{DONE}}$ bit
- Set the ADIF flag bit
- Update the ADRESH:ADRESL registers with new conversion result

9.2.3 TERMINATING A CONVERSION

If a conversion must be terminated before completion, the GO/$\overline{\text{DONE}}$ bit can be cleared in software. The ADRESH:ADRESL registers will not be updated with the partially complete Analog-to-Digital conversion sample. Instead, the ADRESH:ADRESL register pair will retain the value of the previous conversion. Additionally, a 2 T_{AD} delay is required before another acquisition can be initiated. Following this delay, an input acquisition is automatically started on the selected channel.

> **Note:** A device Reset forces all registers to their Reset state. Thus, the ADC module is turned off and any pending conversion is terminated.

9.2.4 ADC OPERATION DURING SLEEP

The ADC module can operate during Sleep. This requires the ADC clock source to be set to the F_{RC} option. When the F_{RC} clock source is selected, the ADC waits one additional instruction before starting the conversion. This allows the `SLEEP` instruction to be executed, which can reduce system noise during the conversion. If the ADC interrupt is enabled, the device will wake-up from Sleep when the conversion completes. If the ADC interrupt is disabled, the ADC module is turned off after the conversion completes, although the ADON bit remains set.

When the ADC clock source is something other than F_{RC}, a `SLEEP` instruction causes the present conversion to be aborted and the ADC module is turned off, although the ADON bit remains set.

9.2.5 SPECIAL EVENT TRIGGER

The ECCP Special Event Trigger allows periodic ADC measurements without software intervention. When this trigger occurs, the GO/$\overline{\text{DONE}}$ bit is set by hardware and the Timer1 counter resets to zero.

Using the Special Event Trigger does not assure proper ADC timing. It is the user's responsibility to ensure that the ADC timing requirements are met.

See **Section 11.0 "Capture/Compare/PWM Modules (CCP1 and CCP2)"** for more information.

9.2.6 A/D CONVERSION PROCEDURE

This is an example procedure for using the ADC to perform an Analog-to-Digital conversion:

1. Configure Port:
   - Disable pin output driver (See TRIS register)
   - Configure pin as analog
2. Configure the ADC module:
   - Select ADC conversion clock
   - Configure voltage reference
   - Select ADC input channel
   - Select result format
   - Turn on ADC module
3. Configure ADC interrupt (optional):
   - Clear ADC interrupt flag
   - Enable ADC interrupt
   - Enable peripheral interrupt
   - Enable global interrupt[(1)]
4. Wait the required acquisition time[(2)].
5. Start conversion by setting the GO/$\overline{\text{DONE}}$ bit.
6. Wait for ADC conversion to complete by one of the following:
   - Polling the GO/$\overline{\text{DONE}}$ bit
   - Waiting for the ADC interrupt (interrupts enabled)
7. Read ADC Result
8. Clear the ADC interrupt flag (required if interrupt is enabled).

> **Note 1:** The global interrupt can be disabled if the user is attempting to wake-up from Sleep and resume in-line code execution.
>
> **2:** See **Section 9.3 "A/D Acquisition Requirements"**.

EXAMPLE 9-1: A/D CONVERSION

```
;This code block configures the ADC
;for polling, Vdd and Vss as reference, Frc
clock and AN0 input.
;
;Conversion start & polling for completion
; are included.
;
BANKSEL   ADCON1      ;
MOVLW     B'10000000' ;right justify
MOVWF     ADCON1      ;Vdd and Vss as Vref
BANKSEL   TRISA       ;
BSF       TRISA,0     ;Set RA0 to input
BANKSEL   ANSEL       ;
BSF       ANSEL,0     ;Set RA0 to analog
BANKSEL   ADCON0      ;
MOVLW     B'11000001' ;ADC Frc clock,
MOVWF     ADCON0      ;AN0, On
CALL      SampleTime  ;Acquisiton delay
BSF       ADCON0,GO   ;Start conversion
BTFSC     ADCON0,GO   ;Is conversion done?
GOTO      $-1         ;No, test again
BANKSEL   ADRESH      ;
MOVF      ADRESH,W    ;Read upper 2 bits
MOVWF     RESULTHI    ;store in GPR space
BANKSEL   ADRESL      ;
MOVF      ADRESL,W    ;Read lower 8 bits
MOVWF     RESULTLO    ;Store in GPR space
```

PIC16F882/883/884/886/887

9.2.7 ADC REGISTER DEFINITIONS

The following registers are used to control the operation of the ADC.

> **Note:** For ANSEL and ANSELH registers, see Register 3-3 and Register 3-4, respectively.

REGISTER 9-1: ADCON0: A/D CONTROL REGISTER 0

R/W-0	R/W-0	R/W-0	R/W-0	R/W-0	R/W-0	R/W-0	R/W-0
ADCS1	ADCS0	CHS3	CHS2	CHS1	CHS0	GO/$\overline{\text{DONE}}$	ADON
bit 7							bit 0

Legend:			
R = Readable bit	W = Writable bit	U = Unimplemented bit, read as '0'	
-n = Value at POR	'1' = Bit is set	'0' = Bit is cleared	x = Bit is unknown

bit 7-6 **ADCS<1:0>:** A/D Conversion Clock Select bits
- `00` = FOSC/2
- `01` = FOSC/8
- `10` = FOSC/32
- `11` = FRC (clock derived from a dedicated internal oscillator = 500 kHz max)

bit 5-2 **CHS<3:0>: Analog Channel Select bits**
- `0000` = AN0
- `0001` = AN1
- `0010` = AN2
- `0011` = AN3
- `0100` = AN4
- `0101` = AN5
- `0110` = AN6
- `0111` = AN7
- `1000` = AN8
- `1001` = AN9
- `1010` = AN10
- `1011` = AN11
- `1100` = AN12
- `1101` = AN13
- `1110` = CVREF
- `1111` = Fixed Ref (0.6V fixed voltage reference)

bit 1 **GO/$\overline{\text{DONE}}$:** A/D Conversion Status bit
- `1` = A/D conversion cycle in progress. Setting this bit starts an A/D conversion cycle. This bit is automatically cleared by hardware when the A/D conversion has completed.
- `0` = A/D conversion completed/not in progress

bit 0 **ADON:** ADC Enable bit
- `1` = ADC is enabled
- `0` = ADC is disabled and consumes no operating current

PIC16F882/883/884/886/887

REGISTER 9-2: ADCON1: A/D CONTROL REGISTER 1

R/W-0	U-0	R/W-0	R/W-0	U-0	U-0	U-0	U-0
ADFM	—	VCFG1	VCFG0	—	—	—	—
bit 7							bit 0

Legend:			
R = Readable bit	W = Writable bit	U = Unimplemented bit, read as '0'	
-n = Value at POR	'1' = Bit is set	'0' = Bit is cleared	x = Bit is unknown

bit 7	**ADFM:** A/D Conversion Result Format Select bit 1 = Right justified 0 = Left justified
bit 6	**Unimplemented:** Read as '0'
bit 5	**VCFG1:** Voltage Reference bit 1 = VREF- pin 0 = VSS
bit 4	**VCFG0:** Voltage Reference bit 1 = VREF+ pin 0 = VDD
bit 3-0	**Unimplemented:** Read as '0'

PIC16F882/883/884/886/887

REGISTER 9-3: ADRESH: ADC RESULT REGISTER HIGH (ADRESH) ADFM = 0

R/W-x	R/W-x	R/W-x	R/W-x	R/W-x	R/W-x	R/W-x	R/W-x
ADRES9	ADRES8	ADRES7	ADRES6	ADRES5	ADRES4	ADRES3	ADRES2
bit 7							bit 0

Legend:			
R = Readable bit	W = Writable bit	U = Unimplemented bit, read as '0'	
-n = Value at POR	'1' = Bit is set	'0' = Bit is cleared	x = Bit is unknown

bit 7-0 **ADRES<9:2>**: ADC Result Register bits
Upper 8 bits of 10-bit conversion result

REGISTER 9-4: ADRESL: ADC RESULT REGISTER LOW (ADRESL) ADFM = 0

R/W-x	R/W-x	R/W-x	R/W-x	R/W-x	R/W-x	R/W-x	R/W-x
ADRES1	ADRES0	—	—	—	—	—	—
bit 7							bit 0

Legend:			
R = Readable bit	W = Writable bit	U = Unimplemented bit, read as '0'	
-n = Value at POR	'1' = Bit is set	'0' = Bit is cleared	x = Bit is unknown

bit 7-6 **ADRES<1:0>**: ADC Result Register bits
Lower 2 bits of 10-bit conversion result

bit 5-0 **Reserved**: Do not use.

REGISTER 9-5: ADRESH: ADC RESULT REGISTER HIGH (ADRESH) ADFM = 1

R/W-x	R/W-x	R/W-x	R/W-x	R/W-x	R/W-x	R/W-x	R/W-x
—	—	—	—	—	—	ADRES9	ADRES8
bit 7							bit 0

Legend:			
R = Readable bit	W = Writable bit	U = Unimplemented bit, read as '0'	
-n = Value at POR	'1' = Bit is set	'0' = Bit is cleared	x = Bit is unknown

bit 7-2 **Reserved**: Do not use.

bit 1-0 **ADRES<9:8>**: ADC Result Register bits
Upper 2 bits of 10-bit conversion result

REGISTER 9-6: ADRESL: ADC RESULT REGISTER LOW (ADRESL) ADFM = 1

R/W-x	R/W-x	R/W-x	R/W-x	R/W-x	R/W-x	R/W-x	R/W-x
ADRES7	ADRES6	ADRES5	ADRES4	ADRES3	ADRES2	ADRES1	ADRES0
bit 7							bit 0

Legend:			
R = Readable bit	W = Writable bit	U = Unimplemented bit, read as '0'	
-n = Value at POR	'1' = Bit is set	'0' = Bit is cleared	x = Bit is unknown

bit 7-0 **ADRES<7:0>**: ADC Result Register bits
Lower 8 bits of 10-bit conversion result

9.3 A/D Acquisition Requirements

For the ADC to meet its specified accuracy, the charge holding capacitor (CHOLD) must be allowed to fully charge to the input channel voltage level. The Analog Input model is shown in Figure 9-4. The source impedance (RS) and the internal sampling switch (RSS) impedance directly affect the time required to charge the capacitor CHOLD. The sampling switch (RSS) impedance varies over the device voltage (VDD), see Figure 9-4. **The maximum recommended impedance for analog sources is 10 kΩ.** As the source impedance is decreased, the acquisition time may be decreased. After the analog input channel is selected (or changed), an A/D acquisition must be done before the conversion can be started. To calculate the minimum acquisition time, Equation 9-1 may be used. This equation assumes that 1/2 LSb error is used (1024 steps for the ADC). The 1/2 LSb error is the maximum error allowed for the ADC to meet its specified resolution.

EQUATION 9-1: ACQUISITION TIME EXAMPLE

Assumptions: *Temperature* = 50°C and external impedance of 10kΩ 5.0V VDD

$$T_{ACQ} = \text{Amplifier Settling Time} + \text{Hold Capacitor Charging Time} + \text{Temperature Coefficient}$$
$$= T_{AMP} + T_C + T_{COFF}$$
$$= 2\mu s + T_C + [(\text{Temperature} - 25°C)(0.05\mu s/°C)]$$

The value for TC can be approximated with the following equations:

$$V_{APPLIED}\left(1 - \frac{1}{(2^{n+1}) - 1}\right) = V_{CHOLD} \qquad ;[1]\ V_{CHOLD} \text{ charged to within 1/2 lsb}$$

$$V_{APPLIED}\left(1 - e^{\frac{-T_C}{RC}}\right) = V_{CHOLD} \qquad ;[2]\ V_{CHOLD} \text{ charge response to } V_{APPLIED}$$

$$V_{APPLIED}\left(1 - e^{\frac{-T_C}{RC}}\right) = V_{APPLIED}\left(1 - \frac{1}{(2^{n+1}) - 1}\right) \qquad ;\text{combining [1] and [2]}$$

Solving for TC:

$$T_C = -C_{HOLD}(R_{IC} + R_{SS} + R_S)\ \ln(1/2047)$$
$$= -10pF(1k\Omega + 7k\Omega + 10k\Omega)\ \ln(0.0004885)$$
$$= 1.37\mu s$$

Therefore:

$$T_{ACQ} = 2MS + 1.37MS + [(50°C - 25°C)(0.05MS/°C)]$$
$$= 4.67MS$$

Note 1: The reference voltage (VREF) has no effect on the equation, since it cancels itself out.
2: The charge holding capacitor (CHOLD) is not discharged after each conversion.
3: The maximum recommended impedance for analog sources is 10 kΩ. This is required to meet the pin leakage specification.

FIGURE 9-4: ANALOG INPUT MODEL

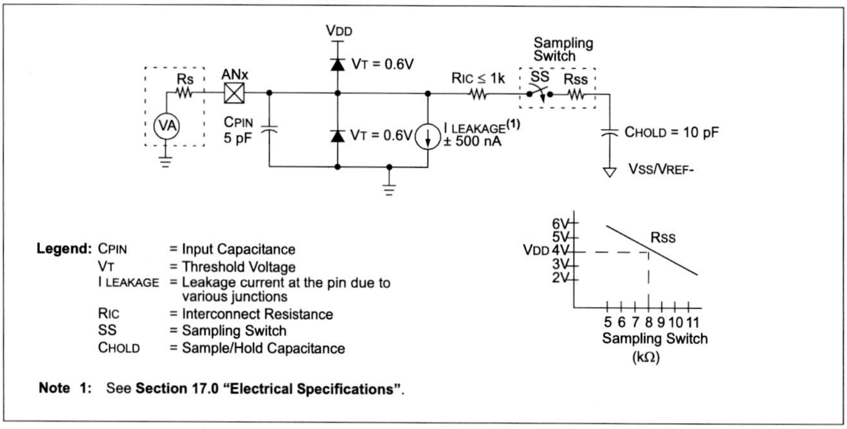

FIGURE 9-5: ADC TRANSFER FUNCTION

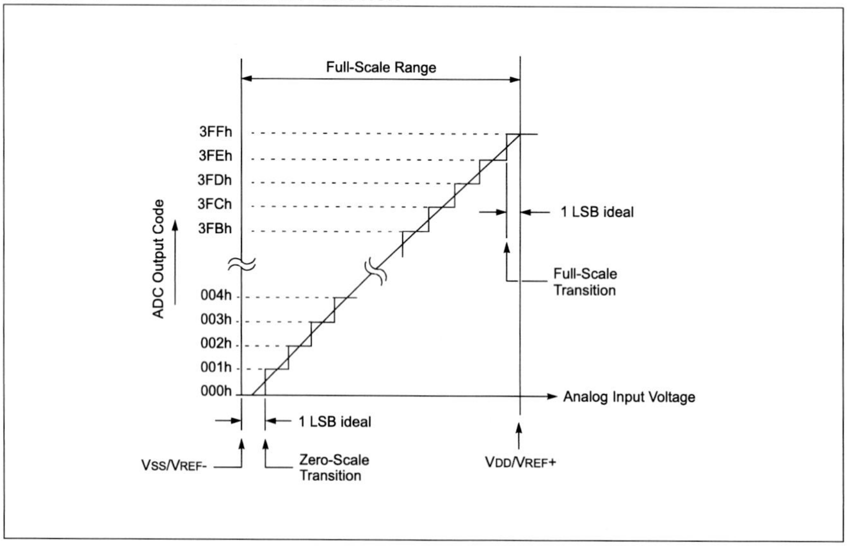

TABLE 9-2: SUMMARY OF ASSOCIATED ADC REGISTERS

Name	Bit 7	Bit 6	Bit 5	Bit 4	Bit 3	Bit 2	Bit 1	Bit 0	Value on POR, BOR	Value on all other Resets
ADCON0	ADCS1	ADCS0	CHS3	CHS2	CHS1	CHS0	GO/$\overline{\text{DONE}}$	ADON	0000 0000	0000 0000
ADCON1	ADFM	—	VCFG1	VCFG0	—	—	—	—	0-00 ----	-000 ----
ANSEL	ANS7	ANS6	ANS5	ANS4	ANS3	ANS2	ANS1	ANS0	1111 1111	1111 1111
ANSELH	—	—	ANS13	ANS12	ANS11	ANS10	ANS9	ANS8	--11 1111	--11 1111
ADRESH	A/D Result Register High Byte								xxxx xxxx	uuuu uuuu
ADRESL	A/D Result Register Low Byte								xxxx xxxx	uuuu uuuu
INTCON	GIE	PEIE	T0IE	INTE	RBIE	T0IF	INTF	RBIF	0000 000x	0000 000x
PIE1	—	ADIE	RCIE	TXIE	SSPIE	CCP1IE	TMR2IE	TMR1IE	-000 0000	-000 0000
PIR1	—	ADIF	RCIF	TXIF	SSPIF	CCP1IF	TMR2IF	TMR1IF	-000 0000	-000 0000
PORTA	RA7	RA6	RA5	RA4	RA3	RA2	RA1	RA0	xxxx xxxx	uuuu uuuu
PORTB	RB7	RB6	RB5	RB4	RB3	RB2	RB1	RB0	xxxx xxxx	uuuu uuuu
PORTE	—	—	—	—	RE3	RE2	RE1	RE0	---- xxxx	---- uuuu
TRISA	TRISA7	TRISA6	TRISA5	TRISA4	TRISA3	TRISA2	TRISA1	TRISA0	1111 1111	1111 1111
TRISB	TRISB7	TRISB6	TRISB5	TRISB4	TRISB3	TRISB2	TRISB1	TRISB0	1111 1111	1111 111
TRISE	—	—	—	—	TRISE3	TRISE2	TRISE1	TRISE0	---- 1111	---- 111

Legend: x = unknown, u = unchanged, – = unimplemented read as '0'. Shaded cells are not used for ADC module.

NOTES:

13.0 MASTER SYNCHRONOUS SERIAL PORT (MSSP) MODULE

13.1 Master SSP (MSSP) Module Overview

The Master Synchronous Serial Port (MSSP) module is a serial interface useful for communicating with other peripheral or microcontroller devices. These peripheral devices may be Serial EEPROMs, shift registers, display drivers, A/D converters, etc. The MSSP module can operate in one of two modes:

- Serial Peripheral Interface (SPI)
- Inter-Integrated Circuit™ (I^2C™)
  - Full Master mode
  - Slave mode (with general address call).

The I^2C interface supports the following modes in hardware:

- Master mode
- Multi-Master mode
- Slave mode.

13.2 Control Registers

The MSSP module has three associated registers. These include a STATUS register and two control registers.

Register 13-1 shows the MSSP STATUS register (SSPSTAT), Register 13-2 shows the MSSP Control Register 1 (SSPCON), and Register 13-3 shows the MSSP Control Register 2 (SSPCON2).

PIC16F882/883/884/886/887

REGISTER 13-1: SSPSTAT: SSP STATUS REGISTER

R/W-0	R/W-0	R-0	R-0	R-0	R-0	R-0	R-0
SMP	CKE	$D/\overline{A}$	P	S	$R/\overline{W}$	UA	BF
bit 7							bit 0

Legend:			
R = Readable bit	W = Writable bit	U = Unimplemented bit, read as '0'	
-n = Value at POR	'1' = Bit is set	'0' = Bit is cleared	x = Bit is unknown

bit 7 **SMP:** Sample bit

SPI Master mode:
1 = Input data sampled at end of data output time
0 = Input data sampled at middle of data output time

SPI Slave mode:
SMP must be cleared when SPI is used in Slave mode

In I^2C Master or Slave mode:
1 = Slew rate control disabled for standard speed mode (100 kHz and 1 MHz)
0 = Slew rate control enabled for high speed mode (400 kHz)

bit 6 **CKE:** SPI Clock Edge Select bit

CKP = 0:
1 = Data transmitted on rising edge of SCK
0 = Data transmitted on falling edge of SCK

CKP = 1:
1 = Data transmitted on falling edge of SCK
0 = Data transmitted on rising edge of SCK

bit 5 **$D/\overline{A}$:** Data/$\overline{\text{Address}}$ bit (I^2C mode only)
1 = Indicates that the last byte received or transmitted was data
0 = Indicates that the last byte received or transmitted was address

bit 4 **P:** Stop bit
(I^2C mode only. This bit is cleared when the MSSP module is disabled, SSPEN is cleared.)
1 = Indicates that a Stop bit has been detected last (this bit is '0' on Reset)
0 = Stop bit was not detected last

bit 3 **S:** Start bit
(I^2C mode only. This bit is cleared when the MSSP module is disabled, SSPEN is cleared.)
1 = Indicates that a Start bit has been detected last (this bit is '0' on Reset)
0 = Start bit was not detected last

bit 2 **$R/\overline{W}$:** Read/Write bit information (I^2C mode only)
This bit holds the $R/\overline{W}$ bit information following the last address match. This bit is only valid from the address match to the next Start bit, Stop bit, or not $\overline{\text{ACK}}$ bit.

In I^2C Slave mode:
1 = Read
0 = Write

In I^2C Master mode:
1 = Transmit is in progress
0 = Transmit is not in progress
OR-ing this bit with SEN, RSEN, PEN, RCEN, or ACKEN will indicate if the MSSP is in Idle mode.

bit 1 **UA:** Update Address bit (10-bit I^2C mode only)
1 = Indicates that the user needs to update the address in the SSPADD register
0 = Address does not need to be updated

bit 0 **BF:** Buffer Full Status bit

Receive (SPI and I^2C modes):
1 = Receive complete, SSPBUF is full
0 = Receive not complete, SSPBUF is empty

Transmit (I^2C mode only):
1 = Data transmit in progress (does not include the $\overline{\text{ACK}}$ and Stop bits), SSPBUF is full
0 = Data transmit complete (does not include the $\overline{\text{ACK}}$ and Stop bits), SSPBUF is empty

REGISTER 13-2: SSPCON: SSP CONTROL REGISTER 1

R/W-0	R/W-0	R/W-0	R/W-0	R/W-0	R/W-0	R/W-0	R/W-0
WCOL	SSPOV	SSPEN	CKP	SSPM3	SSPM2	SSPM1	SSPM0
bit 7							bit 0

Legend:			
R = Readable bit	W = Writable bit	U = Unimplemented bit, read as '0'	
-n = Value at POR	'1' = Bit is set	'0' = Bit is cleared	x = Bit is unknown

bit 7 **WCOL:** Write Collision Detect bit

Master mode:
1 = A write to the SSPBUF register was attempted while the I^2C conditions were not valid for a transmission to be started
0 = No collision

Slave mode:
1 = The SSPBUF register is written while it is still transmitting the previous word (must be cleared in software)
0 = No collision

bit 6 **SSPOV:** Receive Overflow Indicator bit

In SPI mode:
1 = A new byte is received while the SSPBUF register is still holding the previous data. In case of overflow, the data in SSPSR is lost. Overflow can only occur in Slave mode. In Slave mode, the user must read the SSPBUF, even if only transmitting data, to avoid setting overflow. In Master mode, the overflow bit is not set since each new reception (and transmission) is initiated by writing to the SSPBUF register (must be cleared in software).
0 = No overflow

In I^2C mode:
1 = A byte is received while the SSPBUF register is still holding the previous byte. SSPOV is a "don't care" in Transmit mode (must be cleared in software).
0 = No overflow

bit 5 **SSPEN:** Synchronous Serial Port Enable bit

In both modes, when enabled, these pins must be properly configured as input or output

In SPI mode:
1 = Enables serial port and configures SCK, SDO, SDI and $\overline{SS}$ as the source of the serial port pins
0 = Disables serial port and configures these pins as I/O port pins

In I^2C mode:
1 = Enables the serial port and configures the SDA and SCL pins as the source of the serial port pins
0 = Disables serial port and configures these pins as I/O port pins

bit 4 **CKP:** Clock Polarity Select bit

In SPI mode:
1 = Idle state for clock is a high level
0 = Idle state for clock is a low level

In I^2C Slave mode:
SCK release control
1 = Enable clock
0 = Holds clock low (clock stretch). (Used to ensure data setup time.)

In I^2C Master mode:
Unused in this mode

bit 3-0 **SSPM<3:0>:** Synchronous Serial Port Mode Select bits

0000 = SPI Master mode, clock = FOSC/4
0001 = SPI Master mode, clock = FOSC/16
0010 = SPI Master mode, clock = FOSC/64
0011 = SPI Master mode, clock = TMR2 output/2
0100 = SPI Slave mode, clock = SCK pin, $\overline{SS}$ pin control enabled
0101 = SPI Slave mode, clock = SCK pin, $\overline{SS}$ pin control disabled, $\overline{SS}$ can be used as I/O pin
0110 = I^2C Slave mode, 7-bit address
0111 = I^2C Slave mode, 10-bit address
1000 = I^2C Master mode, clock = FOSC / (4 * (SSPADD+1))
1001 = Load Mask function
1010 = Reserved
1011 = I^2C firmware controlled Master mode (Slave idle)
1100 = Reserved
1101 = Reserved
1110 = I^2C Slave mode, 7-bit address with Start and Stop bit interrupts enabled
1111 = I^2C Slave mode, 10-bit address with Start and Stop bit interrupts enabled

PIC16F882/883/884/886/887

REGISTER 13-3: SSPCON2: SSP CONTROL REGISTER 2

R/W-0	R-0	R/W-0	R/W-0	R/W-0	R/W-0	R/W-0	R/W-0
GCEN	ACKSTAT	ACKDT	ACKEN	RCEN	PEN	RSEN	SEN
bit 7							bit 0

Legend:			
R = Readable bit	W = Writable bit	U = Unimplemented bit, read as '0'	
-n = Value at POR	'1' = Bit is set	'0' = Bit is cleared	x = Bit is unknown

bit 7 **GCEN:** General Call Enable bit (in I^2C Slave mode only)
1 = Enable interrupt when a general call address (0000h) is received in the SSPSR
0 = General call address disabled

bit 6 **ACKSTAT:** Acknowledge Status bit (in I^2C Master mode only)
In Master Transmit mode:
1 = Acknowledge was not received from slave
0 = Acknowledge was received from slave

bit 5 **ACKDT:** Acknowledge Data bit (in I^2C Master mode only)
In Master Receive mode:
Value transmitted when the user initiates an Acknowledge sequence at the end of a receive
1 = Not Acknowledge
0 = Acknowledge

bit 4 **ACKEN:** Acknowledge Sequence Enable bit (in I^2C Master mode only)
In Master Receive mode:
1 = Initiate Acknowledge sequence on SDA and SCL pins, and transmit ACKDT data bit. Automatically cleared by hardware.
0 = Acknowledge sequence idle

bit 3 **RCEN:** Receive Enable bit (in I^2C Master mode only)
1 = Enables Receive mode for I^2C
0 = Receive idle

bit 2 **PEN:** Stop Condition Enable bit (in I^2C Master mode only)
SCK Release Control:
1 = Initiate Stop condition on SDA and SCL pins. Automatically cleared by hardware.
0 = Stop condition Idle

bit 1 **RSEN:** Repeated Start Condition Enabled bit (in I^2C Master mode only)
1 = Initiate Repeated Start condition on SDA and SCL pins. Automatically cleared by hardware.
0 = Repeated Start condition Idle

bit 0 **SEN:** Start Condition Enabled bit (in I^2C Master mode only)
In Master mode:
1 = Initiate Start condition on SDA and SCL pins. Automatically cleared by hardware.
0 = Start condition Idle
In Slave mode:
1 = Clock stretching is enabled for both slave transmit and slave receive (stretch enabled)
0 = Clock stretching is disabled

Note 1: For bits ACKEN, RCEN, PEN, RSEN, SEN: If the I^2C module is not in the Idle mode, this bit may not be set (no spooling) and the SSPBUF may not be written (or writes to the SSPBUF are disabled).

PIC16F882/883/884/886/887

13.3 SPI Mode

The SPI mode allows 8 bits of data to be synchronously transmitted and received, simultaneously. All four modes of SPI are supported. To accomplish communication, typically three pins are used:

- Serial Data Out (SDO) – RC5/SDO
- Serial Data In (SDI) – RC4/SDI/SDA
- Serial Clock (SCK) – RC3/SCK/SCL

Additionally, a fourth pin may be used when in any Slave mode of operation:

- Slave Select ($\overline{SS}$) – RA5/$\overline{SS}$/AN4

13.3.1 OPERATION

When initializing the SPI, several options need to be specified. This is done by programming the appropriate control bits SSPCON<5:0> and SSPSTAT<7:6>. These control bits allow the following to be specified:

- Master mode (SCK is the clock output)
- Slave mode (SCK is the clock input)
- Clock polarity (Idle state of SCK)
- Data input sample phase (middle or end of data output time)
- Clock edge (output data on rising/falling edge of SCK)
- Clock rate (Master mode only)
- Slave Select mode (Slave mode only)

Figure 13-1 shows the block diagram of the MSSP module, when in SPI mode.

FIGURE 13-1: MSSP BLOCK DIAGRAM (SPI MODE)

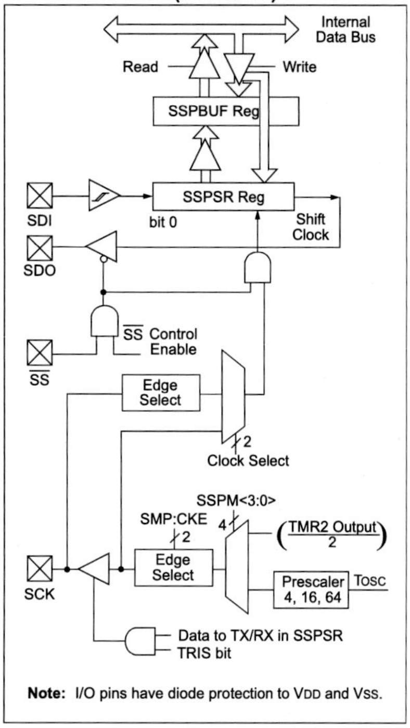

The MSSP consists of a transmit/receive shift register (SSPSR) and a buffer register (SSPBUF). The SSPSR shifts the data in and out of the device, MSb first. The SSPBUF holds the data that was written to the SSPSR, until the received data is ready. Once the 8 bits of data have been received, that byte is moved to the SSPBUF register. Then, the buffer full-detect bit BF of the SSPSTAT register and the interrupt flag bit SSPIF of the PIR1 register are set. This double buffering of the received data (SSPBUF) allows the next byte to start reception before reading the data that was just received. Any write to the SSPBUF register during transmission/reception of data will be ignored, and the write collision detect bit WCOL of the SSPCON register will be set. User software must clear the WCOL bit so that it can be determined if the following write(s) to the SSPBUF register completed successfully.

When the application software is expecting to receive valid data, the SSPBUF should be read before the next byte of data to transfer is written to the SSPBUF. The buffer full bit BF of the SSPSTAT register indicates when SSPBUF has been loaded with the received data (transmission is complete). When the SSPBUF is read, the BF bit is cleared. This data may be irrelevant if the SPI is only a transmitter. Generally, the MSSP Interrupt is used to determine when the transmission/reception has completed. The SSPBUF must be read and/or written. If the interrupt method is not going to be used, then software polling can be done to ensure that a write collision does not occur. Example 13-1 shows the loading of the SSPBUF (SSPSR) for data transmission.

The SSPSR is not directly readable or writable, and can only be accessed by addressing the SSPBUF register. Additionally, the MSSP STATUS register (SSPSTAT register) indicates the various status conditions.

13.3.2 ENABLING SPI I/O

To enable the serial port, SSP Enable bit SSPEN of the SSPCON register must be set. To reset or reconfigure SPI mode, clear the SSPEN bit, re-initialize the SSPCON registers, and then set the SSPEN bit. This configures the SDI, SDO, SCK and $\overline{SS}$ pins as serial port pins. For the pins to behave as the serial port function, some must have their data direction bits (in the TRIS register) appropriately programmed. That is:

- SDI is automatically controlled by the SPI module
- SDO must have TRISC<5> bit cleared
- SCK (Master mode) must have TRISC<3> bit cleared
- SCK (Slave mode) must have TRISC<3> bit set
- $\overline{SS}$ must have TRISA<5> bit set

Any serial port function that is not desired may be overridden by programming the corresponding data direction (TRIS) register to the opposite value.

EXAMPLE 13-1: LOADING THE SSPBUF (SSPSR) REGISTER

```
LOOP BTFSS SSPSTAT, BF    ;Has data been received (transmit complete)?
     GOTO  LOOP           ;No
     MOVF  SSPBUF, W      ;WREG reg = contents of SSPBUF
     MOVWF RXDATA         ;Save in user RAM, if data is meaningful
     MOVF  TXDATA, W      ;W reg = contents of TXDATA
     MOVWF SSPBUF         ;New data to xmit
```

PIC16F882/883/884/886/887

13.3.3 MASTER MODE

The master can initiate the data transfer at any time because it controls the SCK. The master determines when the slave is to broadcast data by the software protocol.

In Master mode, the data is transmitted/received as soon as the SSPBUF register is written to. If the SPI is only going to receive, the SDO output could be disabled (programmed as an input). The SSPSR register will continue to shift in the signal present on the SDI pin at the programmed clock rate. As each byte is received, it will be loaded into the SSPBUF register as a normal received byte (interrupts and Status bits appropriately set). This could be useful in receiver applications as a "Line Activity Monitor" mode.

The clock polarity is selected by appropriately programming the CKP bit of the SSPCON register. This, then, would give waveforms for SPI communication as shown in Figure 13-2, Figure 13-4 and Figure 13-5, where the MSb is transmitted first. In Master mode, the SPI clock rate (bit rate) is user programmable to be one of the following:

- FOSC/4 (or TCY)
- FOSC/16 (or 4 • TCY)
- FOSC/64 (or 16 • TCY)
- Timer2 output/2

This allows a maximum data rate (at 40 MHz) of 10.00 Mbps.

Figure 13-2 shows the waveforms for Master mode. When the CKE bit of the SSPSTAT register is set, the SDO data is valid before there is a clock edge on SCK. The change of the input sample is shown based on the state of the SMP bit of the SSPSTAT register. The time when the SSPBUF is loaded with the received data is shown.

FIGURE 13-2: SPI MODE WAVEFORM (MASTER MODE)

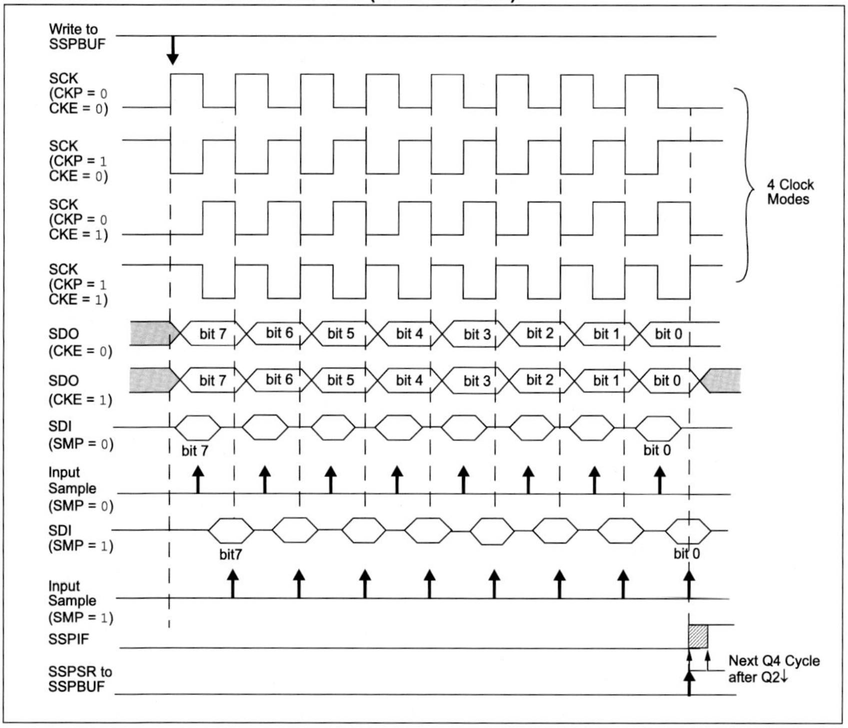

PIC16F882/883/884/886/887

13.3.4 SLAVE MODE

In Slave mode, the data is transmitted and received as the external clock pulses appear on SCK. When the last bit is latched, the SSPIF interrupt flag bit of the PIR1 register is set.

While in Slave mode, the external clock is supplied by the external clock source on the SCK pin. This external clock must meet the minimum high and low times, as specified in the electrical specifications.

While in Sleep mode, the slave can transmit/receive data. When a byte is received, the device will wake-up from Sleep.

13.3.5 SLAVE SELECT SYNCHRONIZATION

The $\overline{SS}$ pin allows a Synchronous Slave mode. The SPI must be in Slave mode with $\overline{SS}$ pin control enabled (SSPCON<3:0> = 04h). The pin must not be driven low for the $\overline{SS}$ pin to function as an input. The Data Latch must be high. When the $\overline{SS}$ pin is low, transmission and reception are enabled and the SDO pin is driven. When the $\overline{SS}$ pin goes high, the SDO pin is no longer driven, even if in the middle of a transmitted byte, and becomes a floating output. External pull-up/pull-down resistors may be desirable, depending on the application.

> **Note 1:** When the SPI is in Slave mode with $\overline{SS}$ pin control enabled (SSPCON<3:0> = 0100), the SPI module will reset if the $\overline{SS}$ pin is set to VDD.
>
> **2:** If the SPI is used in Slave mode with CKE set (SSPSTAT register), then the $\overline{SS}$ pin control must be enabled.

When the SPI module resets, the bit counter is forced to '0'. This can be done by either forcing the $\overline{SS}$ pin to a high level, or clearing the SSPEN bit.

To emulate two-wire communication, the SDO pin can be connected to the SDI pin. When the SPI needs to operate as a receiver, the SDO pin can be configured as an input. This disables transmissions from the SDO. The SDI can always be left as an input (SDI function), since it cannot create a bus conflict.

FIGURE 13-3: SLAVE SYNCHRONIZATION WAVEFORM

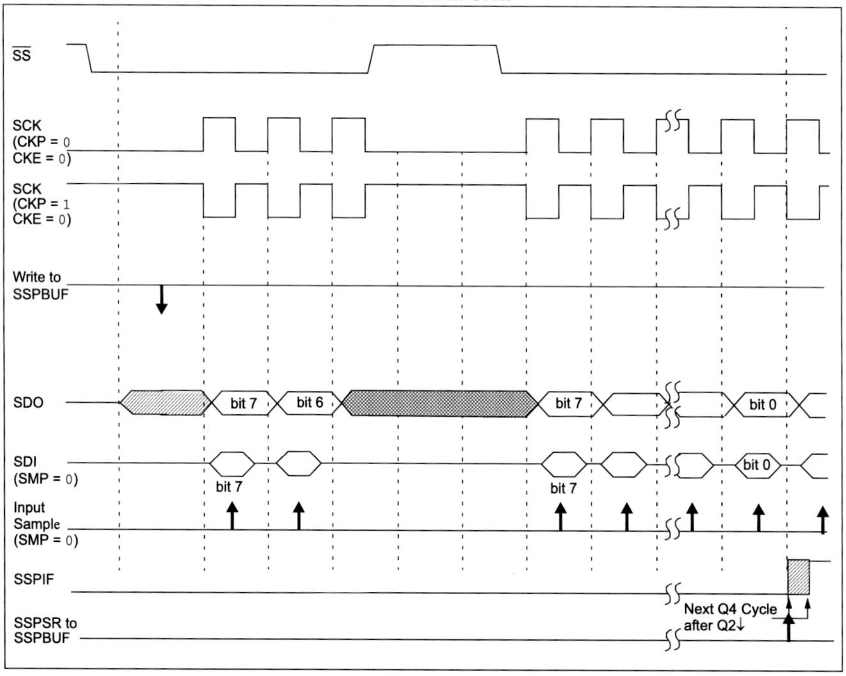

PIC16F882/883/884/886/887

FIGURE 13-4: SPI MODE WAVEFORM (SLAVE MODE WITH CKE = 0)

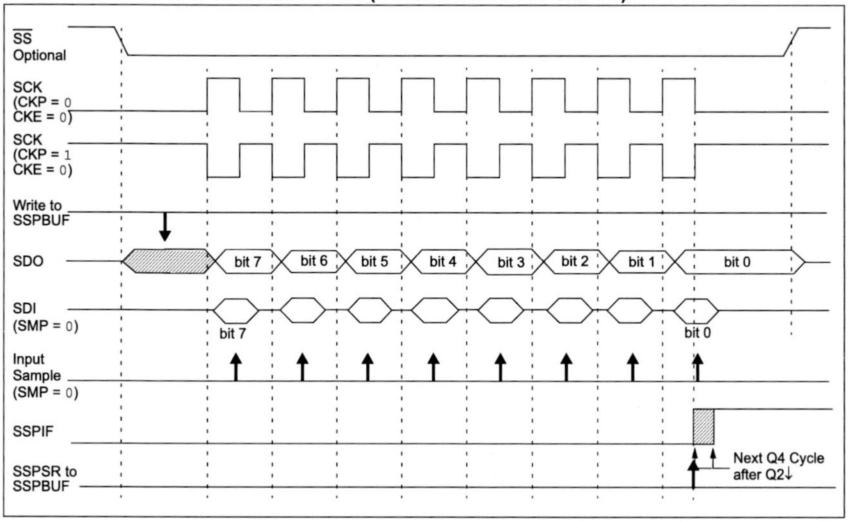

FIGURE 13-5: SPI MODE WAVEFORM (SLAVE MODE WITH CKE = 1)

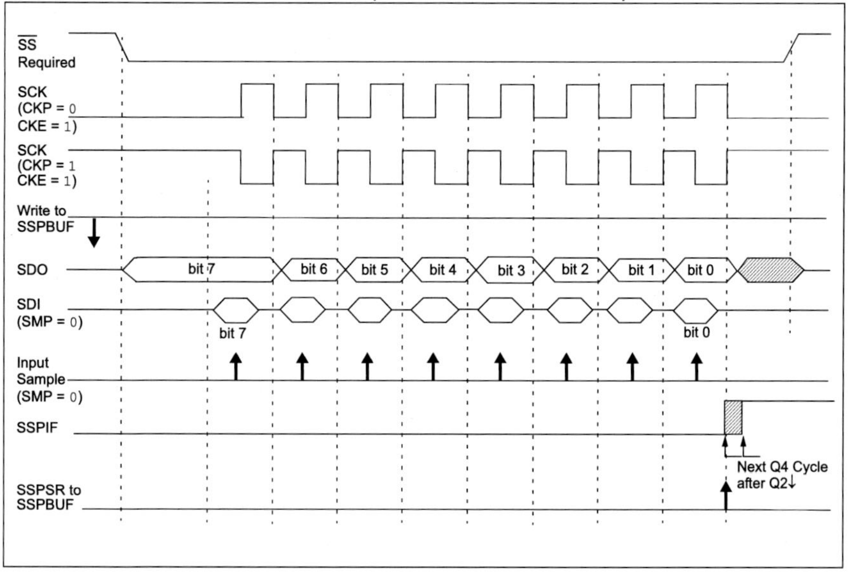

PIC16F882/883/884/886/887

13.3.6 SLEEP OPERATION

In Master mode, all module clocks are halted, and the transmission/reception will remain in that state until the device wakes from Sleep. After the device returns to normal mode, the module will continue to transmit/receive data.

In Slave mode, the SPI transmit/receive shift register operates asynchronously to the device. This allows the device to be placed in Sleep mode and data to be shifted into the SPI transmit/receive shift register. When all eight bits have been received, the MSSP interrupt flag bit will be set and, if enabled, will wake the device from Sleep.

13.3.7 EFFECTS OF A RESET

A Reset disables the MSSP module and terminates the current transfer.

13.3.8 BUS MODE COMPATIBILITY

Table 13-1 shows the compatibility between the standard SPI modes and the states of the CKP and CKE control bits.

TABLE 13-1: SPI BUS MODES

Standard SPI Mode Terminology	Control Bits State	
	CKP	CKE
0, 0	0	1
0, 1	0	0
1, 0	1	1
1, 1	1	0

There is also a SMP bit that controls when the data will be sampled.

TABLE 13-2: REGISTERS ASSOCIATED WITH SPI OPERATION

Name	Bit 7	Bit 6	Bit 5	Bit 4	Bit 3	Bit 2	Bit 1	Bit 0	Value on POR, BOR	Value on all other RESETS
INTCON	GIE/GIEH	PEIE/GIEL	T0IE	INTE	RBIE	T0IF	INTF	RBIF	0000 000x	0000 000u
PIE1	—	ADIE	RCIE	TXIE	SSPIE	CCP1IE	TMR2IE	TMR1IE	0000 0000	0000 0000
PIR1	—	ADIF	RCIF	TXIF	SSPIF	CCP1IF	TMR2IF	TMR1IF	-000 0000	0000 0000
SSPBUF	Synchronous Serial Port Receive Buffer/Transmit Register								xxxx xxxx	uuuu uuuu
SSPCON	WCOL	SSPOV	SSPEN	CKP	SSPM3	SSPM2	SSPM1	SSPM0	0000 0000	0000 0000
SSPSTAT	SMP	CKE	D/$\overline{A}$	P	S	R/$\overline{W}$	UA	BF	0000 0000	0000 0000
TRISA	TRISA7	TRISA6	TRISA5	TRISA4	TRISA3	TRISA2	TRISA1	TRISA0	1111 1111	1111 1111
TRISC	TRISC7	TRISC6	TRISC5	TRISC4	TRISC3	TRISC2	TRISC1	TRISC0	1111 1111	1111 1111

Legend: x = unknown, u = unchanged, – = unimplemented, read as '0'. Shaded cells are not used by the MSSP in SPI mode.
Note 1: Bit 6 of PORTA, LATA and TRISA are enabled in ECIO and RCIO Oscillator modes only. In all other oscillator modes, they are disabled and read '0'.

13.4 MSSP I²C Operation

The MSSP module in I²C mode, fully implements all master and slave functions (including general call support) and provides interrupts on Start and Stop bits in hardware, to determine a free bus (Multi-Master mode). The MSSP module implements the standard mode specifications, as well as 7-bit and 10-bit addressing.

Two pins are used for data transfer. These are the RC3/SCK/SCL pin, which is the clock (SCL), and the RC4/SDI/SDA pin, which is the data (SDA). The user must configure these pins as inputs or outputs through the TRISC<4:3> bits.

The MSSP module functions are enabled by setting MSSP Enable bit SSPEN of the SSPCON register.

FIGURE 13-6: MSSP BLOCK DIAGRAM (I²C MODE)

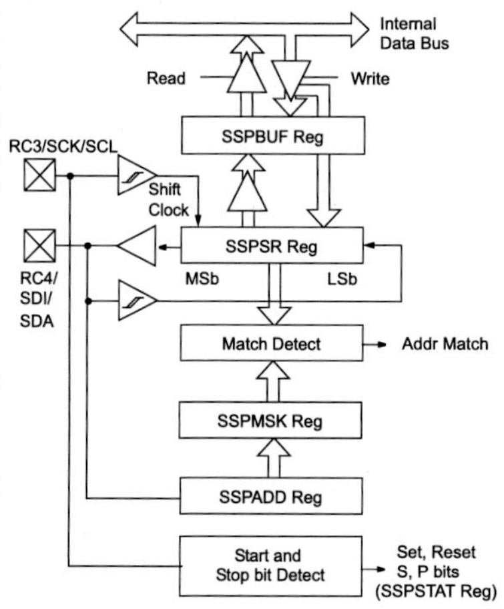

Note: I/O pins have diode protection to VDD and VSS.

The MSSP module has these six registers for I²C operation:

- MSSP Control Register 1 (SSPCON)
- MSSP Control Register 2 (SSPCON2)
- MSSP STATUS register (SSPSTAT)
- Serial Receive/Transmit Buffer (SSPBUF)
- MSSP Shift Register (SSPSR) – Not directly accessible
- MSSP Address register (SSPADD)
- MSSP Mask register (SSPMSK)

The SSPCON register allows control of the I²C operation. The SSPM<3:0> mode selection bits (SSPCON register) allow one of the following I²C modes to be selected:

- I²C Master mode, clock = OSC/4 (SSPADD +1)
- I²C Slave mode (7-bit address)
- I²C Slave mode (10-bit address)
- I²C Slave mode (7-bit address), with Start and Stop bit interrupts enabled
- I²C Slave mode (10-bit address), with Start and Stop bit interrupts enabled
- I²C firmware controlled master operation, slave is idle

Selection of any I²C mode with the SSPEN bit set, forces the SCL and SDA pins to be open drain, provided these pins are programmed to be inputs by setting the appropriate TRISC bits.

13.4.1 SLAVE MODE

In Slave mode, the SCL and SDA pins must be configured as inputs (TRISC<4:3> set). The MSSP module will override the input state with the output data when required (slave-transmitter).

When an address is matched, or the data transfer after an address match is received, the hardware automatically will generate the Acknowledge ($\overline{ACK}$) pulse and load the SSPBUF register with the received value currently in the SSPSR register.

If either or both of the following conditions are true, the MSSP module will not give this $\overline{ACK}$ pulse:

a) The buffer full bit BF (SSPCON register) was set before the transfer was received.

b) The overflow bit SSPOV (SSPCON register) was set before the transfer was received.

In this event, the SSPSR register value is not loaded into the SSPBUF, but bit SSPIF of the PIR1 register is set. The BF bit is cleared by reading the SSPBUF register, while bit SSPOV is cleared through software.

The SCL clock input must have a minimum high and low for proper operation. The high and low times of the I²C specification, as well as the requirement of the MSSP module, are shown in timing parameter #100 and parameter #101.

13.4.1.1 Addressing

Once the MSSP module has been enabled, it waits for a Start condition to occur. Following the Start condition, the eight bits are shifted into the SSPSR register. All incoming bits are sampled with the rising edge of the clock (SCL) line. The value of register SSPSR<7:1> is compared to the value of the SSPADD register. The address is compared on the falling edge of the eighth clock (SCL) pulse. If the addresses match, and the BF and SSPOV bits are clear, the following events occur:

a) The SSPSR register value is loaded into the SSPBUF register.
b) The buffer full bit BF is set.
c) An $\overline{\text{ACK}}$ pulse is generated.
d) MSSP interrupt flag bit, SSPIF of the PIR1 register, is set on the falling edge of the ninth SCL pulse (interrupt is generated, if enabled).

In 10-bit address mode, two address bytes need to be received by the slave. The five Most Significant bits (MSb) of the first address byte specify if this is a 10-bit address. The R/$\overline{\text{W}}$ bit (SSPSTAT register) must specify a write so the slave device will receive the second address byte. For a 10-bit address, the first byte would equal `'1111 0 A9 A8 0'`, where `A9` and `A8` are the two MSb's of the address.

The sequence of events for 10-bit addressing is as follows, with steps 7-9 for slave-transmitter:

1. Receive first (high) byte of address (bit SSPIF of the PIR1 register and bits BF and UA of the SSPSTAT register are set).
2. Update the SSPADD register with second (low) byte of address (clears bit UA and releases the SCL line).
3. Read the SSPBUF register (clears bit BF) and clear flag bit SSPIF.
4. Receive second (low) byte of address (bits SSPIF, BF, and UA are set).
5. Update the SSPADD register with the first (high) byte of address. If match releases SCL line, this will clear bit UA.
6. Read the SSPBUF register (clears bit BF) and clear flag bit SSPIF.
7. Receive Repeated Start condition.
8. Receive first (high) byte of address (bits SSPIF and BF are set).
9. Read the SSPBUF register (clears bit BF) and clear flag bit SSPIF.

13.4.1.2 Reception

When the R/$\overline{\text{W}}$ bit of the address byte is clear and an address match occurs, the R/$\overline{\text{W}}$ bit of the SSPSTAT register is cleared. The received address is loaded into the SSPBUF register.

When the address byte overflow condition exists, then no Acknowledge ($\overline{\text{ACK}}$) pulse is given. An overflow condition is defined as either bit BF (SSPSTAT register) is set, or bit SSPOV (SSPCON register) is set.

An MSSP interrupt is generated for each data transfer byte. Flag bit SSPIF of the PIR1 register must be cleared in software. The SSPSTAT register is used to determine the status of the byte.

13.4.1.3 Transmission

When the R/$\overline{\text{W}}$ bit of the incoming address byte is set and an address match occurs, the R/$\overline{\text{W}}$ bit of the SSPSTAT register is set. The received address is loaded into the SSPBUF register. The $\overline{\text{ACK}}$ pulse will be sent on the ninth bit and pin RC3/SCK/SCL is held low. The transmit data must be loaded into the SSPBUF register, which also loads the SSPSR register. Then pin RC3/SCK/SCL should be enabled by setting bit CKP (SSPCON register). The master must monitor the SCL pin prior to asserting another clock pulse. The slave devices may be holding off the master by stretching the clock. The eight data bits are shifted out on the falling edge of the SCL input. This ensures that the SDA signal is valid during the SCL high time (Figure 13-8).

An MSSP interrupt is generated for each data transfer byte. The SSPIF bit must be cleared in software and the SSPSTAT register is used to determine the status of the byte. The SSPIF bit is set on the falling edge of the ninth clock pulse.

As a slave-transmitter, the $\overline{\text{ACK}}$ pulse from the master-receiver is latched on the rising edge of the ninth SCL input pulse. If the SDA line is high (not $\overline{\text{ACK}}$), then the data transfer is complete. When the $\overline{\text{ACK}}$ is latched by the slave, the slave logic is reset and the slave monitors for another occurrence of the Start bit. If the SDA line was low ($\overline{\text{ACK}}$), the transmit data must be loaded into the SSPBUF register, which also loads the SSPSR register. Pin RC3/SCK/SCL should be enabled by setting bit CKP.

FIGURE 13-7: I²C™ SLAVE MODE WAVEFORMS FOR RECEPTION (7-BIT ADDRESS)

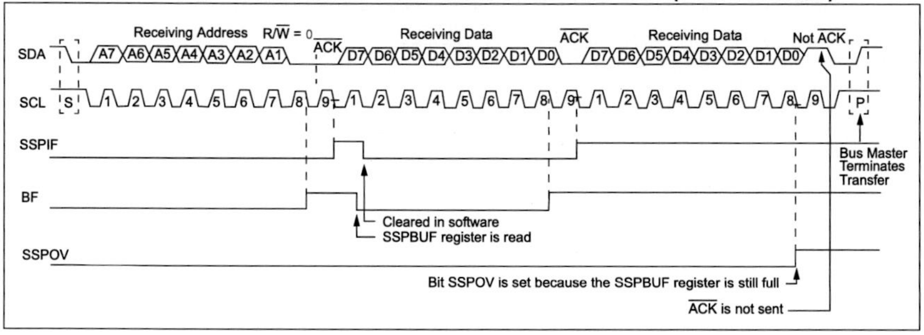

FIGURE 13-8: I²C™ SLAVE MODE WAVEFORMS FOR TRANSMISSION (7-BIT ADDRESS)

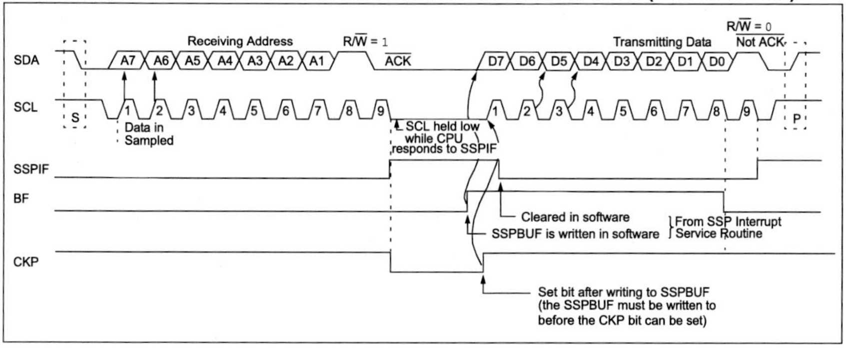

PIC16F882/883/884/886/887

13.4.2 GENERAL CALL ADDRESS SUPPORT

The addressing procedure for the I^2C bus is such that, the first byte after the Start condition usually determines which device will be the slave addressed by the master. The exception is the general call address, which can address all devices. When this address is used, all devices should, in theory, respond with an Acknowledge.

The general call address is one of eight addresses reserved for specific purposes by the I^2C protocol. It consists of all 0's with $R/\overline{W}$ = 0.

The general call address is recognized (enabled) when the General Call Enable (GCEN) bit is set (SSPCON2 register). Following a Start bit detect, eight bits are shifted into the SSPSR and the address is compared against the SSPADD. It is also compared to the general call address and fixed in hardware.

If the general call address matches, the SSPSR is transferred to the SSPBUF, the BF bit is set (eighth bit), and on the falling edge of the ninth bit ($\overline{ACK}$ bit), the SSPIF interrupt flag bit is set.

When the interrupt is serviced, the source for the interrupt can be checked by reading the contents of the SSPBUF. The value can be used to determine if the address was device specific or a general call address.

In 10-bit mode, the SSPADD is required to be updated for the second half of the address to match, and the UA bit is set (SSPSTAT register). If the general call address is sampled when the GCEN bit is set, and while the slave is configured in 10-bit address mode, then the second half of the address is not necessary. The UA bit will not be set, and the slave will begin receiving data after the Acknowledge (Figure 13-9).

FIGURE 13-9: SLAVE MODE GENERAL CALL ADDRESS SEQUENCE (7 OR 10-BIT ADDRESS)

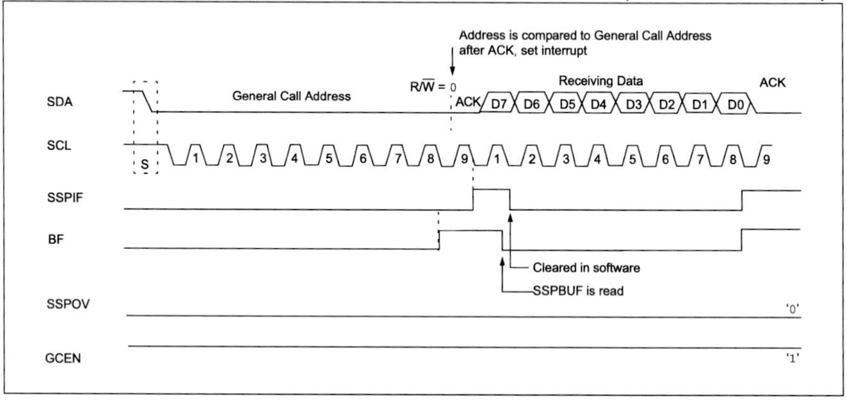

PIC16F882/883/884/886/887

13.4.3 MASTER MODE

Master mode of operation is supported by interrupt generation on the detection of the Start and Stop conditions. The Stop (P) and Start (S) bits are cleared from a Reset, or when the MSSP module is disabled. Control of the I^2C bus may be taken when the P bit is set, or the bus is idle, with both the S and P bits clear.

In Master mode, the SCL and SDA lines are manipulated by the MSSP hardware.

The following events will cause SSP Interrupt Flag bit, SSPIF, to be set (SSP Interrupt if enabled):

- Start condition
- Stop condition
- Data transfer byte transmitted/received
- Acknowledge transmit
- Repeated Start condition

13.4.4 I^2C™ MASTER MODE SUPPORT

Master mode is enabled by setting and clearing the appropriate SSPM bits in SSPCON and by setting the SSPEN bit. Once Master mode is enabled, the user has the following six options:

1. Assert a Start condition on SDA and SCL.
2. Assert a Repeated Start condition on SDA and SCL.
3. Write to the SSPBUF register initiating transmission of data/address.
4. Generate a Stop condition on SDA and SCL.
5. Configure the I^2C port to receive data.
6. Generate an Acknowledge condition at the end of a received byte of data.

Note: The MSSP module, when configured in I^2C Master mode, does not allow queueing of events. For instance, the user is not allowed to initiate a Start condition and immediately write the SSPBUF register to imitate transmission, before the Start condition is complete. In this case, the SSPBUF will not be written to and the WCOL bit will be set, indicating that a write to the SSPBUF did not occur.

FIGURE 13-10: MSSP BLOCK DIAGRAM (I^2C™ MASTER MODE)

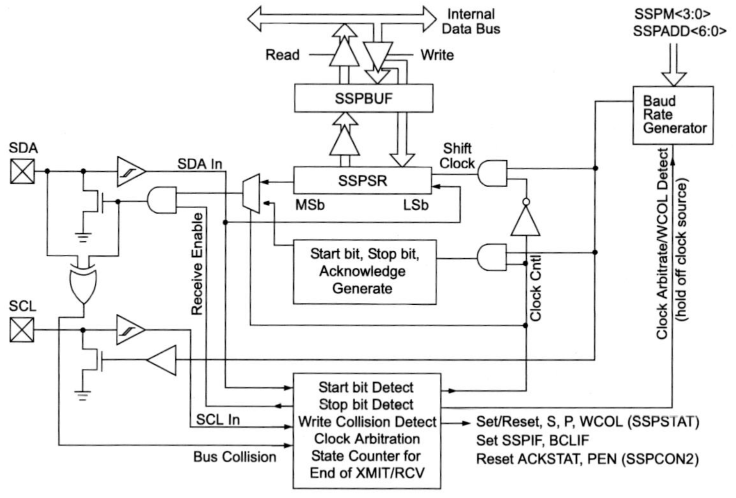

Note: I/O pins have diode protection to VDD and VSS.

PIC16F882/883/884/886/887

13.4.4.1 I²C™ Master Mode Operation

The master device generates all of the serial clock pulses and the Start and Stop conditions. A transfer is ended with a Stop condition or with a Repeated Start condition. Since the Repeated Start condition is also the beginning of the next serial transfer, the I²C bus will not be released.

In Master Transmitter mode, serial data is output through SDA, while SCL outputs the serial clock. The first byte transmitted contains the slave address of the receiving device (7 bits) and the Read/Write ($R/\overline{W}$) bit. In this case, the $R/\overline{W}$ bit will be logic '0'. Serial data is transmitted eight bits at a time. After each byte is transmitted, an Acknowledge bit is received. Start and Stop conditions are output to indicate the beginning and the end of a serial transfer.

In Master Receive mode, the first byte transmitted contains the slave address of the transmitting device (7 bits) and the $R/\overline{W}$ bit. In this case, the $R/\overline{W}$ bit will be logic '1'. Thus, the first byte transmitted is a 7-bit slave address followed by a '1' to indicate receive bit. Serial data is received via SDA, while SCL outputs the serial clock. Serial data is received eight bits at a time. After each byte is received, an Acknowledge bit is transmitted. Start and Stop conditions indicate the beginning and end of transmission.

The Baud Rate Generator used for the SPI mode operation is now used to set the SCL clock frequency for either 100 kHz, 400 kHz, or 1 MHz I²C operation. The Baud Rate Generator reload value is contained in the lower 7 bits of the SSPADD register. The Baud Rate Generator will automatically begin counting on a write to the SSPBUF. Once the given operation is complete (i.e., transmission of the last data bit is followed by ACK), the internal clock will automatically stop counting and the SCL pin will remain in its last state.

A typical transmit sequence would go as follows:

a) The user generates a Start condition by setting the Start Enable (SEN) bit (SSPCON2 register).
b) SSPIF is set. The MSSP module will wait the required start time before any other operation takes place.
c) The user loads the SSPBUF with the address to transmit.
d) Address is shifted out the SDA pin until all eight bits are transmitted.
e) The MSSP module shifts in the ACK bit from the slave device and writes its value into the ACKSTAT bit (SSPCON2 register).
f) The MSSP module generates an interrupt at the end of the ninth clock cycle by setting the SSPIF bit.
g) The user loads the SSPBUF with eight bits of data.
h) Data is shifted out the SDA pin until all eight bits are transmitted.
i) The MSSP module shifts in the ACK bit from the slave device and writes its value into the ACKSTAT bit (SSPCON2 register).
j) The MSSP module generates an interrupt at the end of the ninth clock cycle by setting the SSPIF bit.
k) The user generates a Stop condition by setting the Stop Enable bit PEN (SSPCON2 register).
l) Interrupt is generated once the Stop condition is complete.

13.4.5 BAUD RATE GENERATOR

In I^2C Master mode, the reload value for the BRG is located in the lower 7 bits of the SSPADD register (Figure 13-11). When the BRG is loaded with this value, the BRG counts down to 0 and stops until another reload has taken place. The BRG count is decremented twice per instruction cycle (TCY) on the Q2 and Q4 clocks. In I^2C Master mode, the BRG is reloaded automatically. If clock arbitration is taking place, for instance, the BRG will be reloaded when the SCL pin is sampled high (Figure 13-12).

FIGURE 13-11: BAUD RATE GENERATOR BLOCK DIAGRAM

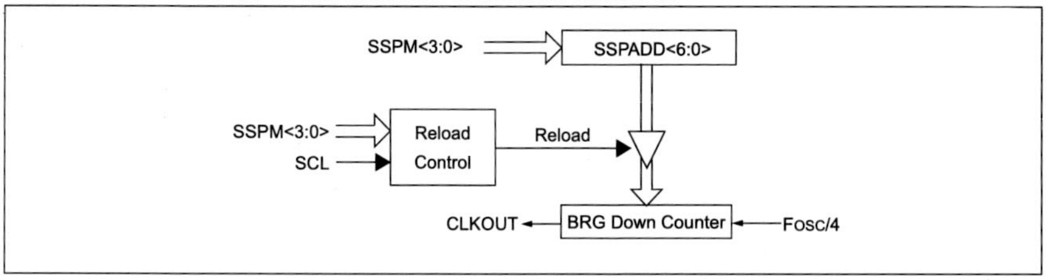

FIGURE 13-12: BAUD RATE GENERATOR TIMING WITH CLOCK ARBITRATION

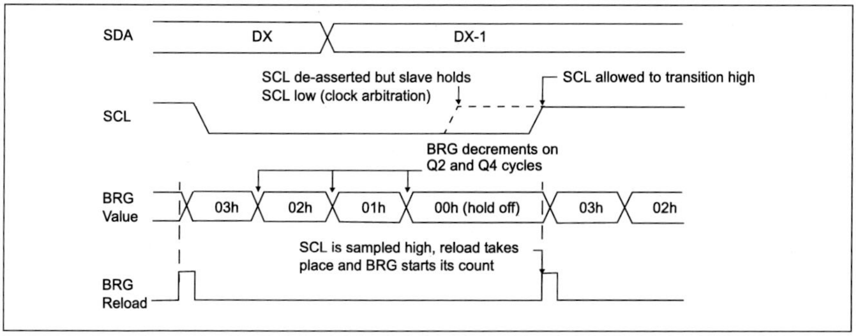

PIC16F882/883/884/886/887

13.4.6 I²C™ MASTER MODE START CONDITION TIMING

To initiate a Start condition, the user sets the Start Condition Enable bit SEN of the SSPCON2 register. If the SDA and SCL pins are sampled high, the Baud Rate Generator is reloaded with the contents of SSPADD<6:0> and starts its count. If SCL and SDA are both sampled high when the Baud Rate Generator times out (TBRG), the SDA pin is driven low. The action of the SDA being driven low, while SCL is high, is the Start condition, and causes the S bit of the SSPSTAT register to be set. Following this, the Baud Rate Generator is reloaded with the contents of SSPADD<6:0> and resumes its count. When the Baud Rate Generator times out (TBRG), the SEN bit of the SSPCON2 register will be automatically cleared by hardware, the Baud Rate Generator is suspended leaving the SDA line held low and the Start condition is complete.

Note:	If, at the beginning of the Start condition, the SDA and SCL pins are already sampled low, or if during the Start condition the SCL line is sampled low before the SDA line is driven low, a bus collision occurs, the Bus Collision Interrupt Flag, BCLIF, is set, the Start condition is aborted, and the I²C module is reset into its Idle state.

13.4.6.1 WCOL Status Flag

If the user writes the SSPBUF when a Start sequence is in progress, the WCOL is set and the contents of the buffer are unchanged (the write doesn't occur).

Note:	Because queueing of events is not allowed, writing to the lower 5 bits of SSPCON2 is disabled until the Start condition is complete.

FIGURE 13-13: FIRST START BIT TIMING

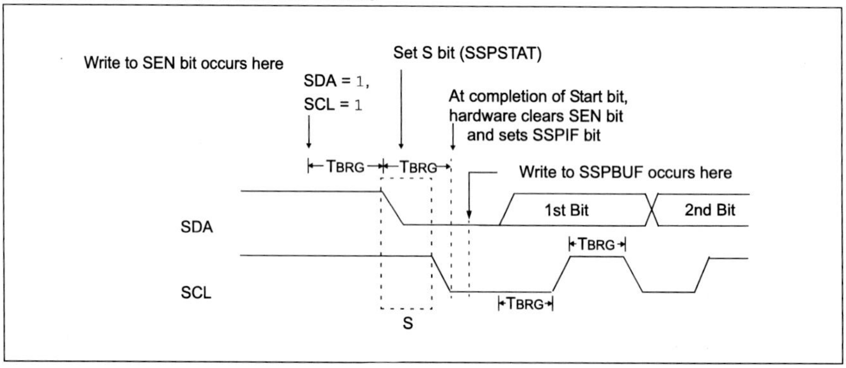

13.4.7 I^{2}C™ MASTER MODE REPEATED START CONDITION TIMING

A Repeated Start condition occurs when the RSEN bit (SSPCON2 register) is programmed high and the I^{2}C logic module is in the Idle state. When the RSEN bit is set, the SCL pin is asserted low. When the SCL pin is sampled low, the Baud Rate Generator is loaded with the contents of SSPADD<5:0> and begins counting. The SDA pin is released (brought high) for one Baud Rate Generator count (TBRG). When the Baud Rate Generator times out, if SDA is sampled high, the SCL pin will be de-asserted (brought high). When SCL is sampled high, the Baud Rate Generator is reloaded with the contents of SSPADD<6:0> and begins counting. SDA and SCL must be sampled high for one TBRG. This action is then followed by assertion of the SDA pin (SDA = 0) for one TBRG, while SCL is high. Following this, the RSEN bit (SSPCON2 register) will be automatically cleared and the Baud Rate Generator will not be reloaded, leaving the SDA pin held low. As soon as a Start condition is detected on the SDA and SCL pins, the S bit (SSPSTAT register) will be set. The SSPIF bit will not be set until the Baud Rate Generator has timed out.

Note 1:	If RSEN is programmed while any other event is in progress, it will not take effect.
2:	A bus collision during the Repeated Start condition occurs if: • SDA is sampled low when SCL goes from low-to-high. • SCL goes low before SDA is asserted low. This may indicate that another master is attempting to transmit a data "1".

Immediately following the SSPIF bit getting set, the user may write the SSPBUF with the 7-bit address in 7-bit mode, or the default first address in 10-bit mode. After the first eight bits are transmitted and an ACK is received, the user may then transmit an additional eight bits of address (10-bit mode), or eight bits of data (7-bit mode).

13.4.7.1 WCOL Status Flag

If the user writes the SSPBUF when a Repeated Start sequence is in progress, the WCOL is set and the contents of the buffer are unchanged (the write doesn't occur).

Note:	Because queueing of events is not allowed, writing of the lower 5 bits of SSPCON2 is disabled until the Repeated Start condition is complete.

FIGURE 13-14: REPEAT START CONDITION WAVEFORM

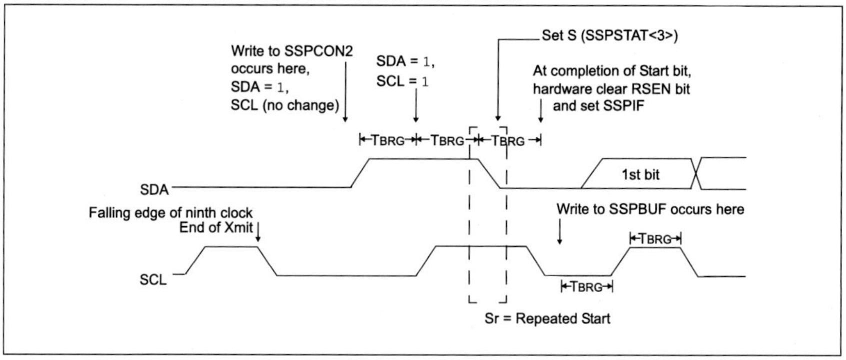

PIC16F882/883/884/886/887

13.4.8 I²C™ MASTER MODE TRANSMISSION

Transmission of a data byte, a 7-bit address, or the other half of a 10-bit address, is accomplished by simply writing a value to the SSPBUF register. This action will set the Buffer Full bit, BF, and allow the Baud Rate Generator to begin counting and start the next transmission. Each bit of address/data will be shifted out onto the SDA pin after the falling edge of SCL is asserted (see data hold time specification, parameter 106). SCL is held low for one Baud Rate Generator rollover count (TBRG). Data should be valid before SCL is released high (see data setup time specification, parameter 107). When the SCL pin is released high, it is held that way for TBRG. The data on the SDA pin must remain stable for that duration and some hold time after the next falling edge of SCL. After the eighth bit is shifted out (the falling edge of the eighth clock), the BF bit is cleared and the master releases SDA, allowing the slave device being addressed to respond with an $\overline{ACK}$ bit during the ninth bit time, if an address match occurs, or if data was received properly. The status of $\overline{ACK}$ is written into the ACKDT bit on the falling edge of the ninth clock. If the master receives an Acknowledge, the Acknowledge Status bit, ACKSTAT, is cleared. If not, the bit is set. After the ninth clock, the SSPIF bit is set and the master clock (Baud Rate Generator) is suspended until the next data byte is loaded into the SSPBUF, leaving SCL low and SDA unchanged (Figure 13-15).

After the write to the SSPBUF, each bit of the address will be shifted out on the falling edge of SCL, until all seven address bits and the R/$\overline{W}$ bit, are completed. On the falling edge of the eighth clock, the master will de-assert the SDA pin, allowing the slave to respond with an Acknowledge. On the falling edge of the ninth clock, the master will sample the SDA pin to see if the address was recognized by a slave. The status of the ACK bit is loaded into the ACKSTAT Status bit (SSPCON2 register). Following the falling edge of the ninth clock transmission of the address, the SSPIF is set, the BF bit is cleared and the Baud Rate Generator is turned off, until another write to the SSPBUF takes place, holding SCL low and allowing SDA to float.

13.4.8.1 BF Status Flag

In Transmit mode, the BF bit (SSPSTAT register) is set when the CPU writes to SSPBUF, and is cleared when all eight bits are shifted out.

13.4.8.2 WCOL Status Flag

If the user writes the SSPBUF when a transmit is already in progress (i.e., SSPSR is still shifting out a data byte), the WCOL is set and the contents of the buffer are unchanged (the write doesn't occur). WCOL must be cleared in software.

13.4.8.3 ACKSTAT Status Flag

In Transmit mode, the ACKSTAT bit (SSPCON2 register) is cleared when the slave has sent an Acknowledge ($\overline{ACK}$ = 0), and is set when the slave does not Acknowledge ($\overline{ACK}$ = 1). A slave sends an Acknowledge when it has recognized its address (including a general call), or when the slave has properly received its data.

13.4.9 I²C™ MASTER MODE RECEPTION

Master mode reception is enabled by programming the Receive Enable bit, RCEN (SSPCON2 register).

Note:	The MSSP module must be in an Idle state before the RCEN bit is set, or the RCEN bit will be disregarded.

The Baud Rate Generator begins counting, and on each rollover, the state of the SCL pin changes (high-to-low/low-to-high) and data is shifted into the SSPSR. After the falling edge of the eighth clock, the RCEN bit is automatically cleared, the contents of the SSPSR are loaded into the SSPBUF, the BF bit is set, the SSPIF flag bit is set and the Baud Rate Generator is suspended from counting, holding SCL low. The MSSP is now in Idle state, awaiting the next command. When the buffer is read by the CPU, the BF bit is automatically cleared. The user can then send an Acknowledge bit at the end of reception, by setting the Acknowledge Sequence Enable bit ACKEN (SSPCON2 register).

13.4.9.1 BF Status Flag

In receive operation, the BF bit is set when an address or data byte is loaded into SSPBUF from SSPSR. It is cleared when the SSPBUF register is read.

13.4.9.2 SSPOV Status Flag

In receive operation, the SSPOV bit is set when eight bits are received into the SSPSR and the BF bit is already set from a previous reception.

13.4.9.3 WCOL Status Flag

If the user writes the SSPBUF when a receive is already in progress (i.e., SSPSR is still shifting in a data byte), the WCOL bit is set and the contents of the buffer are unchanged (the write doesn't occur).

PIC16F882/883/884/886/887

FIGURE 13-15: I²C™ MASTER MODE WAVEFORM (TRANSMISSION, 7 OR 10-BIT ADDRESS)

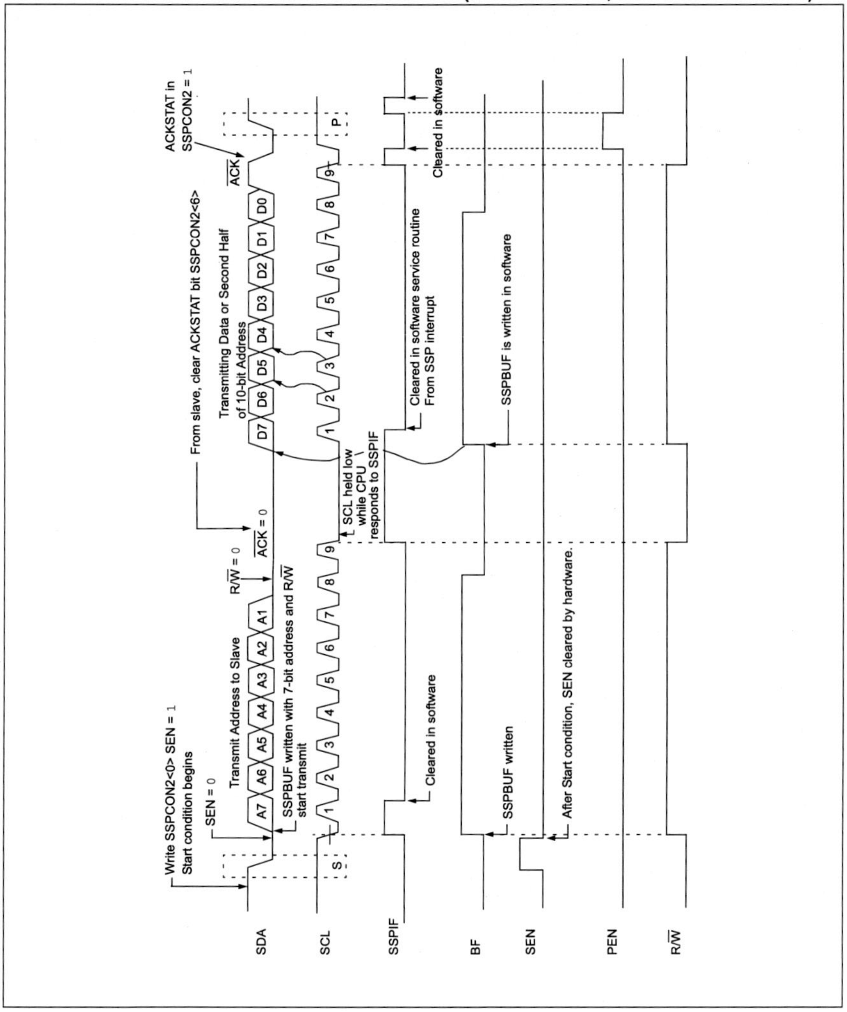

PIC16F882/883/884/886/887

FIGURE 13-16: I^{2}C™ MASTER MODE WAVEFORM (RECEPTION, 7-BIT ADDRESS)

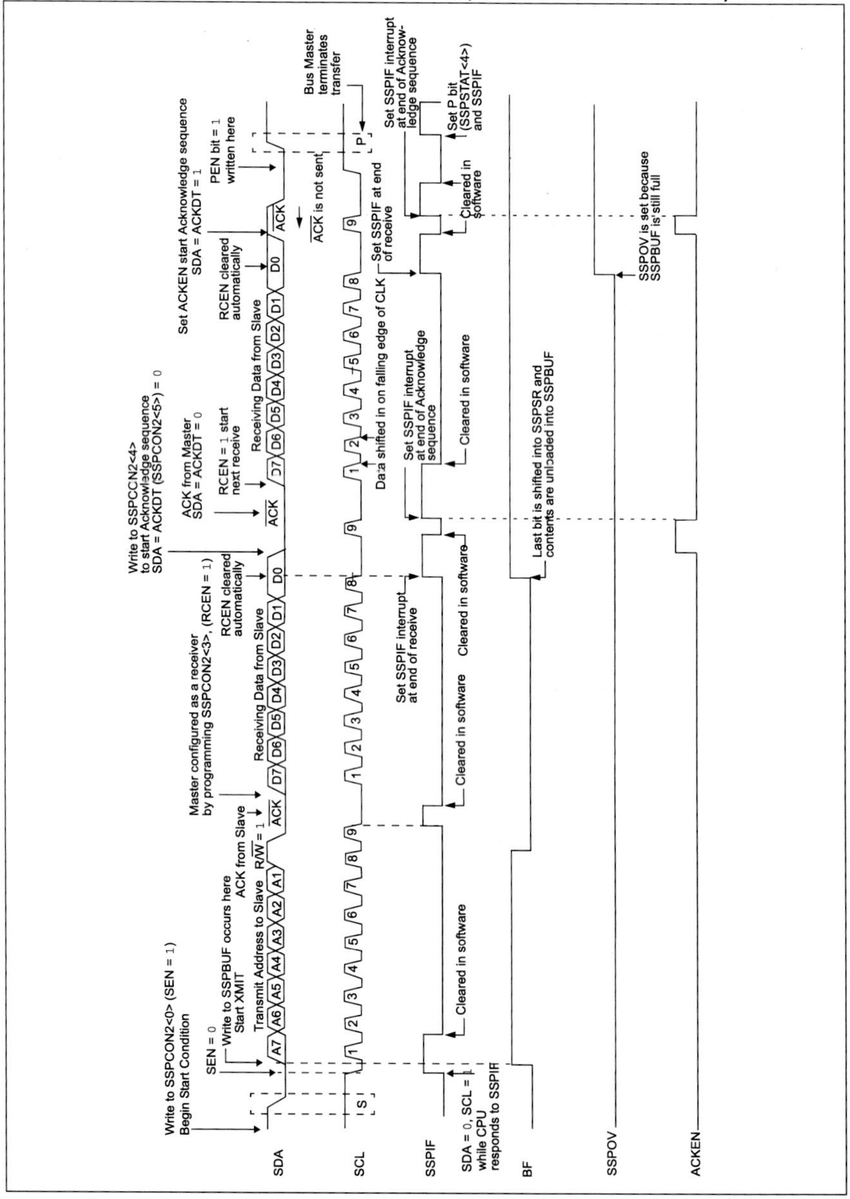

13.4.10 ACKNOWLEDGE SEQUENCE TIMING

An Acknowledge sequence is enabled by setting the Acknowledge Sequence Enable bit, ACKEN (SSPCON2 register). When this bit is set, the SCL pin is pulled low and the contents of the Acknowledge Data bit (ACKDT) is presented on the SDA pin. If the user wishes to generate an Acknowledge, then the ACKDT bit should be cleared. If not, the user should set the ACKDT bit before starting an Acknowledge sequence. The Baud Rate Generator then counts for one rollover period (TBRG) and the SCL pin is de-asserted (pulled high). When the SCL pin is sampled high (clock arbitration), the Baud Rate Generator counts for TBRG. The SCL pin is then pulled low. Following this, the ACKEN bit is automatically cleared, the Baud Rate Generator is turned off and the MSSP module then goes into Idle mode (Figure 13-17).

13.4.10.1 WCOL Status Flag

If the user writes the SSPBUF when an Acknowledge sequence is in progress, then WCOL is set and the contents of the buffer are unchanged (the write doesn't occur).

13.4.11 STOP CONDITION TIMING

A Stop bit is asserted on the SDA pin at the end of a receive/transmit by setting the Stop Sequence Enable bit, PEN (SSPCON2 register). At the end of a receive/transmit, the SCL line is held low after the falling edge of the ninth clock. When the PEN bit is set, the master will assert the SDA line low. When the SDA line is sampled low, the Baud Rate Generator is reloaded and counts down to 0. When the Baud Rate Generator times out, the SCL pin will be brought high, and one TBRG (Baud Rate Generator rollover count) later, the SDA pin will be de-asserted. When the SDA pin is sampled high while SCL is high, the P bit (SSPSTAT register) is set. A TBRG later, the PEN bit is cleared and the SSPIF bit is set (Figure 13-18).

13.4.11.1 WCOL Status Flag

If the user writes the SSPBUF when a Stop sequence is in progress, then the WCOL bit is set and the contents of the buffer are unchanged (the write doesn't occur).

FIGURE 13-17: ACKNOWLEDGE SEQUENCE WAVEFORM

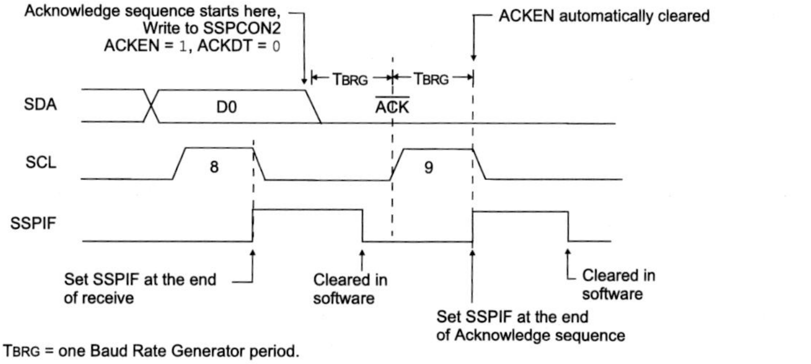

Note: TBRG = one Baud Rate Generator period.

PIC16F882/883/884/886/887

FIGURE 13-18: STOP CONDITION RECEIVE OR TRANSMIT MODE

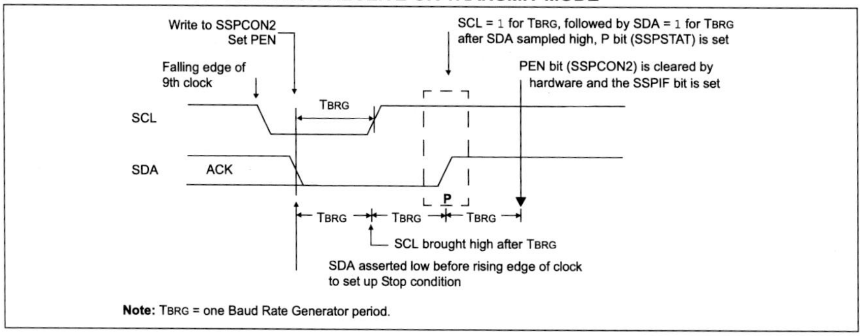

13.4.12 CLOCK ARBITRATION

Clock arbitration occurs when the master, during any receive, transmit or Repeated Start/Stop condition, de-asserts the SCL pin (SCL allowed to float high). When the SCL pin is allowed to float high, the Baud Rate Generator (BRG) is suspended from counting until the SCL pin is actually sampled high. When the SCL pin is sampled high, the Baud Rate Generator is reloaded with the contents of SSPADD<6:0> and begins counting. This ensures that the SCL high time will always be at least one BRG rollover count, in the event that the clock is held low by an external device (Figure 13-19).

13.4.13 SLEEP OPERATION

While in Sleep mode, the I²C module can receive addresses or data, and when an address match or complete byte transfer occurs, wake the processor from Sleep (if the MSSP interrupt is enabled).

13.4.14 EFFECT OF A RESET

A Reset disables the MSSP module and terminates the current transfer.

FIGURE 13-19: CLOCK ARBITRATION TIMING IN MASTER TRANSMIT MODE

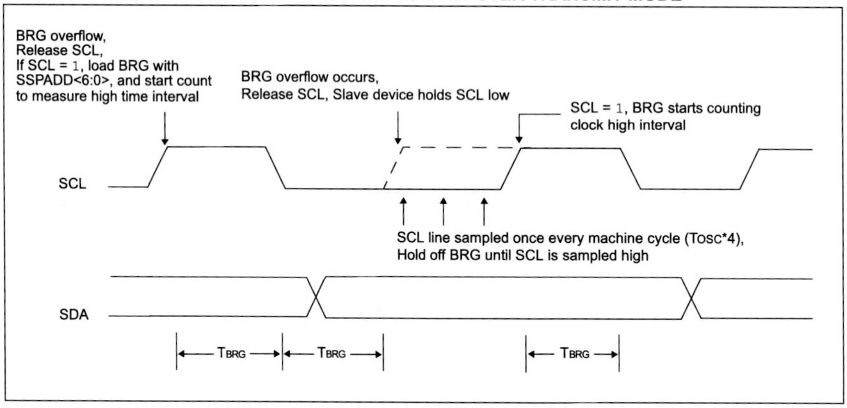

13.4.15 MULTI-MASTER MODE

In Multi-Master mode, the interrupt generation on the detection of the Start and Stop conditions allows the determination of when the bus is free. The Stop (P) and Start (S) bits are cleared from a Reset, or when the MSSP module is disabled. Control of the I^2C bus may be taken when the P bit (SSPSTAT register) is set, or the bus is idle with both the S and P bits clear. When the bus is busy, enabling the SSP Interrupt will generate the interrupt when the Stop condition occurs.

In Multi-Master operation, the SDA line must be monitored for arbitration, to see if the signal level is the expected output level. This check is performed in hardware, with the result placed in the BCLIF bit.

Arbitration can be lost in the following states:

- Address transfer
- Data transfer
- A Start condition
- A Repeated Start condition
- An Acknowledge condition

13.4.16 MULTI -MASTER COMMUNICATION, BUS COLLISION, AND BUS ARBITRATION

Multi-Master mode support is achieved by bus arbitration. When the master outputs address/data bits onto the SDA pin, arbitration takes place when the master outputs a '1' on SDA, by letting SDA float high and another master asserts a '0'. When the SCL pin floats high, data should be stable. If the expected data on SDA is a '1' and the data sampled on the SDA pin = 0, then a bus collision has taken place. The master will set the Bus Collision Interrupt Flag (BCLIF) and reset the I^2C port to its Idle state (Figure 13-20).

If a transmit was in progress when the bus collision occurred, the transmission is halted, the BF bit is cleared, the SDA and SCL lines are de-asserted, and the SSPBUF can be written to. When the user services the bus collision interrupt service routine, and if the I^2C bus is free, the user can resume communication by asserting a Start condition.

If a Start, Repeated Start, Stop, or Acknowledge condition was in progress when the bus collision occurred, the condition is aborted, the SDA and SCL lines are de-asserted, and the respective control bits in the SSPCON2 register are cleared. When the user services the bus collision interrupt service routine, and if the I^2C bus is free, the user can resume communication by asserting a Start condition.

The master will continue to monitor the SDA and SCL pins. If a Stop condition occurs, the SSPIF bit will be set.

A write to the SSPBUF will start the transmission of data at the first data bit, regardless of where the transmitter left off when the bus collision occurred.

In Multi-Master mode, the interrupt generation on the detection of Start and Stop conditions allows the determination of when the bus is free. Control of the I^2C bus can be taken when the P bit is set in the SSPSTAT register, or the bus is idle and the S and P bits are cleared.

FIGURE 13-20: BUS COLLISION TIMING FOR TRANSMIT AND ACKNOWLEDGE

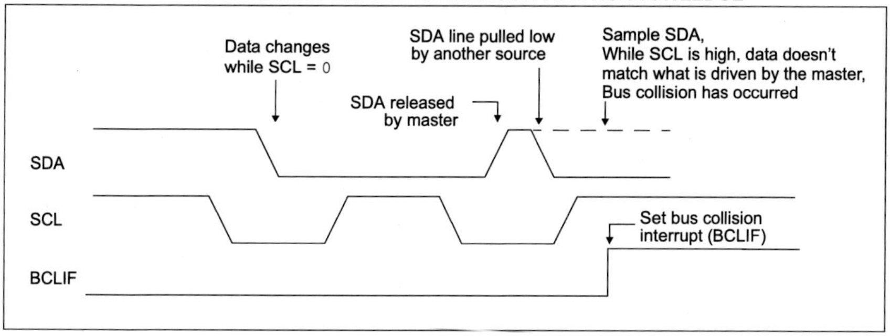

13.4.16.1 Bus Collision During a Start Condition

During a Start condition, a bus collision occurs if:

a) SDA or SCL are sampled low at the beginning of the Start condition (Figure 13-21).

b) SCL is sampled low before SDA is asserted low (Figure 13-22).

During a Start condition, both the SDA and the SCL pins are monitored, if:

the SDA pin is already low,
or the SCL pin is already low,

then:

> the Start condition is aborted,
> <u>and</u> the BCLIF flag is set,
> <u>and</u> the MSSP module is reset to its Idle state (Figure 13-21).

The Start condition begins with the SDA and SCL pins de-asserted. When the SDA pin is sampled high, the Baud Rate Generator is loaded from SSPADD<6:0> and counts down to 0. If the SCL pin is sampled low while SDA is high, a bus collision occurs, because it is assumed that another master is attempting to drive a data '1' during the Start condition.

If the SDA pin is sampled low during this count, the BRG is reset and the SDA line is asserted early (Figure 13-23). If, however, a '1' is sampled on the SDA pin, the SDA pin is asserted low at the end of the BRG count. The Baud Rate Generator is then reloaded and counts down to 0, and during this time, if the SCL pin is sampled as '0', a bus collision does not occur. At the end of the BRG count, the SCL pin is asserted low.

> **Note:** The reason that bus collision is not a factor during a Start condition, is that no two bus masters can assert a Start condition at the exact same time. Therefore, one master will always assert SDA before the other. This condition does not cause a bus collision, because the two masters must be allowed to arbitrate the first address following the Start condition. If the address is the same, arbitration must be allowed to continue into the data portion, Repeated Start or Stop conditions.

FIGURE 13-21: BUS COLLISION DURING START CONDITION (SDA ONLY)

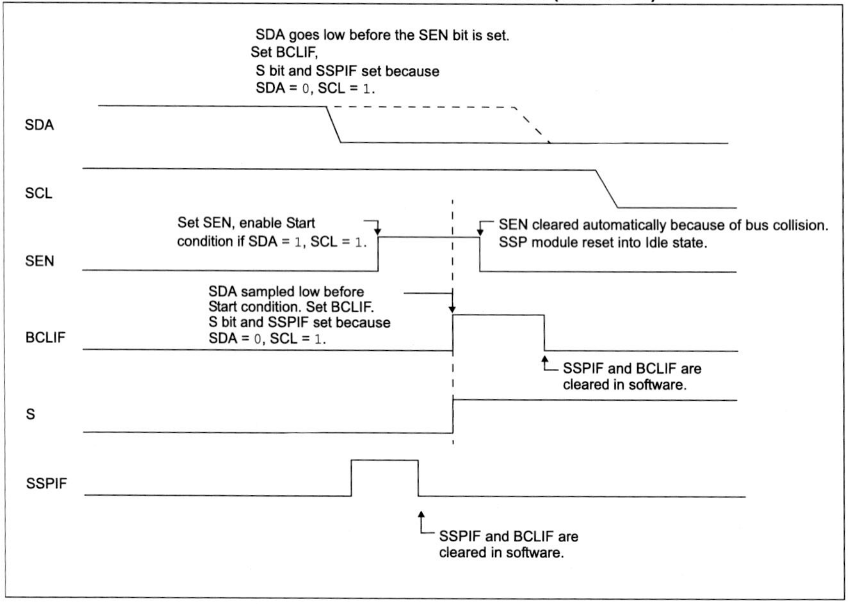

PIC16F882/883/884/886/887

FIGURE 13-22: BUS COLLISION DURING START CONDITION (SCL = 0)

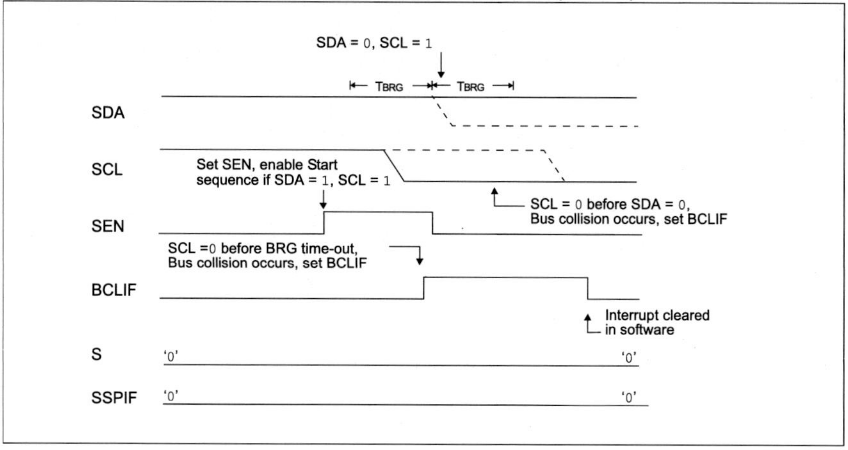

FIGURE 13-23: BRG RESET DUE TO SDA ARBITRATION DURING START CONDITION

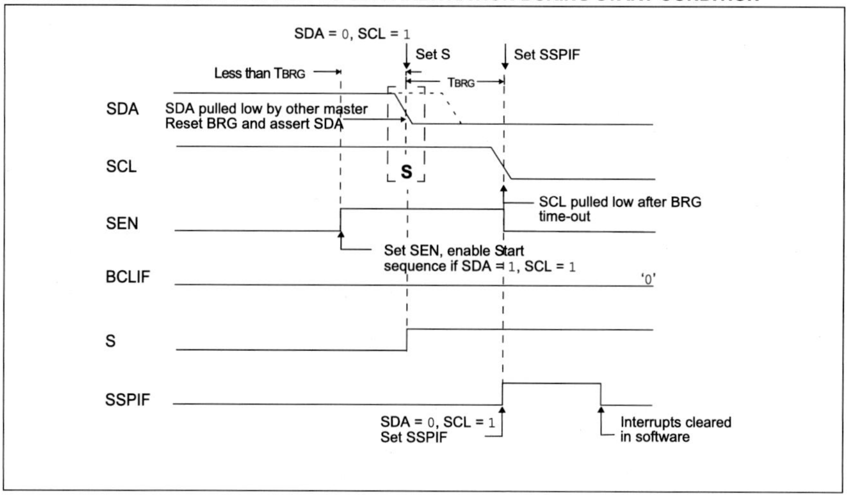

PIC16F882/883/884/886/887

13.4.16.2 Bus Collision During a Repeated Start Condition

During a Repeated Start condition, a bus collision occurs if:

a) A low level is sampled on SDA when SCL goes from low level to high level.
b) SCL goes low before SDA is asserted low, indicating that another master is attempting to transmit a data '1'.

When the user de-asserts SDA and the pin is allowed to float high, the BRG is loaded with SSPADD<6:0> and counts down to 0. The SCL pin is then de-asserted, and when sampled high, the SDA pin is sampled.

If SDA is low, a bus collision has occurred (i.e, another master is attempting to transmit a data '0', see Figure 13-24). If SDA is sampled high, the BRG is reloaded and begins counting. If SDA goes from high-to-low before the BRG times out, no bus collision occurs because no two masters can assert SDA at exactly the same time.

If SCL goes from high-to-low before the BRG times out and SDA has not already been asserted, a bus collision occurs. In this case, another master is attempting to transmit a data '1' during the Repeated Start condition (Figure 13-25).

If at the end of the BRG time-out, both SCL and SDA are still high, the SDA pin is driven low and the BRG is reloaded and begins counting. At the end of the count, regardless of the status of the SCL pin, the SCL pin is driven low and the Repeated Start condition is complete.

FIGURE 13-24: BUS COLLISION DURING A REPEATED START CONDITION (CASE 1)

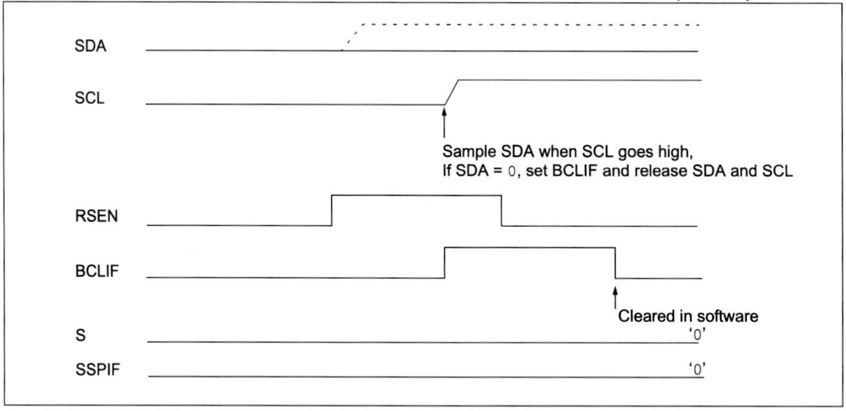

FIGURE 13-25: BUS COLLISION DURING REPEATED START CONDITION (CASE 2)

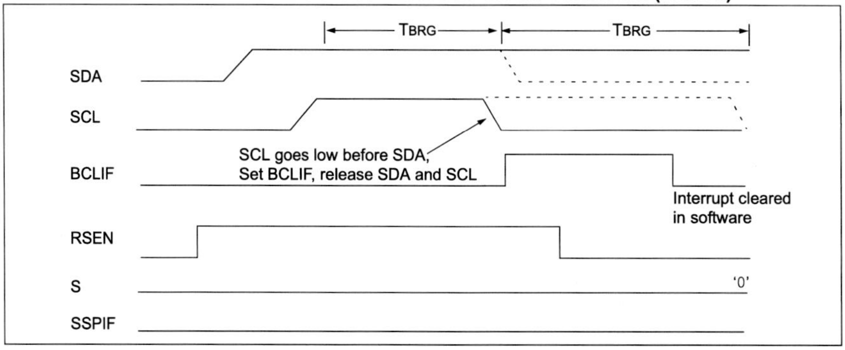

13.4.16.3 Bus Collision During a Stop Condition

Bus collision occurs during a Stop condition if:

a) After the SDA pin has been de-asserted and allowed to float high, SDA is sampled low after the BRG has timed out.
b) After the SCL pin is de-asserted, SCL is sampled low before SDA goes high.

The Stop condition begins with SDA asserted low. When SDA is sampled low, the SCL pin is allowed to float. When the pin is sampled high (clock arbitration), the Baud Rate Generator is loaded with SSPADD<6:0> and counts down to 0. After the BRG times out, SDA is sampled. If SDA is sampled low, a bus collision has occurred. This is due to another master attempting to drive a data '0' (Figure 13-26). If the SCL pin is sampled low before SDA is allowed to float high, a bus collision occurs. This is another case of another master attempting to drive a data '0' (Figure 13-27).

FIGURE 13-26: BUS COLLISION DURING A STOP CONDITION (CASE 1)

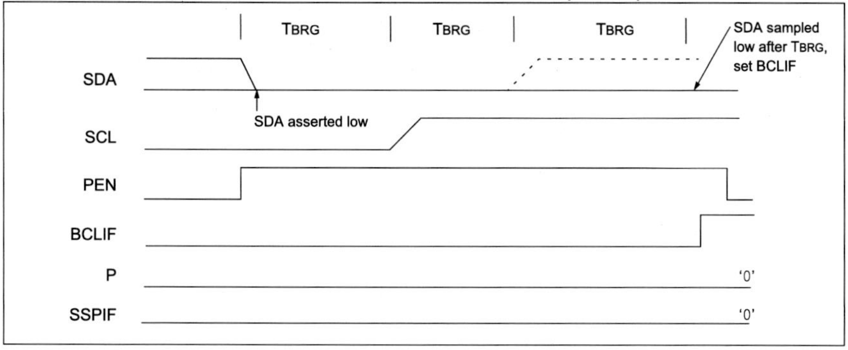

FIGURE 13-27: BUS COLLISION DURING A STOP CONDITION (CASE 2)

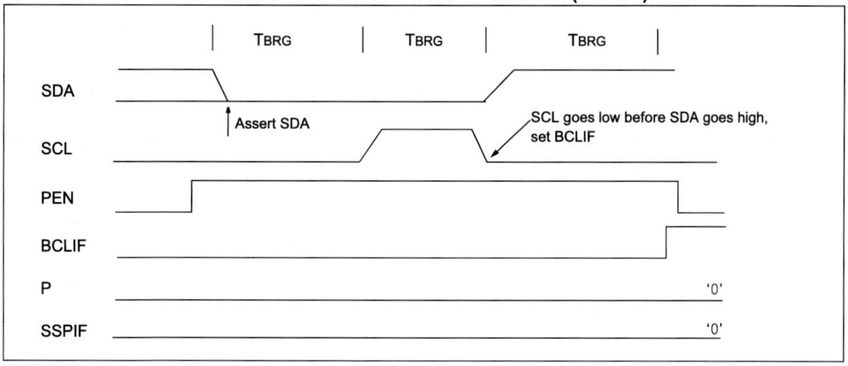

13.4.17 SSP MASK REGISTER

An SSP Mask (SSPMSK) register is available in I^2C Slave mode as a mask for the value held in the SSPSR register during an address comparison operation. A zero ('0') bit in the SSPMSK register has the effect of making the corresponding bit in the SSPSR register a "don't care".

This register is reset to all '1's upon any Reset condition and, therefore, has no effect on standard SSP operation until written with a mask value.

This register must be initiated prior to setting SSPM<3:0> bits to select the I^2C Slave mode (7-bit or 10-bit address).

This register can only be accessed when the appropriate mode is selected by bits (SSPM<3:0> of SSPCON).

The SSP Mask register is active during:

- 7-bit Address mode: address compare of A<7:1>.
- 10-bit Address mode: address compare of A<7:0> only. The SSP mask has no effect during the reception of the first (high) byte of the address.

REGISTER 13-4: SSPMSK: SSP MASK REGISTER(1)

R/W-1	R/W-1	R/W-1	R/W-1	R/W-1	R/W-1	R/W-1	R/W-1
MSK7	MSK6	MSK5	MSK4	MSK3	MSK2	MSK1	MSK0(2)
bit 7							bit 0

Legend:			
R = Readable bit	W = Writable bit	U = Unimplemented bit, read as '0'	
-n = Value at POR	'1' = Bit is set	'0' = Bit is cleared	x = Bit is unknown

bit 7-1 **MSK<7:1>:** Mask bits
1 = The received address bit n is compared to SSPADD<n> to detect I^2C address match
0 = The received address bit n is not used to detect I^2C address match

bit 0 **MSK<0>:** Mask bit for I^2C Slave mode, 10-bit Address(2)
I^2C Slave mode, 10-bit Address (SSPM<3:0> = 0111):
1 = The received address bit 0 is compared to SSPADD<0> to detect I^2C address match
0 = The received address bit 0 is not used to detect I^2C address match

Note 1: When SSPCON bits SSPM<3:0> = 1001, any reads or writes to the SSPADD SFR address are accessed through the SSPMSK register.

2: In all other SSP modes, this bit has no effect.

PIC16F882/883/884/886/887

TABLE 15-2: PIC16F882/883/884/886/887 INSTRUCTION SET

Mnemonic, Operands		Description	Cycles	14-Bit Opcode				Status Affected	Notes
				MSb			LSb		
BYTE-ORIENTED FILE REGISTER OPERATIONS									
ADDWF	f, d	Add W and f	1	00	0111	dfff	ffff	C, DC, Z	1, 2
ANDWF	f, d	AND W with f	1	00	0101	dfff	ffff	Z	1, 2
CLRF	f	Clear f	1	00	0001	lfff	ffff	Z	2
CLRW	–	Clear W	1	00	0001	0xxx	xxxx	Z	
COMF	f, d	Complement f	1	00	1001	dfff	ffff	Z	1, 2
DECF	f, d	Decrement f	1	00	0011	dfff	ffff	Z	1, 2
DECFSZ	f, d	Decrement f, Skip if 0	1(2)	00	1011	dfff	ffff		1, 2, 3
INCF	f, d	Increment f	1	00	1010	dfff	ffff	Z	1, 2
INCFSZ	f, d	Increment f, Skip if 0	1(2)	00	1111	dfff	ffff		1, 2, 3
IORWF	f, d	Inclusive OR W with f	1	00	0100	dfff	ffff	Z	1, 2
MOVF	f, d	Move f	1	00	1000	dfff	ffff	Z	1, 2
MOVWF	f	Move W to f	1	00	0000	lfff	ffff		
NOP	–	No Operation	1	00	0000	0xx0	0000		
RLF	f, d	Rotate Left f through Carry	1	00	1101	dfff	ffff	C	1, 2
RRF	f, d	Rotate Right f through Carry	1	00	1100	dfff	ffff	C	1, 2
SUBWF	f, d	Subtract W from f	1	00	0010	dfff	ffff	C, DC, Z	1, 2
SWAPF	f, d	Swap nibbles in f	1	00	1110	dfff	ffff		1, 2
XORWF	f, d	Exclusive OR W with f	1	00	0110	dfff	ffff	Z	1, 2
BIT-ORIENTED FILE REGISTER OPERATIONS									
BCF	f, b	Bit Clear f	1	01	00bb	bfff	ffff		1, 2
BSF	f, b	Bit Set f	1	01	01bb	bfff	ffff		1, 2
BTFSC	f, b	Bit Test f, Skip if Clear	1 (2)	01	10bb	bfff	ffff		3
BTFSS	f, b	Bit Test f, Skip if Set	1 (2)	01	11bb	bfff	ffff		3
LITERAL AND CONTROL OPERATIONS									
ADDLW	k	Add literal and W	1	11	111x	kkkk	kkkk	C, DC, Z	
ANDLW	k	AND literal with W	1	11	1001	kkkk	kkkk	Z	
CALL	k	Call Subroutine	2	10	0kkk	kkkk	kkkk		
CLRWDT	–	Clear Watchdog Timer	1	00	0000	0110	0100	$\overline{TO}$, $\overline{PD}$	
GOTO	k	Go to address	2	10	1kkk	kkkk	kkkk		
IORLW	k	Inclusive OR literal with W	1	11	1000	kkkk	kkkk	Z	
MOVLW	k	Move literal to W	1	11	00xx	kkkk	kkkk		
RETFIE	–	Return from interrupt	2	00	0000	0000	1001		
RETLW	k	Return with literal in W	2	11	01xx	kkkk	kkkk		
RETURN	–	Return from Subroutine	2	00	0000	0000	1000		
SLEEP	–	Go into Standby mode	1	00	0000	0110	0011	$\overline{TO}$, $\overline{PD}$	
SUBLW	k	Subtract W from literal	1	11	110x	kkkk	kkkk	C, DC, Z	
XORLW	k	Exclusive OR literal with W	1	11	1010	kkkk	kkkk	Z	

Note 1: When an I/O register is modified as a function of itself (e.g., `MOVF GPIO, 1`), the value used will be that value present on the pins themselves. For example, if the data latch is '1' for a pin configured as input and is driven low by an external device, the data will be written back with a '0'.

2: If this instruction is executed on the TMR0 register (and where applicable, d = 1), the prescaler will be cleared if assigned to the Timer0 module.

3: If the Program Counter (PC) is modified, or a conditional test is true, the instruction requires two cycles. The second cycle is executed as a `NOP`.

15.2 Instruction Descriptions

ADDLW	**Add literal and W**
Syntax:	[*label*] ADDLW k
Operands:	0 ≤ k ≤ 255
Operation:	(W) + k → (W)
Status Affected:	C, DC, Z
Description:	The contents of the W register are added to the eight-bit literal ‘k’ and the result is placed in the W register.

ADDWF	**Add W and f**
Syntax:	[*label*] ADDWF f,d
Operands:	0 ≤ f ≤ 127 d ∈ [0,1]
Operation:	(W) + (f) → (destination)
Status Affected:	C, DC, Z
Description:	Add the contents of the W register with register ‘f’. If ‘d’ is ‘0’, the result is stored in the W register. If ‘d’ is ‘1’, the result is stored back in register ‘f’.

ANDLW	**AND literal with W**
Syntax:	[*label*] ANDLW k
Operands:	0 ≤ k ≤ 255
Operation:	(W) .AND. (k) → (W)
Status Affected:	Z
Description:	The contents of W register are AND’ed with the eight-bit literal ‘k’. The result is placed in the W register.

ANDWF	**AND W with f**
Syntax:	[*label*] ANDWF f,d
Operands:	0 ≤ f ≤ 127 d ∈ [0,1]
Operation:	(W) .AND. (f) → (destination)
Status Affected:	Z
Description:	AND the W register with register ‘f’. If ‘d’ is ‘0’, the result is stored in the W register. If ‘d’ is ‘1’, the result is stored back in register ‘f’.

BCF	**Bit Clear f**
Syntax:	[*label*] BCF f,b
Operands:	0 ≤ f ≤ 127 0 ≤ b ≤ 7
Operation:	0 → (f<b>)
Status Affected:	None
Description:	Bit ‘b’ in register ‘f’ is cleared.

BSF	**Bit Set f**
Syntax:	[*label*] BSF f,b
Operands:	0 ≤ f ≤ 127 0 ≤ b ≤ 7
Operation:	1 → (f<b>)
Status Affected:	None
Description:	Bit ‘b’ in register ‘f’ is set.

BTFSC	**Bit Test f, Skip if Clear**
Syntax:	[*label*] BTFSC f,b
Operands:	0 ≤ f ≤ 127 0 ≤ b ≤ 7
Operation:	skip if (f<b>) = 0
Status Affected:	None
Description:	If bit ‘b’ in register ‘f’ is ‘1’, the next instruction is executed. If bit ‘b’ in register ‘f’ is ‘0’, the next instruction is discarded, and a NOP is executed instead, making this a two-cycle instruction.

PIC16F882/883/884/886/887

BTFSS	Bit Test f, Skip if Set
Syntax:	[*label*] BTFSS f,b
Operands:	$0 \le f \le 127$ $0 \le b < 7$
Operation:	skip if (f<b>) = 1
Status Affected:	None
Description:	If bit 'b' in register 'f' is '0', the next instruction is executed. If bit 'b' is '1', then the next instruction is discarded and a NOP is executed instead, making this a two-cycle instruction.

CALL	Call Subroutine
Syntax:	[*label*] CALL k
Operands:	$0 \le k \le 2047$
Operation:	(PC)+ 1→ TOS, k → PC<10:0>, (PCLATH<4:3>) → PC<12:11>
Status Affected:	None
Description:	Call Subroutine. First, return address (PC + 1) is pushed onto the stack. The eleven-bit immediate address is loaded into PC bits <10:0>. The upper bits of the PC are loaded from PCLATH. CALL is a two-cycle instruction.

CLRF	Clear f
Syntax:	[*label*] CLRF f
Operands:	$0 \le f \le 127$
Operation:	00h → (f) 1 → Z
Status Affected:	Z
Description:	The contents of register 'f' are cleared and the Z bit is set.

CLRW	Clear W
Syntax:	[*label*] CLRW
Operands:	None
Operation:	00h → (W) 1 → Z
Status Affected:	Z
Description:	W register is cleared. Zero bit (Z) is set.

CLRWDT	Clear Watchdog Timer
Syntax:	[*label*] CLRWDT
Operands:	None
Operation:	00h → WDT 0 → WDT prescaler, 1 → $\overline{TO}$ 1 → $\overline{PD}$
Status Affected:	$\overline{TO}$, $\overline{PD}$
Description:	CLRWDT instruction resets the Watchdog Timer. It also resets the prescaler of the WDT. Status bits $\overline{TO}$ and $\overline{PD}$ are set.

COMF	Complement f
Syntax:	[*label*] COMF f,d
Operands:	$0 \le f \le 127$ $d \in [0,1]$
Operation:	$(\bar{f})$ → (destination)
Status Affected:	Z
Description:	The contents of register 'f' are complemented. If 'd' is '0', the result is stored in W. If 'd' is '1', the result is stored back in register 'f'.

DECF	Decrement f
Syntax:	[*label*] DECF f,d
Operands:	$0 \le f \le 127$ $d \in [0,1]$
Operation:	(f) - 1 → (destination)
Status Affected:	Z
Description:	Decrement register 'f'. If 'd' is '0', the result is stored in the W register. If 'd' is '1', the result is stored back in register 'f'.

DECFSZ	Decrement f, Skip if 0
Syntax:	[*label*] DECFSZ f,d
Operands:	$0 \le f \le 127$ $d \in [0,1]$
Operation:	(f) - 1 → (destination); skip if result = 0
Status Affected:	None
Description:	The contents of register 'f' are decremented. If 'd' is '0', the result is placed in the W register. If 'd' is '1', the result is placed back in register 'f'. If the result is '1', the next instruction is executed. If the result is '0', then a NOP is executed instead, making it a two-cycle instruction.

GOTO	Unconditional Branch
Syntax:	[*label*] GOTO k
Operands:	$0 \le k \le 2047$
Operation:	k → PC<10:0> PCLATH<4:3> → PC<12:11>
Status Affected:	None
Description:	GOTO is an unconditional branch. The eleven-bit immediate value is loaded into PC bits <10:0>. The upper bits of PC are loaded from PCLATH<4:3>. GOTO is a two-cycle instruction.

INCF	Increment f
Syntax:	[*label*] INCF f,d
Operands:	$0 \le f \le 127$ $d \in [0,1]$
Operation:	(f) + 1 → (destination)
Status Affected:	Z
Description:	The contents of register 'f' are incremented. If 'd' is '0', the result is placed in the W register. If 'd' is '1', the result is placed back in register 'f'.

INCFSZ	Increment f, Skip if 0
Syntax:	[*label*] INCFSZ f,d
Operands:	$0 \le f \le 127$ $d \in [0,1]$
Operation:	(f) + 1 → (destination), skip if result = 0
Status Affected:	None
Description:	The contents of register 'f' are incremented. If 'd' is '0', the result is placed in the W register. If 'd' is '1', the result is placed back in register 'f'. If the result is '1', the next instruction is executed. If the result is '0', a NOP is executed instead, making it a two-cycle instruction.

IORLW	Inclusive OR literal with W
Syntax:	[*label*] IORLW k
Operands:	$0 \le k \le 255$
Operation:	(W) .OR. k → (W)
Status Affected:	Z
Description:	The contents of the W register are OR'ed with the eight-bit literal 'k'. The result is placed in the W register.

IORWF	Inclusive OR W with f
Syntax:	[*label*] IORWF f,d
Operands:	$0 \le f \le 127$ $d \in [0,1]$
Operation:	(W) .OR. (f) → (destination)
Status Affected:	Z
Description:	Inclusive OR the W register with register 'f'. If 'd' is '0', the result is placed in the W register. If 'd' is '1', the result is placed back in register 'f'.

PIC16F882/883/884/886/887

MOVF	**Move f**
Syntax:	[*label*] MOVF f,d
Operands:	$0 \le f \le 127$ $d \in [0,1]$
Operation:	(f) → (dest)
Status Affected:	Z
Description:	The contents of register 'f' is moved to a destination dependent upon the status of 'd'. If d = 0, destination is W register. If d = 1, the destination is file register 'f' itself. d = 1 is useful to test a file register since status flag Z is affected.
Words:	1
Cycles:	1
Example:	`MOVF FSR, 0` After Instruction W = value in FSR register Z = 1

MOVWF	**Move W to f**
Syntax:	[*label*] MOVWF f
Operands:	$0 \le f \le 127$
Operation:	(W) → (f)
Status Affected:	None
Description:	Move data from W register to register 'f'.
Words:	1
Cycles:	1
Example:	`MOVW OPTION F` Before Instruction OPTION = 0xFF W = 0x4F After Instruction OPTION = 0x4F W = 0x4F

MOVLW	**Move literal to W**
Syntax:	[*label*] MOVLW k
Operands:	$0 \le k \le 255$
Operation:	k → (W)
Status Affected:	None
Description:	The eight-bit literal 'k' is loaded into W register. The "don't cares" will assemble as '0's.
Words:	1
Cycles:	1
Example:	`MOVLW 0x5A` After Instruction W = 0x5A

NOP	**No Operation**
Syntax:	[*label*] NOP
Operands:	None
Operation:	No operation
Status Affected:	None
Description:	No operation.
Words:	1
Cycles:	1
Example:	`NOP`

RETFIE	**Return from Interrupt**
Syntax:	[*label*] RETFIE
Operands:	None
Operation:	TOS → PC, 1 → GIE
Status Affected:	None
Description:	Return from Interrupt. Stack is POPed and Top-of-Stack (TOS) is loaded in the PC. Interrupts are enabled by setting Global Interrupt Enable bit, GIE (INTCON<7>). This is a two-cycle instruction.
Words:	1
Cycles:	2
<u>Example</u>:	`RETFIE` After Interrupt PC = TOS GIE = 1

RETLW	**Return with literal in W**
Syntax:	[*label*] RETLW k
Operands:	$0 \leq k \leq 255$
Operation:	k → (W); TOS → PC
Status Affected:	None
Description:	The W register is loaded with the eight-bit literal 'k'. The program counter is loaded from the top of the stack (the return address). This is a two-cycle instruction.
Words:	1
Cycles:	2

<u>Example</u>:

```
          CALL TABLE;W contains
          table
                  ;offset value
TABLE     •       ;W now has
          •       ;table value
          •
          •
          ADDWF PC ;W = offset
          RETLW k1 ;Begin table
          RETLW k2 ;
          •
          •
          •
          RETLW kn ;End of table
```

Before Instruction
W = 0x07
After Instruction
W = value of k8

RETURN	**Return from Subroutine**
Syntax:	[*label*] RETURN
Operands:	None
Operation:	TOS → PC
Status Affected:	None
Description:	Return from subroutine. The stack is POPed and the top of the stack (TOS) is loaded into the program counter. This is a two-cycle instruction.

PIC16F882/883/884/886/887

RLF — **Rotate Left f through Carry**

Syntax: [*label*] RLF f,d

Operands: $0 \le f \le 127$
$d \in [0,1]$

Operation: See description below

Status Affected: C

Description: The contents of register 'f' are rotated one bit to the left through the Carry flag. If 'd' is '0', the result is placed in the W register. If 'd' is '1', the result is stored back in register 'f'.

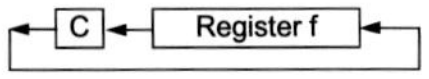

Words: 1

Cycles: 1

Example:

```
RLF     REG1,0
```

Before Instruction

```
REG1  =  1110 0110
C     =  0
```

After Instruction

```
REG1  =  1110 0110
W     =  1100 1100
C     =  1
```

SLEEP — **Enter Sleep mode**

Syntax: [*label*] SLEEP

Operands: None

Operation: 00h → WDT,
0 → WDT prescaler,
1 → $\overline{TO}$,
0 → $\overline{PD}$

Status Affected: $\overline{TO}$, $\overline{PD}$

Description: The power-down Status bit, $\overline{PD}$ is cleared. Time-out Status bit, $\overline{TO}$ is set. Watchdog Timer and its prescaler are cleared.
The processor is put into Sleep mode with the oscillator stopped.

RRF — **Rotate Right f through Carry**

Syntax: [*label*] RRF f,d

Operands: $0 \le f \le 127$
$d \in [0,1]$

Operation: See description below

Status Affected: C

Description: The contents of register 'f' are rotated one bit to the right through the Carry flag. If 'd' is '0', the result is placed in the W register. If 'd' is '1', the result is placed back in register 'f'.

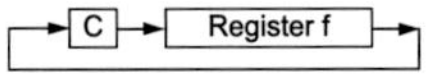

SUBLW — **Subtract W from literal**

Syntax: [*label*] SUBLW k

Operands: $0 \le k \le 255$

Operation: k - (W) → (W)

Status Affected: C, DC, Z

Description: The W register is subtracted (2's complement method) from the eight-bit literal 'k'. The result is placed in the W register.

C = 0	W > k
C = 1	W ≤ k
DC = 0	W<3:0> > k<3:0>
DC = 1	W<3:0> ≤ k<3:0>

SUBWF	Subtract W from f
Syntax:	[*label*] SUBWF f,d
Operands:	0 ≤ f ≤ 127 d ∈ [0,1]
Operation:	(f) - (W) → (destination)
Status Affected:	C, DC, Z
Description:	Subtract (2's complement method) W register from register 'f'. If 'd' is '0', the result is stored in the W register. If 'd' is '1', the result is stored back in register 'f'.

C = 0	W > f
C = 1	W ≤ f
DC = 0	W<3:0> > f<3:0>
DC = 1	W<3:0> ≤ f<3:0>

SWAPF	Swap Nibbles in f
Syntax:	[*label*] SWAPF f,d
Operands:	0 ≤ f ≤ 127 d ∈ [0,1]
Operation:	(f<3:0>) → (destination<7:4>), (f<7:4>) → (destination<3:0>)
Status Affected:	None
Description:	The upper and lower nibbles of register 'f' are exchanged. If 'd' is '0', the result is placed in the W register. If 'd' is '1', the result is placed in register 'f'.

XORLW	Exclusive OR literal with W
Syntax:	[*label*] XORLW k
Operands:	0 ≤ k ≤ 255
Operation:	(W) .XOR. k → (W)
Status Affected:	Z
Description:	The contents of the W register are XOR'ed with the eight-bit literal 'k'. The result is placed in the W register.

XORWF	Exclusive OR W with f
Syntax:	[*label*] XORWF f,d
Operands:	0 ≤ f ≤ 127 d ∈ [0,1]
Operation:	(W) .XOR. (f) → (destination)
Status Affected:	Z
Description:	Exclusive OR the contents of the W register with register 'f'. If 'd' is '0', the result is stored in the W register. If 'd' is '1', the result is stored back in register 'f'.

NOTES:

16.0 DEVELOPMENT SUPPORT

The PIC® microcontrollers are supported with a full range of hardware and software development tools:

- Integrated Development Environment
  - MPLAB® IDE Software
- Assemblers/Compilers/Linkers
  - MPASM™ Assembler
  - MPLAB C18 and MPLAB C30 C Compilers
  - MPLINK™ Object Linker/ MPLIB™ Object Librarian
  - MPLAB ASM30 Assembler/Linker/Library
- Simulators
  - MPLAB SIM Software Simulator
- Emulators
  - MPLAB ICE 2000 In-Circuit Emulator
  - MPLAB REAL ICE™ In-Circuit Emulator
- In-Circuit Debugger
  - MPLAB ICD 2
- Device Programmers
  - PICSTART® Plus Development Programmer
  - MPLAB PM3 Device Programmer
  - PICkit™ 2 Development Programmer
- Low-Cost Demonstration and Development Boards and Evaluation Kits

16.1 MPLAB Integrated Development Environment Software

The MPLAB IDE software brings an ease of software development previously unseen in the 8/16-bit microcontroller market. The MPLAB IDE is a Windows® operating system-based application that contains:

- A single graphical interface to all debugging tools
  - Simulator
  - Programmer (sold separately)
  - Emulator (sold separately)
  - In-Circuit Debugger (sold separately)
- A full-featured editor with color-coded context
- A multiple project manager
- Customizable data windows with direct edit of contents
- High-level source code debugging
- Visual device initializer for easy register initialization
- Mouse over variable inspection
- Drag and drop variables from source to watch windows
- Extensive on-line help
- Integration of select third party tools, such as HI-TECH Software C Compilers and IAR C Compilers

The MPLAB IDE allows you to:

- Edit your source files (either assembly or C)
- One touch assemble (or compile) and download to PIC MCU emulator and simulator tools (automatically updates all project information)
- Debug using:
  - Source files (assembly or C)
  - Mixed assembly and C
  - Machine code

MPLAB IDE supports multiple debugging tools in a single development paradigm, from the cost-effective simulators, through low-cost in-circuit debuggers, to full-featured emulators. This eliminates the learning curve when upgrading to tools with increased flexibility and power.

16.2 MPASM Assembler

The MPASM Assembler is a full-featured, universal macro assembler for all PIC MCUs.

The MPASM Assembler generates relocatable object files for the MPLINK Object Linker, Intel® standard HEX files, MAP files to detail memory usage and symbol reference, absolute LST files that contain source lines and generated machine code and COFF files for debugging.

The MPASM Assembler features include:

- Integration into MPLAB IDE projects
- User-defined macros to streamline assembly code
- Conditional assembly for multi-purpose source files
- Directives that allow complete control over the assembly process

16.3 MPLAB C18 and MPLAB C30 C Compilers

The MPLAB C18 and MPLAB C30 Code Development Systems are complete ANSI C compilers for Microchip's PIC18 and PIC24 families of microcontrollers and the dsPIC30 and dsPIC33 family of digital signal controllers. These compilers provide powerful integration capabilities, superior code optimization and ease of use not found with other compilers.

For easy source level debugging, the compilers provide symbol information that is optimized to the MPLAB IDE debugger.

16.4 MPLINK Object Linker/ MPLIB Object Librarian

The MPLINK Object Linker combines relocatable objects created by the MPASM Assembler and the MPLAB C18 C Compiler. It can link relocatable objects from precompiled libraries, using directives from a linker script.

The MPLIB Object Librarian manages the creation and modification of library files of precompiled code. When a routine from a library is called from a source file, only the modules that contain that routine will be linked in with the application. This allows large libraries to be used efficiently in many different applications.

The object linker/library features include:

- Efficient linking of single libraries instead of many smaller files
- Enhanced code maintainability by grouping related modules together
- Flexible creation of libraries with easy module listing, replacement, deletion and extraction

16.5 MPLAB ASM30 Assembler, Linker and Librarian

MPLAB ASM30 Assembler produces relocatable machine code from symbolic assembly language for dsPIC30F devices. MPLAB C30 C Compiler uses the assembler to produce its object file. The assembler generates relocatable object files that can then be archived or linked with other relocatable object files and archives to create an executable file. Notable features of the assembler include:

- Support for the entire dsPIC30F instruction set
- Support for fixed-point and floating-point data
- Command line interface
- Rich directive set
- Flexible macro language
- MPLAB IDE compatibility

16.6 MPLAB SIM Software Simulator

The MPLAB SIM Software Simulator allows code development in a PC-hosted environment by simulating the PIC MCUs and dsPIC® DSCs on an instruction level. On any given instruction, the data areas can be examined or modified and stimuli can be applied from a comprehensive stimulus controller. Registers can be logged to files for further run-time analysis. The trace buffer and logic analyzer display extend the power of the simulator to record and track program execution, actions on I/O, most peripherals and internal registers.

The MPLAB SIM Software Simulator fully supports symbolic debugging using the MPLAB C18 and MPLAB C30 C Compilers, and the MPASM and MPLAB ASM30 Assemblers. The software simulator offers the flexibility to develop and debug code outside of the hardware laboratory environment, making it an excellent, economical software development tool.

16.7 MPLAB ICE 2000 High-Performance In-Circuit Emulator

The MPLAB ICE 2000 In-Circuit Emulator is intended to provide the product development engineer with a complete microcontroller design tool set for PIC microcontrollers. Software control of the MPLAB ICE 2000 In-Circuit Emulator is advanced by the MPLAB Integrated Development Environment, which allows editing, building, downloading and source debugging from a single environment.

The MPLAB ICE 2000 is a full-featured emulator system with enhanced trace, trigger and data monitoring features. Interchangeable processor modules allow the system to be easily reconfigured for emulation of different processors. The architecture of the MPLAB ICE 2000 In-Circuit Emulator allows expansion to support new PIC microcontrollers.

The MPLAB ICE 2000 In-Circuit Emulator system has been designed as a real-time emulation system with advanced features that are typically found on more expensive development tools. The PC platform and Microsoft® Windows® 32-bit operating system were chosen to best make these features available in a simple, unified application.

16.8 MPLAB REAL ICE In-Circuit Emulator System

MPLAB REAL ICE In-Circuit Emulator System is Microchip's next generation high-speed emulator for Microchip Flash DSC® and MCU devices. It debugs and programs PIC® and dsPIC® Flash microcontrollers with the easy-to-use, powerful graphical user interface of the MPLAB Integrated Development Environment (IDE), included with each kit.

The MPLAB REAL ICE probe is connected to the design engineer's PC using a high-speed USB 2.0 interface and is connected to the target with either a connector compatible with the popular MPLAB ICD 2 system (RJ11) or with the new high speed, noise tolerant, low-voltage differential signal (LVDS) interconnection (CAT5).

MPLAB REAL ICE is field upgradeable through future firmware downloads in MPLAB IDE. In upcoming releases of MPLAB IDE, new devices will be supported, and new features will be added, such as software breakpoints and assembly code trace. MPLAB REAL ICE offers significant advantages over competitive emulators including low-cost, full-speed emulation, real-time variable watches, trace analysis, complex breakpoints, a ruggedized probe interface and long (up to three meters) interconnection cables.

16.9 MPLAB ICD 2 In-Circuit Debugger

Microchip's In-Circuit Debugger, MPLAB ICD 2, is a powerful, low-cost, run-time development tool, connecting to the host PC via an RS-232 or high-speed USB interface. This tool is based on the Flash PIC MCUs and can be used to develop for these and other PIC MCUs and dsPIC DSCs. The MPLAB ICD 2 utilizes the in-circuit debugging capability built into the Flash devices. This feature, along with Microchip's In-Circuit Serial Programming™ (ICSP™) protocol, offers cost-effective, in-circuit Flash debugging from the graphical user interface of the MPLAB Integrated Development Environment. This enables a designer to develop and debug source code by setting breakpoints, single stepping and watching variables, and CPU status and peripheral registers. Running at full speed enables testing hardware and applications in real time. MPLAB ICD 2 also serves as a development programmer for selected PIC devices.

16.10 MPLAB PM3 Device Programmer

The MPLAB PM3 Device Programmer is a universal, CE compliant device programmer with programmable voltage verification at VDDMIN and VDDMAX for maximum reliability. It features a large LCD display (128 x 64) for menus and error messages and a modular, detachable socket assembly to support various package types. The ICSP™ cable assembly is included as a standard item. In Stand-Alone mode, the MPLAB PM3 Device Programmer can read, verify and program PIC devices without a PC connection. It can also set code protection in this mode. The MPLAB PM3 connects to the host PC via an RS-232 or USB cable. The MPLAB PM3 has high-speed communications and optimized algorithms for quick programming of large memory devices and incorporates an SD/MMC card for file storage and secure data applications.

PIC16F882/883/884/886/887

16.11 PICSTART Plus Development Programmer

The PICSTART Plus Development Programmer is an easy-to-use, low-cost, prototype programmer. It connects to the PC via a COM (RS-232) port. MPLAB Integrated Development Environment software makes using the programmer simple and efficient. The PICSTART Plus Development Programmer supports most PIC devices in DIP packages up to 40 pins. Larger pin count devices, such as the PIC16C92X and PIC17C76X, may be supported with an adapter socket. The PICSTART Plus Development Programmer is CE compliant.

16.12 PICkit 2 Development Programmer

The PICkit™ 2 Development Programmer is a low-cost programmer and selected Flash device debugger with an easy-to-use interface for programming many of Microchip's baseline, mid-range and PIC18F families of Flash memory microcontrollers. The PICkit 2 Starter Kit includes a prototyping development board, twelve sequential lessons, software and HI-TECH's PICC™ Lite C compiler, and is designed to help get up to speed quickly using PIC® microcontrollers. The kit provides everything needed to program, evaluate and develop applications using Microchip's powerful, mid-range Flash memory family of microcontrollers.

16.13 Demonstration, Development and Evaluation Boards

A wide variety of demonstration, development and evaluation boards for various PIC MCUs and dsPIC DSCs allows quick application development on fully functional systems. Most boards include prototyping areas for adding custom circuitry and provide application firmware and source code for examination and modification.

The boards support a variety of features, including LEDs, temperature sensors, switches, speakers, RS-232 interfaces, LCD displays, potentiometers and additional EEPROM memory.

The demonstration and development boards can be used in teaching environments, for prototyping custom circuits and for learning about various microcontroller applications.

In addition to the PICDEM™ and dsPICDEM™ demonstration/development board series of circuits, Microchip has a line of evaluation kits and demonstration software for analog filter design, KEELOQ® security ICs, CAN, IrDA®, PowerSmart® battery management, SEEVAL® evaluation system, Sigma-Delta ADC, flow rate sensing, plus many more.

Check the Microchip web page (www.microchip.com) and the latest *"Product Selector Guide"* (DS00148) for the complete list of demonstration, development and evaluation kits.

THE MICROCHIP WEB SITE

Microchip provides online support via our WWW site at www.microchip.com. This web site is used as a means to make files and information easily available to customers. Accessible by using your favorite Internet browser, the web site contains the following information:

- **Product Support** – Data sheets and errata, application notes and sample programs, design resources, user's guides and hardware support documents, latest software releases and archived software
- **General Technical Support** – Frequently Asked Questions (FAQ), technical support requests, online discussion groups, Microchip consultant program member listing
- **Business of Microchip** – Product selector and ordering guides, latest Microchip press releases, listing of seminars and events, listings of Microchip sales offices, distributors and factory representatives

CUSTOMER CHANGE NOTIFICATION SERVICE

Microchip's customer notification service helps keep customers current on Microchip products. Subscribers will receive e-mail notification whenever there are changes, updates, revisions or errata related to a specified product family or development tool of interest.

To register, access the Microchip web site at www.microchip.com, click on Customer Change Notification and follow the registration instructions.

CUSTOMER SUPPORT

Users of Microchip products can receive assistance through several channels:

- Distributor or Representative
- Local Sales Office
- Field Application Engineer (FAE)
- Technical Support
- Development Systems Information Line

Customers should contact their distributor, representative or field application engineer (FAE) for support. Local sales offices are also available to help customers. A listing of sales offices and locations is included in the back of this document.

Technical support is available through the web site at: http://support.microchip.com

PIC16F882/883/884/886/887

READER RESPONSE

It is our intention to provide you with the best documentation possible to ensure successful use of your Microchip product. If you wish to provide your comments on organization, clarity, subject matter, and ways in which our documentation can better serve you, please FAX your comments to the Technical Publications Manager at (480) 792-4150.

Please list the following information, and use this outline to provide us with your comments about this document.

To: Technical Publications Manager   Total Pages Sent ________

RE: Reader Response

From: Name ______________________________

Company ______________________________

Address ______________________________

City / State / ZIP / Country ______________________________

Telephone: (_______) ________ - ________   FAX: (______) ________ - ________

Application (optional):

Would you like a reply?____Y ____N

Device: PIC16F882/883/884/886/887   Literature Number: DS41291F

Questions:

1. What are the best features of this document?

____

____

2. How does this document meet your hardware and software development needs?

____

____

3. Do you find the organization of this document easy to follow? If not, why?

____

____

4. What additions to the document do you think would enhance the structure and subject?

____

____

5. What deletions from the document could be made without affecting the overall usefulness?

____

____

6. Is there any incorrect or misleading information (what and where)?

____

____

7. How would you improve this document?

____

____

PRODUCT IDENTIFICATION SYSTEM

To order or obtain information, e.g., on pricing or delivery, refer to the factory or the listed sales office.

PART NO.	X	/XX	XXX
Device	Temperature Range	Package	Pattern

Device:	PIC16F883(1), PIC16F883T(1, 2), PIC16F884(1), PIC16F884T(1, 2), PIC16F886(1), PIC16F886T(1, 2), PIC16F887(1), PIC16F887T(1, 2) VDD range 2.0V to 5.5V
Temperature Range:	I = -40°C to +85°C (Industrial) E = -40°C to +125°C (Extended)
Package:	ML = Quad Flat No Leads (QFN) P = Plastic DIP PT = Plastic Thin-Quad Flatpack (TQFP) SO = Plastic Small Outline (SOIC) (7.50 mm) SP = Skinny Plastic DIP SS = Plastic Shrink Small Outline
Pattern:	QTP, SQTP, Code or Special Requirements (blank otherwise)

Examples:

a) PIC16F883-E/P 301 = Extended Temp., PDIP package, 20 MHz, QTP pattern #301

b) PIC16F883-I/SO = Industrial Temp., SOIC package, 20 MHz

Note 1: T = In tape and reel SSOP, SOIC and QFN packages only.

Index

Note: Page numbers followed by *f* indicate figures and *t* indicate tables.

I

J

L

M

X

Printed in the United States
By Bookmasters